Plant Protection, Diversity and Conservation

Plant Protection, Diversity and Conservation

— Volume 2 —

Editors

Dr. D. Prasad

Division of Nematology,
L.B.S. Centre, IARI, New Delhi – 110 012

Dr. Satya Kumar

Department of Plant Pathology
PRBC, Khanna Farm
Bhagwanpur, U.S. Nagar

2011

BIOTECH BOOKS®

Delhi - 110 002

ISBN 978-81-7622-238-9 (SET)
ISBN 978-81-7622-239-6 (Vol. 1)
ISBN 978-81-7622-240-2 (Vol. 2)

The views expressed in various articles are those of the authors and not of editor or publisher of the book.

Published by : **BIOTECH BOOKS®**
4762-63/23, Ansari Road, Darya Ganj,
New Delhi - 110 002
Phone: +91-011-23262132
E-mail: biotechbooks@yahoo.co.in
website: www.dayabooks.com

Laser Typesetting : **Classic Computer Services**
Delhi - 110 035

Printed at : **Chawla Offset Printers**
Delhi - 110 052

PRINTED IN INDIA

Preface

The global estimates for biodiversity indicate occurrence of about 13-14 million species including one million species of fungi, bacteria and nematodes. At present, nearly 1.7 million species have been described so far. The plant species number is around 2.5 lakh of which about 3000 plant species have been utilized by human beings. Most of our needs are being met by just 30 species which provide for 90 per cent of world's food. The number of endemic species is estimated to be close to 3470 in Himalayas, 2000 in Peninsular India and 240 in Andaman and Nicobar Islands. It is reported that the Indian gene centre has about 4000 species of medical values, 500 fibre yielding species, 100 aromatic and essential oil yielding species and 400 fodder yielding species. The rich wealth of biodiversity has evolved over the millennia through evolutionary processes such as selection, mutation, recombination, genetic drift and migration. We must conserve it for sustainable use and benefit of posterity. Agricultural systems based on large number of different crop species and varieties traditionally bestowed a high degree of genetic heterogeneity on crop habitats. Research on breeding for commercial crops also change the habits of feeding of insect pests and subsequently adaptation for natural enemies such as parasitoids, predators etc need to be matched with cropping systems. These need to be conserve to enhance the sustainability in organic farming.

In recent years, cultivation of transgenic cotton in larger areas has resulted in increase of thrips incidence along with mosaic virus. The intensive agriculture especially the new high yielding genotypes has become susceptible to the pests and pathogens and their races already present in the country. The changing cropping patterns including cultivation in non-traditional area have resulted in a spurt of biotic stresses. It is not only that new pest problem has emerged but the minor pests assumed the status of key pests. In the pursuit of faster economic growth, man has been over exploiting the natural resources without any serious consideration of the ecological consequences. Effective and timely management of a vast array of insect pests, diseases and nematodes of crops of agricultural importance by the use of bioagents and biopesticides in Integrated Pest Management recognizes the importance of eco-friendly insects to the judiciously conserved promoters.

The fifty two chapters covered in two volumes deal with various facets of plant protection. The book begins with chapters on Allelopathy in action, Natural enemies conservations, Nanotechnology in Plant Protection, Bacterial antagonists, Plant growth regulation in nematodes and entomopathogenic nematodes as bioinsecticides. *Plant Protection, Diversity and Conservation* also deals with the chapters like chemical hazards, Nematode Fusarium Wilt interactions, Novel nematicides, bacterial and fungal diseases, Root-knot nematode on rice, Micro-metabolites, Environmental pollution on crop production, IPM technology on various crops. Some of the knowledge based biotech chapters are being included such as Biopesticides from bioenergy sector, Food application of lac dye, Various application of lac, Potential of biopesticides, Mode of action of bioagents and biopesticides, Insects friends to man, Microbial toxin and Potentiality of Rhizobacteria and Products development from water hyacinth. Few topics pertaining to Key pest complexes on Maize, Management Strategies, IPM in Pigeon pea, chick pea, Major Crops in A. P., INM, IDM in Onion and Garlic. Pests complex in Cotton, Pest of Jute, Eco-friendly management of nematodes in vegetables, Potential of allelopathic plants for weed control, Herbicide resistant Rodents, Eco-friendly management and Electronic wastes. Two chapters from Andaman & Nicobar Islands are of great values due to the unexplored area in the field of Plant Protection, these are: Coconut diseases in Bay Islands and Insect pest of vegetables and their future management strategies.

The book *Plant Protection, Diversity and Conservation* in 2 volumes is a compilation of work of various scientists their expertise and experience. There are 52 thought provoking chapters covered on the problems and purposes of Diversity of biotic components and conservation of important biological agents and beneficial insects especially lac. Eco-friendly managements have been discussed with respect to the insect pests, diseases, nematodes and weeds.

The book is timely for the scientists, teachers, students, scholars, administrators and policy makers dealing with insect, disease, nematode and weed management in Canada having very useful information.

We received a tremendous response and cooperation from all the contributors for preparing the manuscripts in tune with the theme of the book for which we express our sincere gratitude to one and all.

We are also grateful to Shri Sunil Mittal of Biotech Books, New Delhi for timely taking up the task of publication of this important book and provided us all technical helps required during the preparation and finalization of the manuscripts. Last but not the least, I am also thankful to my wife, Gita Prasad, who has helped during the finalization of the chapters.

D. Prasad

Satya Kumar

Contents

— Volume 2 —

Plant Protection, Diversity and Conservation (2011) *Pages* **359–365**
Editors: **D. Prasad & Satya Kumar**
Published by: **BIOTECH BOOKS, NEW DELHI**

Chapter 27
Integrated Management of Diseases of Chickpea

Sanjeev Kumar
College of Agriculture, Ganj Basoda (JNKVV-Jabalpur), Vidisha – 464 221, M.P.

Chickpea (*Cicer arietinum*) commonly known as gram or Bengal gram is an important pulse crop of india contributing 40 per cent to the national pulse basket. It is cultivated throughout the country excepting high altitudes and coastal region. However, the major production domain lies in the states of Madhya Pradesh, Rajasthan, Uttar Pradesh, Maharashtra, Karnataka and Haryana which contributes 93 percent towards total hectarage. It is also important for sustainable agriculture through its inherent capacity to fix atmospheric nitrogen in symbiotic association with *Rhizobium* and to harvest the limited available soil moisture because of its deep penetrating root system.

The production of chickpea has remained low or stagnant. The important factor causing low or stagnating production of chickpea in India are cultivation of these crops mostly on marginal and sub marginal lands of poor fertility, low or non adoption of package of improved production technology and susceptibility of most chickpea varieties to disease and pests. The crop has highly unstable performance due to many biotic and abiotic stresses. Through the role of plant protection is well documented in chickpea but there is a great scope to increase the total production through minimizing the losses caused by diseases. Hence there is a urgent need to implement the IDM technology, for effective management of diseases in chickpea. IDM strategy support absorption of traditional knowledge to facilitate reversing the negative impact of chemical pathway and increase the efficiency in crop production which will directly contribute to pesticide residue free food grains and food commodities.

The disease distribution of chickpea varies in different states. Climatic patterns and cropping system also affect their spread and intensity. State wise distribution of key diseases and nematodes is given below.

Table 27.1: Major Diseases of Chickpea and their Distribution

Diseases	*Causal Organism*	*States*
Fusarium wilt	*Fusarium oxysporum f. sp. ciceri*	U.P., Bihar, W.B. and Gujarat, M.P., Rajasthan, Maharashtra, Karnataka and A.P
Dry root rot	*Rhizoctonia bataticola*	U.P., Bihar, W.B. and Assam, M.P., Rajasthan, Maharashtra
Collar rot	*Sclerotium rolfsii*	M.P, Bihar, U.P.
Ascochyta blight	*Ascochyta rabies*	Delhi, Punjab, Haryana, West Rajasthan, West Bengal, U.P.
Botrytis gray mould	*Botrytis cinerea*	Tarai areas of U.P., Tal areas of Bihar
Root-knot nematodes	*Meloidogyne incognita*	Maharashtra, Bihar, Punjab, U.P., Gujarat and Rajasthan
Reniform nematodes	*Rotylenchulus reniformis*	U.P., M.P.

Key Diseases and their Diagnostic Details

Wilt

It is a typical vascular disease causing xylem browning or blackening. The disease is systemic in nature and the pathogen may infect plants at any growth stage. The affected seedlings collapse and lie on the ground. Grown up plants show typical wilting *i.e* drooping of petioles, rachis and leaflets. The leaves turn yellow and then light brown or straw colored. When the stem is split vertically, internal discoloration can be seen. Transverse cut of the stem shows blackening of xylem vessels. The pathogen is both soil and seed borne and survives to serve as source of inoculum in next season.

Collar Rot

The affected plants appear yellowish with drooping leaves. The disease is further characterized by the presence of water soaked lesions on the lower part of the plant in contact with the soil. White cottony mycelia growth on and inside the affected parts later on bears mustard like yellow circular structures, turning later into black hard sclerotia. In extreme cases the leaves dry and turn straw colored prematurely and the plant gets killed.

Dry Root Rot

Yellowing and sudden drying of the plants. Normally, the leaves of the affected plants may be straw colored and occasionally the lower leaves and stem may be brown. The tap root becomes dark brown, quite brittle in dry soil and often shows extensive rooting resulting in the loss of lateral roots. The fungus produces dark brown or black sclerotia on infected dead plant, if the plant are pulled from the soil and examined, the basal stem and the main roots may show dry rot symptoms.

Ascochyta Blight

Initial circular spots appear near the tip of the young shoots and top most leaves. Later on brown spot appear in other aerial parts such as leaves, petioles, flowers, pods, branches and stems. The apical twig, branches and stems often show girdling, lesions on pods are prominent and usually circular with dark margins.

Botrytis Gray Mould (BGM)

Initial symptoms are water soaking and softening of affected plant parts *viz.*, leaves, flowers and tender shoots. On the stem, the grey mould symptoms are gradually replaced by dark grey and black

Figure 27.1: Wilt of Chickpea

Figure 27.2: Dry Root Rot Of Chickpea

Figure 27.3: *Ascochyta* Blight of Chickpea

Figure 27.4: *Botrytis* Gray Mould of Chickpea

sporodochia under humid condition, all the flowers are attacked resulting in the complete failure of the crop. BGM affected pod either do not produce any seed or produce only small shrivelled seed.

Root-knot Nematode

They cause chlorosis, wilting, delay in flowering and fruit formation, reduction in plant size and yield etc. Distinct gall shaped nodules develop around the position where nematode places itself in the root of the plant. The infected portion of the root may also become curved and on the convex side of the curvature, large number of small thin roots may develop above the gall. Eggs are deposited under gelatinous matrix and the nematode completes several generations in a crop season.

Reniform Nematode

The female attains a kidney type shape and is a semi endoparasite of roots. It can complete multiple generations in each crop season. The nematode feed in cortex, pericycle, endodermis and phloem parenchyma and complete its life cycle in 24 to 30 days. It can delay the emergence of seedlings and seedling stands is reduced to 6 to 11 per cent at population of above 1 nematode per gram of soil. The plant succumbs to wilt at an early stage. Many a time nematode invades the roots, suppreses *Rhizobium* root nodulation and nitrogen fixation and can induce physiological disorders resulting in reduced plant vigour and low crop yield. Prasad (2006) worked on management of *R. reniformis* on Chickpea.

Table 27.1: Integrated Management of *Rotylenchus reniformis*

Treatments	*Before Treatments*	*30 Days After Treatments*	*Per cent Decrease Over Control*	*At Harvest*	*Per cent Decrease Over Control*
T_1 = *Trichoderma viride* @ 2 kg/ha	365.3 (19.06)	220.0 (14.83)	39.0	281.3 (16.77)	22.4
T_2 = *Lantana camara* @ 2 q/ha	413.3 (20.33)	, 213.3 (14.60)	54.4	293.3 (17.13)	32.7
T_3 = Phorate 10G @ 1 kg a.i./ha	380.0 (19.49)	180.3 (13.43)	54.5	266.6 (16.33)	31.1
$T_4 = T_1 + T_2$	360.0 (18.97)	198.3 (14.08)	33.3	233.3 (15.27)	23.7
$T_5 = T_1 + T_3$	410.0 (20.25)	224.0 (14.97)	50.8	290.0 (17.03)	32.7
$T_6 = T_2 + T_3$	417.3 (20.43)	194.3 (13.94)	60.9	293.3 (17.13)	33.8
$T_7 = T_1 + T_2 + T_3$	366.6 (19.15)	174.3 (13.20)	54.5	263.3 (16.23)	28.1
T_8 = Control	366.6 (19.15)	396.6 (19.91)	–	410.3 (20.26)	–
CD (P = 0.05)	NS	4.71			2.13

Values in parentheses are $\sqrt{n+1}$ (transformation).

Three components of INM such as a bioagent, *Trichoderma viride* @ 2 kg/ha; a weed, *Lantana camara* @ 2 q/ha and a nematicide, phorate 10G, @ 1 kg/ha were used in well infested field of I.A.R.I., New Delhi (initial pop about 3.1 nema/g). The field was well ploughed, and plots of 3×2m were prepared. Leaves and twigs of *L. camara* were applied one week before sowing (T_1), *T. viride* powder formulation was applied at the time of sowing of Chick pea (T_2) cv. Pusa-267 in lines and then, phorate 10 G was applied one week after germination of seeds (T_3). All the treatments were combined such as $T_1 + T_2(T_4)$; $T_1 + T_3(T_5)$; $T_2 + T_3(T_6)$; $T_1 + T_2 + T_3(T_7)$ and control (T_8), replicated thrice, soil samples were drawn before treatments, after 30 days of treatment and at harvest for estimation of nematode population.

Results on *R. reniformis* revealed that the population was non-significantly when sampled before the treatments. There were significant reduction of nematode population after 30 days of treatment, maximum being in treatments having two components such as *L. camara* + phorate (T_6), which gave 60.9 per cent reduction whereas phorate application three components (*T. viride* + *L. camara* + phorate) resulted 54.5 per cent decrease in both the treatments. *R. reniformis* population at harvest also followed the same trend, maximum reduction was noticed in T_6 treatment followed by T_5 (*T. viride* + phorate) and *L. camara* (T_2) resulted 32.7 per cent decrease of *R. reniformis* population, probably indicating that the toxic principles from *Lantana* released in soil that took more time to kill nematodes. Similarly, phorate immediately after application released toxic molecules to act fast and then, bio-agent took the part to reduce population. Prasad *et al.* (1997) reported that *Calotropis* resulted decrease of *M. arenaria* and increased groundnut yield. Similarly, phorate @ 1.5 kg ai/ha significantly reduced *M. incognita* on soybean (Prasad and Mittal, 2004).

Table 27.2: Environmental Factors Favourable for the Important Diseases

Wilt	✫ Incidence is generally higher in warmer(25ºC) and drier climates
	✫ Early sown crop suffers more than the ones sown late.
	✫ Alkaline soil favours the wilt incidence.
	✫ Disease is more if rotation is not practiced.
Root rot	✫ Minimum ambient temperatures above 20ºC and maximum above 30ºC and moisture stress.
Collar rot	✫ Presence of undecomposed organic matter on soil surface.
	✫ Warm temperatures and high soil moistures
Botrytis grey mould	✫ Dense planting and excessive canopy
	✫ High humidity
	✫ Fogy conditions
Ascochyta blight	✫ Wet weather with strong wind
	✫ Temperature around 22-26ºC and R.H of soil surface around 0-3 per cent
Root-knot nematode	✫ Sandy loam soils
Reniform nematode	✫ Irrigated soils

Development of IPM Package

IPM Components

IPM components include

1. Sowing a crop variety with resistance to a disease.
2. Modified farm management practices that result in reduction of disease damage *viz.* altering sowing date or sowing pattern, inter cropping etc.
3. The enhancement of natural control processes.
4. The judicious application of pesticides only when necessary.

Based on the above components a package can be developed for the management of chickpea diseases detailed below.

Step Wise IPM in Chickpea

Cultural Practices

1. Deep summer ploughing by soil turning plough to expose soil inhabiting, pathogens and nematodes population.
2. Training of the field bunds to destroy the existing rodent burrows
3. Apply recommended dose of NPK.
4. Adjustment of planting and harvesting time for avoidance of diseases. avoid early planting for controlling BGM and *Fusarium* wilt in the plains.
5. Follow crop rotation with non host crops like wheat, barley, linseed, mustard etc to minimize soil and seed borne disease.
6. Inter or mixed cropping of chickpea+ linseed +coriander (1:1) to reduce wilt and dry root rot incidence.

Table 27.3: Grow Resistant/Tolerant Varieties

Diseases	*Resistant/Tolerant Varieties*
For wilt:	Avrodhi, DCP 92-3, JG 74, JG 315, BG 372. Pusa 391, ICCV-10, BG244, Vijay,
For dry root rot:	Bharti, Kranti, Sadhawana, JG30. Avrodhi, JG315, K850,C-104, ICC12337, ICC12269
For *Ascophyta* blight:	BG 256,C 235, Pusa 261, Pusa 408, Pusa 413, Pusa 244, Pusa 372, MCK 54, Gaurav
For BGM:	Pusa 209, SAKI 9516, and Sweta

Seed Treatment with

1. *Trichoderma viride/Trichoderma harzianum* + Vitavax (@4+1 g/kg seed) against wilt and dry root rot disease.
2. Rovaral @ 2.5 g/kg of seed against *Ascochyta* blight and BGM were found effective.
3. *Rhizobium* culture @8 ml/packets/ha.

Mechanical Practices

1. Collect and destroy the affected plant parts after harvest by burning
2. Affected plant should be roughed out.

Bio Control Practices

1. In case of *Ascochyta* blight or *Botrytis* gray mould are observed spray mixed formulation of *Trichoderma harzianum* (TH)+ *Pseudomonas fluorescence* (Psf) (@ 5 g/litre) or carbendazim+ Psf (1+4 g/litre).

Chemical Practices

1. Apply Mancozeb 75 per cent WP @ 2.5 kg/ha for control of *Ascochyta* blight, BGM and rust disease.

Conclusion

With global perspectives of environmental protection and food security, integration of modern tools is must in traditional methods for restoration and use of resources judiciously to revolutionize pulse production. Moreover, issues on patenting warrants an urgent need to develop and exercise efficient tools indigenously by biodiversity rich developing countries. Hence to combat with various disease problem integration of technologies would be a better approach than to opt for a single one. In this context, farmer can play a crucial role in combating various disease problems.

References

Ali, S.S., 1995. *Nematode Problems in Chickpea*. Indian Institue of Pulses Research, Kanpur.

Ali., Masood, Kumar, Shiv and Singh, N.B., 2003. *Chickpea Research in India* (Edited). Indian Institute of Pulses Research, Kanpur, India, 334p.

Anon., 2003. *Annual Report*. National Centre for Integrated Pest Management, New Delhi.

Kumar, J., 2009. *Recent Advances in Plant Disease Management*. CAS, G.B. Pant University of Agriculture and Technology, Pantnagar, Compendium.221p

Gangawane, V.K and Khilare, V.C., 2009. *Crop Diseases: Identification and Management*. Daya Publishing House., New Delhi, 243p.

Kumar., Sanjeev, Upadhyay, J.P. and Roy, S., 2002. Effect of pesticide seed dressing and *Rhizobium* inoculation on nodulation and yield of Chickpea. *J. of Applied Biology*, Patna, 12(1–2): 81–83.

Kumar., Sanjeev and Upadhyay, J.P., 2003. Compatibility of *Rhizobium* and *Trichoderma in vitro* and *in vivo*. *RAU J. of Research,* Pusa, Bihar, 13(1–2): 61–63.

Kumar, Sanjeev and Singh, Y., 2010. Food security in chickpea by integrated disease and pest management. *Indian Farmer Digest*, GBPUA&T University, Pantnagar, p. 5–7.

Mittal, R.K., 2003. Diagnosis and management of diseases of pulse and oil seed crops in Uttarakhand hills. Vivekananda Parvatiya University Anusandhan Sansthan, Almora (Uttarakhand). *Technical Bulletin*, 20.

Prasad, D., Dubey, K.N. and Mittal, A., 1997. An integrated approach for management of plant parasitic nematodes in groundnut. *Ann. Pl. Protec. Sci.*, 5: 217–219.

Prasad, D. and Mittal, A., 2004. Effect of calotropis, oilcakes and phorate on growth of soybean and *Meloidogyne incognita* population. *Ann. Pl. Protec. Sci.*, 12: 234–235.

Singh, Amerika, Sardana, H.R. and Sabir, Naved, 2004. Validated IPM Technologies foe Selected Crops. National Centre for Integrated Pest Management, New Delhi, 253 p.

Plant Protection, Diversity and Conservation (2011) *Pages* **366–371**
Editors: **D. Prasad & Satya Kumar**
Published by: **BIOTECH BOOKS, NEW DELHI**

Chapter 28

Integrated Disease Management in Pigeonpea for Ensuring Food Security

Sanjeev Kumar

College of Agriculture, Ganj Basoda (JNKVV-Jabalpur), Vidisha – 464 221, M.P.

Pigeonpea (*Cajanus cajan*), popularly known as tur or arhar is the second important pulse crop after chickpea in india. It occupies about 15.14 per cent of the total area under pulses contributing about14.91 per cent of the total production. The major pigeonpea growing states are Maharashtra, Uttar Pradesh, Madhya Pradesh, Karnataka, Gujarat, Andhra Pradesh, Tamil Nadu and Bihar. Besides being rich source of protein (22.3 per cent), essential amino acids particularly lysine, this crop also help in maintaining the soil fertility through natural biological nitrogen fixation (Joshi and Srivastava, 2006). The ability of pigeonpea to produce economic yields in soils characterized by moisture deficiency, makes it an important crop of dry land agriculture. Farmers grow it in various production systems as a mixed crop, inter crop and perennial crop using long established traditional practices (Chauhan, 1990). In majority of the cases pigeonpea is seldom or never grown pure except on a very small scale. Pigeonpea as a sole crop is relatively inefficient because of its low initial growth rate and harvest index (Willey *et al.*, 1980).

The other major factor for low production of pigeonpea in India are ecological factors, lack of appropriate pulse production and protection technologies, poor postharvest management technologies, inadequate supply of quality seed to farmers, less thrust on basic research and socio economic constraints etc. Apart from these, the pest and disease problems are the major bottlenecks in realizing the higher yields (Dubey and Sharma, 2002). A wide array of diseases which have been extensively studied over the years are amenable to suggested management strategies but still claim a sizable toll on yield. In this category, vascular wilt, stem blight, sterility mosaic and *Alternaria* blight etc. are a few that can be quoted frequently.

Therefore, apart from facing the above mentioned constraints of pigeon pea production, mitigation of disease damage is one of the possible alternatives to boost the production of the country, especially when other than economy the nutritive value of the crop to the masses is of critical importance. In order to combat the losses caused by disease, farmers often resort to injudicious use pesticides leading to pest resurgence, resistance, pesticides residues and associated environmental problems. This has dictated the need to look for other alternatives like use of botanicals pesticides, bioagents, manipulation of cultural practices and lastly the need based use of eco friendly pesticides. Integrated disease management(IDM) combines a wide array of all the above methods and practices for the control of diseases and at the same time protects the environment in the long run as natural enemies are conserved and pesticide use are minimized. IDM is a systems approach, which is not only ecofriendly but also sustainable to the crop produce.

Key Diseases and their Diagnostic Details

Wilt (*Fusarium udum*)

It is widely distributed soil borne disease which starts appearing on 6 week old plants but more wilt is observed during the flowering and podding stages of crop. Infected seedlings show loss of turgidity, intervenial clearing chlorosis and drying. The old plants show gradual or sudden withering and drying of green foliage. The disease appears in patches in the field. The characteristic symptoms are browning of the xylem vessels in the form of streaks under the bark. In some cases dark purple band can be seen on the surface of stem also. Sometimes, specially at later stages of crop growth, the branches of affected plant may show die bark symptom along with discoloration of xylem vessels.

Phytophthora Blight (*Phytophthora drechsleri f sp. cajani*)

The disease may occur on seedlings and their death may go unnoticed. On grown up plants, brown to dark brown lesions appear on stem and branches. The lesions increase in size, girdle the stem resulting in drying and death of affected part above the girdle. The infected stems or branches may break at lesion site with strong wind or other force. Water soaked, brown circular to irregular lesions may be formed on leaves causing rapid blighting of foliage, if rain and cloudy weather continues.

The symptoms of this disease resemble with wilt due to which it has probably been mistaken for wilt disease in the past. However, both can be differentiated as (a) the leaves of plant affected by wilt turn yellow before drying while those of blight are light green with upward rolling followed by rapid drying,(b) leaves in case of wilt remain attached to plant while they shed in case of blight, and (c) frequent blackening of xylem tissues occurs in wilt and not in blight.

Sterlity Mosaic (Virus Transmitted by *Aceria cajani*)

It is caused by pigeonpea sterility mosaic virus transmitted through an eriophyid mite, *Aceria cajani*. The affected plant remains stunted with profuse branching bearing large number of small light green leaves due to which the plant appears bushy. The leaves show mosaic with dark and light green to yellowish patches. The branches are without flowers partially or completely. The diseased plant is easily spotted out in the field due to its yellowish and bushy appearance. The disease may cause heavy reduction in yield.

Alternaria Blight (*Alternaria tenuissima*)

It is considered as a most serious disease of post-rainy season crop of pigeonpea in north eastern india and epiphytotics of the disease has been reported in part of Bihar. The severe occurrence of the disease has been reported on july sown crop also. The symptoms of the disease appear as small brownish spots on leaves, which later increase in size and coalease each other resulting in blighting and defoliation. Spots may appear on pods and the fungus infects seed also.

Figure 28.1: *Pytophthora* Blight of Pigeonpea

Figure 28.2: Sterility Mosaic of Pigeonpea

Figure 28.3: Wilt of Pigeonpea

Integrated Disease Management (IDM)

Barring the wilt, diseases of pigeonpea are mostly area specific and also more than one diseases occur in the same field/area. For instance, in Uttar Pradesh, Bihar, Maharahtra, Gujarat and some Southern states of india, wilt and sterility mosaic are serious problems while in Madhya Pradesh and West Bengal, wilt and *Phytophthora* blight are common. *Alternaria* blight is the serious problem of Bihar and Eastern Uttar Pradesh. Hence, adoption of Integrated disease management(IDM) approach is pertinent.

IDM Component

It includes:

1. Sowing a crop variety with resistance/tolerant to a disease.
2. Modified farm management practices that result in reduction of disease damage.
3. The enhancement of natural control processes.
4. Need based application of pesticides.

Based on the above components a package can be developed for the management of pigeon pea diseases detailed below.

Stepwise IDM in Pigeonpea

Summer Ploughing

Due to summer ploughing, most of the crop residue which harbor the dormant mycelium/spores of the pathogen get exposed to the temperature above 40° C for several days. Thus most of the inoculums of soil borne pathogens (*Fusarium spp., Rhizoctonia spp, Phytophthora spp* and nematodes) automatically get killed under hot condition and their number are reduced.

Soil Solarization

In this practice, ploughed fields are covered with a thin polythene sheet for a period of 6-7 weeks during April – May, which results in rising of the soil temperature up to 60° C. The resultant high temperature is lethal to soil inhabiting pathogen, nematode and insects. This practice is very effective in controlling the *Heterodera cajani* and *Fusarium udum* provided the soil is irrigated before polythene covering. Solarization in conjunction with neem oil seed meal has been found very effective against wilt pathogen.

Plant Nutrients

Pigeonpea is a legume crop so it is believed that it requires less amount of nitrogen. Therefore, diammonium phosphate(DAP) @ 100 kg/ha is applied in the field which provides nitrogen and phosphorus. Application of 20 kg of K/ha reduces the incidence of *Phytophthora* blight. Manures and oil cakes reduces wilt incidence. It increases the population of beneficial microorganism and reduces the population of fungal and bacterial propagules. Sulphur as micronutrient plays an important role in the metabolism, which in turn increase the vigour and resistance of the plant. Use of sulphur @ 10-15 kg/ha also increased the yield.

Varieties

The improved pigeonpea varieties having good yield potential as well as resistant/tolerant to diseases are mentioned in Table 28.1.

Table 28.1: Disease Resistant/Tolerant Varieties of Pigeonpea

Sl.No.	*Resistant/Tolerant to*	*Improved Varieties*
1.	Wilt	Mukta, Maruti, Sharda BDN 1. 2, C11, JA4
2.	Sterility mosaic(SMD)	Bahar, Narendra Arhar-1, Shard, Jagrati, MA6, Azad
3.	Wilt+SMD	Asha, MA3, BSMR 736, Narendra Arhar-1
4.	*Alternaria* blight+SMD	Shard (DA11), Pusa-9

Seed Treatment

The main objective of IDM is to manage the diseases by minimizing the use hazardous chemicals like fungicides so as to avoid the environmental pollution. Seed treatment requires much less quantity of chemicals with comparatively high benefits. Hence, they are well included in the IDM system. Seed treatment with carbendazim+ thiram(1:2) @ 2.5 g/kg of seed reduces the incidence of wilt and *Alternaria* blight immensely. In *Phytophthora* prone area, metalaxyl should be used for seed treatment @ 2.5 g/kg of seed. Seed treatment with *Trichoderma* formulations @ 5 g/kg of seed reduces the incidence of wilt considerably.

Sowing Time and Season

Date of sowing and season also plays an important role in disease management. *Phytophthora* blight occurs less in pre monsoon sown crops as compared to late sown crops of which early plant stage is caught by cloudy weather and continuous rains. The disease occurs more on plants at the age of 15 to 20 days and the susceptibility decreases with advancing age, being minimum at 120 days. The pre rabi pigeonpea crop escapes the attack of *phytophthora* blight as well as wilt. On the contrary, *Alternaria* blight is serious on pre *rabi* crops as compared to rainy season crop. Early sowing of pigeonpea reduces the sterility mosaic incidence.

Method of Sowing

Ridge sowing helps in the better crop establishment than the flat bed sowing method. Ridges not only provide proper drainage but also avoid the water to come in contact of plants for the longer duration. Otherwise, water stagnated in the field facilitates congenial condition for sporangium formation, germination and liberation of zoospores. It also helps in free swimming of zoospore and infection of seedling to cause *Phytophthora* blight disease.

Crop Geometry

The crop geometry influences the crop growth as well as microclimate of the area. The closely spaced plants are more weak and prone to attack of *Phytophthora* blight and stem rot diseases, while widely spaced plant suffer less. On the other hand, less plant spacing (25cm x10cm) are known to reduce the incidence of wilt and sterility mosaic diseases. The optimum row to row spacing recommended for pigeon pea are 40cm, 50- 60cm and 75 cm for early, medium and late maturing varieties.

Inter/Mixed Cropping

Inter/mixed cropping provides diversity in the field and ultimately creates sequential and temporal discontinuity to the pathogen. Crops like sorghum, urdbean, sunhemp, mungbean, and sesame are grown as inter/mixed crop during cultivation. These crop are of short duration(60 to 65 day) and mature at different intervals and perform better under such conditions.In other words, the total economic gain under mixed crop is more than the monoculture of pigeon pea. Inter-cropping of sorghum with pigeonpea at row ratio of 2:2 has been found to reduce the incidence of early stage mortality, wilt and *phytophthora* blight of pigeonpea. Roots of sorghum exudates hydrogen cyanide(HCN) which inhibit the growth and reproduction of the pathogen.

Crop Rotation

Crop rotation with non host crop helps in reduction of soil borne diseases. Three to four year rotations with tobacco/sorghum and cereal crops reduces intensity of wilt and *Phytophthora* blight diseases.

Sanitation

The pathogens *Phytophthora, Fusarium, Sclerotium* and several others survive in the field on the infected plant debris. Therefore, removal and destruction of the infected plant debris after harvesting or at the time of field preparation, help in reducing the severity of diseases. The mite vector aswell as the virus of SDM survives on the perennial plants of pigeonpea growing in the field bund or near the houses. Such plant should be uprooted and destroyed to eliminate the inoculums. In the standing crop also, the diseased plants may be destroyed as soon as they appear in the field.

Chemical Control

In sterility mosaic prone areas, soil application of phorate@1kg a.i/ha or carbofuran 3 per cent @ 1.2 kg/ha at the time of sowing can protect the pigeonpea plant from infection of sterility mosaic diseases for the period of 75 days after sowing. Control of vector mites by foliar spray of Kelthane 0.15 or Metasystox 0.1 per cent *Alternaria* blight by foliar spray of Carbendazim @ 0.1 per cent reduces the spread of disease considerably.

Conclusion

Of late, pigeonpea production has shifted from a largely subsistence farming to maket oriented commercial production. However, it has not been possible to exploit the full genetic potential of high yielding pigeonpea varieties because of immense losses due to diseases and pests. Mangement of one single disease in isolation would prove counterproductive; hence, it calls for the management of all the diseases in comprehensive manner. An integrated disease management to combat the disease is the right approach.

References

Agrawal, S.C., 2006. Integrated management of diseases of pigeonpea. *Indian Farming*, New Delhi. p. 15–17.

Chauhan, Y.S., 1990. Pigeonpea: Optimum organic management. In: *The Pigeon Pea*, (Eds.) Y.L. Nene S.D. Hal and V.K. Sheila. CAB International, Wallingford, U.K., pp. 157–258.

Joshi, Nitin and Srivastava, C.P., 2006. Yield maximization in pigeon pea through crop protection measures. *Indian Farming*, New Delhi, p. 4–7.

Dorsami, L.S., 1940. The cultivar of pulses in Mysore state. *J. Mysore Agric. Experimental Union*, 19(1–4): 63–68.

Gangawane, V.K. and Khilare, V.C., 2009. *Crop Diseases: Identification and Management*. Daya Publishing House, New Delhi, 243 p.

Kumar, J., 2009. *CAS on Recent Advances in Plant Disease Management*. G.B. Pant University of Agriculture and Technology, Pantnagar, Compendium, 221p.

Kumar, Sanjeev, Upadhyay, J.P. and Rani, Arachana, 2009. Evaluation of *Trichoderma species* against *Fusarium udum* Butler causing wilt of pigeonpea *J. Biol. Control*, 23(3): 329–332.

Singh, Amerika, Sardana, H.R. and Sabir, Naved, 2004. *Validated IPM Technologies foe Selected Crops*. National Centre for Integrated Pest Management,New Delhi, 253 p.

Willey, R.W., Rao, M.R. and Natrajan, M., 1980. Traditional cropping systems with pigeonpea and their improvement. In: *Proc. International Workshop on Pigeonpea* (Vol. 1), held during15 -19 December at ICRISAT Centre, India, pp. 11–25.

Plant Protection, Diversity and Conservation (2011) *Pages* **372–386**
Editors: **D. Prasad & Satya Kumar**
Published by: **BIOTECH BOOKS, NEW DELHI**

Chapter 29

Potentiality of Rhizobacteria to Control Soil Borne Bacterial Diseases

Dinesh Singh

Division of Plant Pathology, Indian Agricultural Research Institute, New Delhi – 110 012

Bacterial Domain is numerically dominant component of most microbial communities on plants. The bacteria can attain densities as high as 10^9 cells/g of plant tissue of roots based on culturing and 10^9 cells/g based on cultivation independent methods. Besides bacteria, eukaryotic micro flora (filamentous fungi, yeast and algae) are also associated to the plants at different densities but lower than the prokaryotes. Plants offer a wide range of habitats that support microbial growth. These include sites that are moist and rich in nutrients and thus ideal for fostering bacterial growth as well as sites that are nutrient- poor or exposed to stressful environmental conditions. The surfaces of seeds, roots, leaves, and fruits are often harbour large, diverse bacterial communities, whereas blossom, stem and vascular tissues and intercellular spaces within plant tissues often are free of microorganism or support only limited bacterial communities.

The rhizosphere, volume of soil surrounding roots and influenced chemically, physically and biologically by plant root. It is a highly favourable habitat for the proliferation of bacteria. It also exerts a potential impact on plant health soil fertility (Sorensen, 1997). The members of the total rhizospheric bacteria that colonize the roots have been termed rhizobacteria by Kloepper and Schroth (1978). Root exudates rich in amino acids, monosaccharide and organic acids, serve as the primary source of nutrients and support the dynamic growth and activities of various microorganisms within vicinity of the roots. These bacteria may be beneficial, harmful or neutral to the plants (Dommergues, 1987) and their diversity remains dynamic with a frequent shift in community structure and species abundance (Kunc and Macura, 1988).

Diversity of Rhizobacteria

The diversity of rhizobacteria depends on its ability to live in a commensal relationship with plants, utilize nutrients exuded by roots and survive environmental stress by inhabiting protected sites provided by the architecture of the root. (Paulsen *et al.*, 2005) and those micro-sites formed by the combined intermingling of root and soil components. The composition of rhizobacterial communities in the rhizosphere is significantly affected by root borne carbohydrates (Newman and Romheld, 2002). It reflects the metabolic versality that allows rhizobacteria to compete with non-rhizosphere bacteria and dominate the rhizosphere environment (Nehl *et al.*, 1997). The rhizosphere inhabitants include a diverse array of rhizobacteria that differ widely in structure or taxonomy and function. Species of *Pseudomonas, Burkholderia, Aeromonas, Agrobacterium, Alcaligenes, Acenitobacter, Athrobacter Azoarcus, Azospirillum, Azotobacter, Bacillus, Chryseomonas, Clostridium, Klebsiella, Enterobacter, Flavobacterium, Gluconacetobacter Stenotrophomonas, Xanthomonas, Lysobacter, Pantoea* and *Serratia* are representative of members of the rhizobacterial community that have been identified in rhizosphere of numerous plants. Because, the similar species of rhizobacteria either may inhabit the rhizosphere, rhizoplane and endorhizal cells where as some rhizobacteria may colonize all three niches.

Root Colonization of Rhizobacteria

Rhizobacteria are aggressive rhizosphere colonizers because their growth is stimulated by root-exuded organic compounds, and root cell lysates and secretions (de Weger *et al.*, 1995; Sommers *et al.*, 2004). Plant exudes high levels of nutrients, many of which act as chemoattractants of bacteria. Indeed, a strong relationship exists between successful root colonization by rhizobacteria and their high growth rates on root derived carbon substrates (sommers *et al.*, 2004). Vital characteristics of rhizobacteria necessary for establishment, persistence, and interaction in the rhizosphere include motility, chemotaxis towards root exudes, and colonizing ability. Attachment to the plant surface is one of the initial steps in root colonization. Flagella and different types of pili are involved in attachment to diverse surfaces. Flagella of *Pseudomonas* are necessary for biofilm formation on different surfaces (O'Toole and Kolter, 1998 b). Pili (fimbriae) can have different functions among different organisms. For example, type 4 pili are known to be involved in twitching motility and the initial contact between bacteria and epithelial cell surfaces. Type 4 pili from *P. fluorescens* WCS365 are involved in efficient attachment and biofilm formation on PVC. It is also involved in competitive colonization of tomato root tip (Camacho, 2001).

Motility

It was reported that flagella- less mutants (no flagella present) do not have the ability to colonize the root tip efficiently (de Weger *et al.*, 1987). On the contrary, Howie *et al.* (1987) and Scher *et al.* (1988) stated that flagella less mutants are not impaired in colonization of wheat and soybean roots respectively. Motility in relation to root colonization has been reported to depend on the soil type, the plant and bacterial strains used (Weller and Thomashow, 1994). However, bacteria need functional flagella to migrate down with the growing root and to reach the deeper root parts such as the root tip.

Chemotaxis

Signal transduction pathways mediated by histidine kinase play a central role in information processing. Histidine kinase allows bacteria, plants and fungi to sense and respond to their environment. These signal transduction pathways are some times referred to as two component systems and they regulate cellular responses like chemotaxis, sporulation or microbial pathogenesis (Appleby *et al.*, 1996). Most histidine kinases are homodimers that use ATP to phosphorelate a specific histidine residue on the flanking submit within the dimer (Swanson *et al.*, 1993). By rotation of their flagella bacteria are able to swim and tumble. The bacterial chemotaxis pathway is encoded by a number of genes and has been best described for *Escherichia coli*. Genes necessary for chemotaxis and their function are listed in Table 29.1. This phosphorylation group is transferred to a specific aspartyl residue on a response regulation domain. The phophorylated response regulator acts directly by modifying effectors, which will turn lead to a change in cellular behaviour. A well studied signaling pathway that uses a two component system is bacterial chemotaxis. Bacterial chemotaxis acts upon sensing of attractants and repellents. Whenever an attractant is sensed by MCPs these membrane proteins are methylated by CheR. In this case, histidine kinase CheA is not able to phosphorylate its histidine residue. There is no interaction with the flagellar motor by response regulator CheY and flagella will rotate counterclockwise (CCW) causing the bacterial cell to swim smoothly and not change its direction of swimming. However, when the bacterial cell senses a repellent the MCPs are demethylated by CheB. In this case, CheA is able to phosphorylate it histidine residue and transfer this phosphoryl group to CheY. Phosphorylated CheY will interact with the flagellar motor causing a conformational change and the flagella will switch from CCW to clockwise (CW) rotation. CW rotation of flagella will result in tumbling of the cell, thereby changing the swimming direction of the cell. Tomato root exudates consist mainly of organic acids, sugars and amino acids (Simons *et al.*, 1997, Lugtenberg *et al.*, 1999). All of these components can function as potential nutrient sources for the bacteria present in the rhizosphere (Gamliel and Katan, 1992).

Utilization of specific nutrients is of importance for rhizosphere establishment of bacteria. It can be hypothesized that a directed movement towards exudates components is necessary for efficient competitive colonizing abilities and that specific components of the root exudates function as chemo-attractants. Flagella- driven chemotaxis and its specific chemo-attraction towards individual root exudates components were shown to be essential for competitive root colonization. Testing compound previously being identified to be present in tomato root exudates showed that WCS365 exhibit a chemotactic response towards some organic acids and some amino acids. The sugar present in the root exudates did not initiate a chemotactic response from the WCS365 cells. Comparision of the minimal concentrations required for a chemotactic response with concentrations estimated to be present in the root exudates suggested that malic acid and citric acid are among major chemo- attractants for *P. fluorescens* WCS365 cells in the tomato rhizosphere (de Weert *et al.*, 2002).

Table 29.1: Genes Involved in Bacterial Chemotaxis and their Mode of Action

Gene	*Protein*	*Mode of action*
cheA	Histidine kinase	Autophosphorylation and transfer of P- group to *CheY*
cheY	Response regulator	Interaction with flagellar motor
cheR	Methyltransferase	Methylation of methyl accepting chemotaxis proteins (MCPs)
cheW	Coupling protein	Coupling of transmembrane receptors to *CheA*
cheB	Methylesterase	Demethylation of MCPs
cheZ	Phosphatase	Dephosphorylation of *CheY*

Soil Borne Diseases in Plants Incited by Rhizobacteria

There are various group of bacteria such as Species of *Agrobacterium, Erwinia, Burkholderia, Ralstonia, Clavibacter* and *Streptomyces* causing diseases in the plants. They are mostly soil inhabitants and survive for the longer period. Some important rhizobactaria causing diseases in plants are given as given in Table 29.2.

Table 29.2: List of Soil Borne Important Bacterial Diseases of Plants

Sl.No.	*Name of Disease*	*Causal Organism*	*Host*
1.	Crown gall	*Agrobacterium* spp.	Dicotylendonous plants, include fruits and nuts crops, grapes ornamentals and landscape plants
2.	Wilt	*Erwinia tracheiphila*	Cucurbits
3.	Corm and pseudobulbrot	*Burkholderia gladioli*	Gladiolus
4.	Wilt and stem rot	*B. caryophylli*	Carnation
5.	Wilt	*Ralstonia solanacearum*	Solanaceous crops
6.	Corky roots	*Rhizomonas suberifaciens*	Lettuce
7.	Wilt and tuber rot (ring rot)	*Clavibacter michiganensis sub sp. sepedonicus*	Potato
8.	Wilt	*C. michiganensis subsp. insidiosus*	Alfalfa
9.	Common scab	*Streptomyces scabiei*	Potato
10.	Soil rot (pox)	*Streptomyces ipomoeae*	Sweet potato

Important Antagonistic Rhizobacteria

Bacillus cereus

Bacillus cereus strain UW85 is an effective biocontrol agent to suppress various diseases under diverse field conditions. It has potential to control bacterial wilt disease of solanaceous crops. This species is common and present in the soils and occurring at approximately 10^4 to 10^5 cfu/g soils. Bacteria are rod shaped, 1.0 – 1.2 x 3.0 – 5.0 µm, gram positive, non capsular, aerobic, motile and in chains. Endospores are central; form in 18–48h and 1.0 x 1.5 µm. Colonies are large, usually rough, flat, and whitish, form wavy arrangement of chains of cells that may develop into whip- like out growth. The mechanism of disease suppression has been extensively studied, and led to discovery of a novel antibiotic, zwittermicin A, that prevents normal development of several plant pathogens.

Bacillus subtilis

One of the most ubiquitous bacterial species found in agricultural systems is *Bacillus subtilis*. These bacteria can be readily cultured from soils, food stuffs, plant and animal tissues. Bacteria are rod shaped and seldom in chains. Endospores are oval, 0.8 x 1.5- 1.8 µm, thin walled, located centrally in the cell but not distend the walls. Cells are relatively small and 0.7- 0.8 x 2.0 – 3.0 µm in size. Flagella are found laterally. Colonies are whitish, round or irregular. Numerous strains of this species have been shown to express activities that suppress plant pathogens and promote plant growth. Several *Bacillus*-containing products are available to growers who are interested in using biological seed treatments or drench applications. Mechanism activities have been offered including direct antagonism of pathogens (*i.e.* through antibiosis and competition for food) and indirect stimulation of plant growth (*i.e.* by promoting nodulation and secreting growth factors). The GB03 strain is the active ingredient in one of the most widely distributed biofungicides (Kodiak, Gustafson LLC) which is applied in cotton and other field crops every year.

Burkholderia cepacia

The *Burkholderia cepacia* complex is a group of closely related, gram-negative, non-fluorescent pigment producing bacteria that include strains useful in biological control and bioremediation. Flagella are polar or multitrichous. Bacteria are straight or curved rods (0.5 – 1.0 x 1.5 – 4.0 μm), catalase positive and accumulate poly- β-hydroxybutyrate. Most notable are the production of antibiotics that are involved in the bacterium's ability to function as a biological control agent against many bacterial pathogens of the plants. Certain members of this group are naturally abundant in the plant rhizosphere and play an important role in managing root health. Antibiotic pyrrolnitrin is produced by *B. cepacia* (Hwang *et al.*, 2002).

Pseudomonas fluorescens

Pseudomonas fluorescens is a gram-negative rod shaped bacteria that inhabit soil, plants, and water surfaces. The Pf-5 strain resides in the plant's rhizosphere and produces a variety of secondary metabolites including antibiotics against soil borne plant pathogens (Paulsen *et al.*, 2005). *Pseudomonas fluorescens* PFO-1 is well adapted to the soil where it was first isolated in agricultural soil. *Pseudomonas fluorescens* strain SBW25 grow on plant leaves and roots where they can contribute to plant growth. Soluble, green fluorescent pigments are produced when the iron concentration is low. Production of secondary metabolites plays an important role in plant disease suppression. Antibiotics such as pyrrolnitrin, pyoluteorin, phinazine and 2, 4-diacetylphloroglucinol are produced by *Pseudomonas fluorescens* Pf-5, which inhibit growth of the pathogen. *Pseudomonas fluorescens* produces hydrogen cyanide and the siderophores like pyocheline and pyoverdine which suppress bacterial pathogens in rhizosphere.

Pantoea agglomerans Strain C9-1

Pantoea agglomerans is a member of the *Enterobacteriaceae* found commonly in association with plants or as saprophytes from soil, water and insects. Since *P. agglomerans* can compete successfully with the indigenous flora of a variety of microenvironments. Some examples are two peptide antibiotics termed herbicolin A and B that inhibit fungi and sterol-requiring mycoplasmas, novel β-lactam antibiotics, and a new class of antibiotics called pantocins that inhibit *Erwinia amylovora*. *P. agglomerans* strain C9-1, secretes at least three antibiotics (herbicolin O, I, and 2C) when grown in a chemically defined medium. C9-1 was isolated from apple tissue and has been used effectively as a biological control of fire blight. Many of the antibiotics produced by *P. agglomerans* are inactive in the presence of amino acids, such as histidine or arginine. Herbicolin O, which is not active in the presence of histidine, inhibits growth of several species of enteric bacteria and *Bacillus cereus*, whereas herbicolin 2C and I inhibit growth of only *E. amylovora* and *Staphylococcus aureus* (Ishimaru *et al.*, 1988, Johnson and Stockwell, 1988)

Agrobacterium radiobacter

The bacterium *Agrobacterium radiobacter* strain K84 occurs naturally in many types of soil and near plant roots. A. radiobacter is small, short rods 0.4- 0.8 x 1.0-3.0 μm, non- spore forming, typically motile by 1 -4 peritrichous flagella and gram –ve. Optimum temperature for growth of bacteria is 25–30°C, minimum 0°C, maximum 37°C and thermal death point is 51°C. As a pesticide active ingredient, it is used in nurseries and greenhouses for controlling the closely related bacterium *Agrobacterium tumefaciens*, which causes crown gall disease in plants. No adverse effects are expected to humans, other non-target organisms, or the environment when pesticidal products containing

Agrobacterium radiobacter strain K84 are applied. It can be applied by dipping or spraying, as appropriate, to germinating seeds, seedlings, cuttings, and roots and stems.

Rhizobactria Used as Antagonists

Rhizobacteria which exert a beneficial effect on the plant being colonized. Some of these these bacteria have antagonistic properties to control soil borne bacterial disease of plants. In most of these cases, biological control results from bacterial production of metabolites which directly inhibit the pathogen, such as antibiotics, hydrogen cyanide, iron chelating, siderophores and cell wall degrading enzymes. Some examples of bacterial antagonist which control plant diseases are given in Table 29.3.

Table 29.3: List of Bacterial Antagonist to Control Soil Borne Bacterial Diseases of Plants

Sl.No.	*Name of Antagonist*	*Mode of Action*	*Name of Pathogen*	*Name of Disease*
1.	*Agrobacterium radiobacter* K 84	Antibiosis	*Agrobacterium* spp.	Crown gall
2.	*Bacillus amyloliquefaciens, Azotobacter chrococcum*	Antibiosis	*Erwinia tracheiphila*	Wilt
3.	*Pseudomonas fluorescens, B. subtilis*	Antibiosis	*Ralstonia solanacearum*	Wilt
4.	*B. subtilis, Pseudomonas fluorescens*	Antibiotics and competition	*Streptomyces scabiei*	Common scab
5.	*B. subtilis, Pseudomonas fluorescens*	Antibiotics and competition	*S. ipomoeae*	Soil rot (pox)

Mechanism of Biocontrol in Rhizospheric Bacteria

An overview of biocontrol mechanism of rhizospheric bacteria are induction of systemic resistance, Niche exclusion and competition for nutrients, parasitism and predation, production of antimicrobial metabolites. The detail of these mechanism are given below:

Induction of Systemic Resistance

Plants show immune response towards a pathogen is known as systemic acquired resistance (SAR). This type of defense is characterized by the accumulation of salicylic acid (SA), which in turn induces a set of genes encoding pathogenesis related (PR) proteins in the plant. Although SAR and ISR both result in resistance against pathogens from the plant. There is no indication that these two pathways have an overlap (van Wees *et al.*, 2000). The ISR response is induced by non – pathogenic rhizobacteria and does not result in SA accumulation or in expression of PR proteins. When ISR is induced the plant produces jasmonic acid and ethylene. It was shown that several *Pseudomonas* spp. are able to induce ISR in a wide range of plants towards different pathogens. For example, *P. fluorescens* strain WCS417 is able to induce ISR in *A. thaliana* ecotype *Columbia* against the pathogen *P. syringae* pv. *tomato*. The signals of different *Pseudomonas* spp. that induce ISR have not yet been completely elucidated but there are indications that bacterial surface structures such as flagella and lipopolysaccharide can be involved (Bakker and Schippers, 1995). Examples of specific rhizobacterial strains in induction of systemic resistance to more than one species of plants as in given in Table 29.4.

Table 29.4: Induction of Systemic Resistance in the Plants Induced by Rhizobacteria.

Bacterial Strains	*Plant Species*
Bacillus amyloliquefaciens IN937a	Arabidopsis, Cucumber, pearl millet, pepper, tomato
B. pumilus INR7	Cucumber, tomato
B. subtilis GB03	Cucumber, pearl millet, pepper, tomato
Pseudomonas aeruginosa 7NSK2	Arabidopsis, bean, tobacco, tomato
P. fluorescens WCS417	Arabidopsis, bean, radish, tomato
Serratia marcescens 90- 166	Arabidopsis, cucumber, tobacco, pine

Niche Exclusion and Competition for Nutrients

The biocontrol agents should have ability to establish themselves in rhizosphere. They must be able to compete with the indigenous microflora, including pathogens. A rapid establishment of the bioagent on the root and competition for the exuded nutrients has been suggested as an important mechanism for biological control. The cell- cell junction are the preferably sites on the root where bioagent as well as pathogens colonise the root system (Bolwerk *et al.*, 2003). These sites are assumed to be the locations where nutrients leak from the root. An overlap of preferably niches for colonization is presumably the basis for competition in the rhizosphere. An efficient and fast colonization is required to be able to prevent the establishment of the pathogen on the root system and its subsequent infection. There is also competition for other components such as iron ions can take place (Buyer and Leong, 1986). Iron is often used as a cofactor in enzymes that are involved in pathways essential for microbial growth. In the rhizosphere, there is often a limitation for soluble Fe^{3+}. Rhizobacteria mostly aerobic and facultative anaerobic respond to iron limitation by producing iron chelating compounds which known as siderophores. Siderophores have fluorescent properties resulting in a yellow- green colour. Siderophores are the extra cellular low molecular weight (500- 1000 Daltons iron (III)- transport agents, which selectively make complex with iron (III) (Fe^{3+} ions) with very high affinity. Examples of siderophores produced by rhizobacteria is given in Table 29.5. The ability to produce siderophores or to take up related siderophores produced by other microorganism in the rhizosphere results in a competitive advantages and an iron deficiency for pathogens. This inhibits the growth of pathogens and eventually lead to the suppression of the disease.

Table 29.5: Distribution of Siderophores in Bacteria

Species of Bacteria	*Siderophore*
Actinomyces sp.	Ferrioxamines
Agrobacterium tumefaciens	Agrobactin
Arthrobacter sp.	Arthrobactin
Bacillus megaterium	Schizokinen
Pseudomonas sp.	Pseudobactin, Pyochelin, Pyoverdine, Terribactin
Mycobacteria	Mycobactins

Parasitism and Predation

In bacterial antagonistic, parasitism and predation mechanism is not much common. However, they produce hydrolytic enzymes which shown to posses biocontrol abilities (Markovich and

Kononova, 2003). For an effective suppression of a plant disease root colonization is important to reach the side to compete with the pathogen.

Production of Antimicrobial Metabolites

Antibiosis is defined as antagonism mediated by specific or non-specific metabolites of microbial origin by lytic agents, enzymes, volatile compounds or other toxic substances. Antibiosis plays an important role in biological control. Antibiosis is a situation where the metabolites are secreted by under ground parts of the plants, soil bacteria, plant residues etc. it occurs when the pathogen is inhibited or killed by metabolic products of the antagonists. For example fluorescent pseudomonads are well known to produce a large diversity of antimicrobial secondary metabolites which are responsible for their biocontrol capacity (Raaijmakers *et al.*, 2002). Antibiotic production by some strains of fluorescent pseudomonads is now recognized as an impotant mechanism in disease control. These antibiotics are phenazines, pyoluteorin, topolone, pyocyanin, 2,4-diacetyl phloroglucinol (DAPG) and pyrrolnitrin. Tropolone have broad spectrum activity against many bacteria. The role of individual antibiotic compounds in suppression of root pathogens has been clearly established using mutation analysis and molecular genetic tools and purified antibiotic compounds. For example, DAPG over producing mutants of *P. fluorescens* offered a better protection against bacterial wilt in tomato, though the root colonization pattern remained the same (Hongyou *et al.*, 2005). Hass and Keel (2003) isolated genes encode the enzymes responsible for synthesis of few above mentioned antibiotics and studied their regulation both at transcriptional and post transcriptional stages to facilitate the genetic improvement of antibiotic production.

Bacteriocins are antibiotic- like compounds with bactericidal specificity, which restrict bacterial strains closely related to the bacteriocin producer. For example avirulent non bactriocin- producing strains of *Ralstonia solanacearum* were unable to protect tomato plants from infection by *R. solanacearum* in *green* house test, where as the bacteriocin producing avirulent strains gave protection against the virulent *R. solanacearum*.

Plant Health: Plant Growth Promoting Rhizobacteria (PGPR)

Rhizospheric bacteria that favorably affect plant growth and yield of many commercially important crops are designated plant growth promoting rhizobacteria (PGPR). PGPR can improve seed germination, root development, mineral nutrition and water utilization (Phytostimulation). The use of PGPR has become rather common practice and is on the rise in many regions of the world. Inoculation with PGPR also aims at reducing the application of potentially polluting chemical fertilizers and pesticides. Various inoculants (bacteria with a suitable carrier) referred to as biofertilizers, phytostimulators, and biopesticides, are commercially available (Nelson, 2004).

Plant growth in agricultural soils is influenced by a myriad of abiotic and biotic factors. While growers routinely use physical and chemical approaches to manage the soil environment to improve crop yields, the application of microbial products for this purpose is less common. The region around the root, the rhizosphere, is relatively rich in nutrients, due to the loss of as much as 40 per cent of plant photosynthates from the roots (Lynch and Whipps, 1991). The importance of rhizosphere microbial populations for maintenance of root health, nutrient uptake, and tolerance of environmental stress is now recognized (Bowen and Rovira, 1999, Cook, 2002). These beneficial microorganisms can be a significant component of management practices to achieve the attainable yield, which has been defined as crop yield limited only by the natural physical environment of the crop and its innate genetic potential (Cook, 2002).

The prospect of manipulating crop rhizosphere microbial populations by inoculation of beneficial bacteria to increase plant growth has shown considerable promise in laboratory and greenhouse studies, but responses have been variable in the field (Bowen and Rovira, 1999). The potential environmental benefits of this approach, leading to a reduction in the use of agricultural chemicals and the fit with sustainable management practices, are driving this technology. Recent progress in our understanding of the biological interactions that occur in the rhizosphere and of the practical requirements for inoculant formulation and delivery should increase the technology's reliability in the field and facilitate its commercial development.

Mechanisms of Action

PGPR enhance plant growth by direct and indirect means, but the specific mechanisms involved have not all been well-characterized (Glick, 1995, Kloepper, 1993). Direct mechanisms of plant growth promotion by PGPR can be demonstrated in the absence of plant pathogens or other rhizosphere microorganisms. While indirect mechanisms involve the ability of PGPR to reduce the deleterious effects of plant pathogens on crop yield. PGPR directly enhance plant growth by a variety of mechanisms: fixation of atmospheric nitrogen that is transferred to the plant, production of siderophores that chelate iron and make it available to the plant root, solubilization of minerals such as phosphorus, and synthesis of phytohormones (Glick, 1995). Direct enhancement of mineral uptake due to increases in specific ion fluxes at the root surface in the presence of PGPR has also been reported (Bashan and Levanony, 1991). PGPR strains may use one or more of these mechanisms in the rhizosphere. Molecular approaches using microbial and plant mutants altered in their ability to synthesize or respond to specific phytohormones have increased our understanding of the role of phytohormone synthesis as a direct mechanism of plant growth enhancement by PGPR (Glick, 1995, Rainey, 1999). PGPR that synthesize auxins and cytokinins or that interfere with plant ethylene synthesis have been identified (Glick, 1995, Rainey, 1999).

Synergistic Action of PGPR, Rhizobia and Arbuscular Mycorhizal (AM) Fungi

Among the different bioinoculants, it is proved as potential alternatives to chemical fertilizers. *Rhizobium* and AM are also important group of rhizobacteria to improve plant health. It became necessary to evaluate the specific functional capability of PGPR with rhozobia and AM fungi to develop multifunctional based bioinoculants. Barea *et al.* (2005) studied the diversity, dynamics and significance of rhizosphere microbial populations and their co-operative activities. In several of these studies a positive co-operation was observed between PGPR, rhizobia and/or AM fungi by various workers as given in Table 29.6.

Challenges in Selection and Characterization of PGPR and Antagonists

One of the challenges in developing PGPR for commercial application is ensuring that an effective selection and screening procedure is in place. Hence, the most promising organisms are identified and brought forward. In the agricultural chemical industry, thousands of prospective compounds are screened annually in efficient high-throughput assays to select the best one or two compounds for further development. Similar approaches are not yet in place for PGPR. Effective strategies for initial selection and screening of rhizobacterial isolates are required. It may be important to consider host plant specificity or adaptation to a particular soil, climatic conditions or pathogen in selecting the isolation conditions, and screening assays (Bowen and Rovira, 1999, Chanway *et al.*, 1989). One approach for selection of organisms with the potential to control soil-borne phytopathogens is to isolate from soils that are suppressive to that pathogen (Weller *et al.*, 2002). Other approaches involve selection based on traits known to be associated with PGPR such as root colonization, 1-aminocyclopropane-1-carboxylate (ACC) deaminase activity (Cattelan *et al.*, 1999, Glick, 1995), and antibiotic and siderophore production (Cattelan *et al.*, 1999). The development of high throughput assay systems and effective bioassays will facilitate selection of superior strains (Mathre *et al.*, 1999,). An important aspect of colonization is the ability to compete with indigenous microorganisms already present in the soil and rhizosphere of the developing plant. Phenotypic and genotypic approaches are now available to characterize rhizobacterial community structure. Phenotypic methods that depend on the ability to culture microorganisms like standard plating methods on selective media, community level physiological profiles (CLPP) using the BIOLOG system (Garland, 1996), phospholipid fatty acid (PLFA) (Tunlid and White, 1992), and fatty acid methyl ester (FAME) profiling (Germida *et al.*, 1998). Culture-independent molecular techniques are based on direct extraction of DNA from soil and 16S-rRNA gene sequence analysis, bacterial artificial chromosome or expression cloning systems (Rondon *et al.*, 1999). These are providing new insight into the diversity of rhizosphere microbial communities, the heterogeneity of the root environment, and the importance of environmental and biological factors in determining community structure (Baudoin *et al.*, 2002, Smalla *et al.*, 2001). These approaches can also be used to determine the impact of inoculation of plant growth-promoting rhizobacteria on the rhizosphere community (Steddom *et al.*, 2002).

Table 29.6: Beneficial Association of PGPR and Rhizobium/AM Fungi on Plant Growth

PGPR Strain	*Rhizobium/AM Strain*	*Benefit as an Increase*	*Reference*
Pseudomonas fluorescens 92rk	*G. mosseae* BEG12	The growth of tomato plants	Gamalero *et al.*, 2004
Pseudomonas striata	*Bradyrhizobium* sp. and *G. fasciculatum*	N_2 and Puptake by green gram	Zaidi *et al.*, 2004
Pseudomonas sp.	*Mesorhizobium* sp. *cicer*	Increase shoot dry weight by 3-4 fold and N_2 uptake	Goel *et al.*, 2002
Azospirillum sp; *Pseudomonas striata*	*Rhizobium* sp.	Nodule number and dry weight, plant height, dry weight and yield in pigeonpea	Devanand *et al.*, 2002
Pseudomonas spp.	*R. leguminosarum* bv. *viceae*	The shoot height, root length and dry weight of pea	Kumar *et al.*, 2001
Bacillus circulans	*Glomus* sp.	N_2 and P uptake and grain and straw yield of wheat	Singh and Kapoor, 1998

Source: Appa Rao and Kishore, 2006.

Challenges in Field Application of PGPR and Antagonists

The application of PGPR for control of bacterial pathogens in greenhouse systems and other control conditions shows considerable promise, due in part to the consistent environmental conditions and high incidence of disease in greenhouses. Achieving consistent performance in the field where there is heterogeneity of abiotic and biotic factors and competition with indigenous organisms is more difficult. Knowledge of these factors can aid in determination of optimal concentration, timing and placement of inoculant, and of soil and crop management strategies to enhance survival and proliferation of the inoculant (Bowen and Rovira, 1999). The concept of engineering or managing the rhizosphere to enhance PGPR function by manipulation of the host plant, substrates for PGPR, or through agronomic practices, is gaining increasing attention (Bowen and Rovira, 1999, Mansouri *et al.*, 2002). Development of better formulations to ensure survival and activity in the field and compatibility with chemical and biological seed treatments is another area of focus; approaches include optimization of growth conditions prior to formulation and development of improved carriers and application technology (Bashan, 1998, Bowen and Rovira, 1999).

Challenges in Commercialization of PGPR and Antagonists

Prior to registration and commercialization of PGPR products, a number of hurdles must be overcome (Fravel *et al.*, 1999). These include scale up and production of the organism under commercial fermentation conditions while maintaining quality, stability, and efficacy of the product. Formulation development must consider factors such as shelf life, compatibility with current application practices, cost, and ease of application. Health and safety testing may be required to address such issues as non-target effects on other organisms including toxigenicity, allergenicity and pathogenicity, persistence in the environment, and potential for horizontal gene transfer. The product claim, whether as a fertilizer supplement or for biological control, will determine to which federal agency applications for registration should be addressed. Some commercially available products of plant growth promoting rhizobacteria in the market as given in Table 29.7.

Table 29.7: Commercially Available PGPR Strains that Essentially Act through Direct Plant Growth Promoting Mechanism

Trade Name	*PGPR Strains*	*Manufacturer*	*Recommended Application*
Bioboost	*Delftia acidovorans*	Brett- young seeds Ltd., Manotoba	Seed treatment in canola
Bioplin	*Azotobacter* spp.	Kumar Krishi Mitra Bioproducts Pvt. Ltd, Pune, India	Soil drenching for sunflower, tomato and other vegetable crops
Bioyield	*Bacillus* spp.	Guftason, LLC, Plano, Tx	Seed treatment in tomato, tobacco, cucumber and pepper
Compete	*Bacillus, Pseudomonas* and *Streptomyces* spp.	Plant health care, BV, CA Vught	Soil drenching for turfgrass, nursery and greenhouse plantations.
Kodiak	*B. subtilis* GB03	Gustafson, LLC, Dallaas, Tx	Seed treatment in fruit and vegetables

Source: Appa Rao and Kishore, 2006.

As described briefly in this article, rhizobacteria can have diverse interactions with plant roots. It is good opportunity to convert this interaction into practical benefits for crops by the introduction of specific PGPR or antagonistic bacteria or by the targeted enhancement of components of the indigenous soil bacterial community. The future challenge will be to integrate these approaches to provide an economically significant level of induced suppressiveness against crop pathogens.

References

Appa Rao, P. and Kishore, G.K., 2006. Plant growth promoting rhizobacteria. In: *Plant Associated Bacteria,* (Ed.) S.S. Gnanamanickam. Springer, The Netherlands.

Appleby, J.L., Parkinson, J.S. and Bourret, R.B., 1996. Signal transduction via the multi-step phosphorelay: not necessarily a road less traveled. *Cell,* 86: 845 – 848.

Bakker, A.W. and Schippers, B., 1995. induction of systemic resistance against Fusarium wilt of radish by lipopolysaccharides of *Pseudomonas fluorescens*. *Phytopathol.*, 85: 1021 – 1027.

Barea, J.M., Pozo, M.J., Azcon, R. and Azcon-Aguilar, C., 2005. Microbial co-operation in the rhizosphere. J. *Exptl. Botany*, 56: 1761– 1778.

Bashan, Y., 1998. Inoculants of plant growth-promoting bacteria for use in agriculture. *Biotechnol. Adv.*, 16: 729–770.

Bashan, Y. and Levanony, H., 1991. Alterations in membrane potential and in proton efflux in plant roots induced by *Azosprillum brasilense. Plant and Soil*, 137: 99–103.

Baudoin, E., Benizri, E., and Guckert, A., 2002. Impact of growth stage on the bacterial community structure along maize roots, as determined by metabolic and genetic fingerprinting. *Appl. Soil Ecol.*, 19: 135–145.

Bolwerk, A., Lagopodi, A.L., Wijfjes, A.H.M., Lamers, G.E.M., Chin-A-Woeng, T.F.C., Lugtenberg, B.J.J. and Bloemberg, G.V., 2003. interactions in the tomato rrhizosphere of two *Pseudomonas* biocontrol strains with the phytopathogenic fungus *Fusarium oxysporum* f. sp. *radicis lycopersici. Mol. Plant Microbe. Interact.*, 16: 983–993.

Bowen, G.D. and Rovira, A.D., 1999. The rhizosphere and its management to improve plant growth. *Adv. Agron.*, 66: 1–102.

Buyer, J.S. and Leong, J., 1986. Iron transport-mediated antagonism between plant growth-promoting and plant-deleterious *Pseudomonas* strains. *J. Biol. Chem.*, 261: 791–794.

Camacho, M.M., 2001. Molecular characterization of the role of Type 4 pili, NDH–1 and PyrR in rhizosphere colonization of *Pseudomonas fluorescens* WCS365. *Thesis*, Leiden, University.

Cattelan, A.J., Hartel, P.G. and Fuhrmann, J.J., 1999. Screening for plant growth-promoting rhizobacteria to promote early soybean growth. *Soil Sci. Soc. Am. J.*, 63: 1670–1680.

Chanway, C.P., Nelson, L.M. and Holl, F.B., 1989. Cultivar-specific growth promotion of spring wheat (*Triticum aestivum* L.) by co-existent *Bacillus* species. *Can. J. Microbiol.*, 34: 925–929.

Cook, R.J., 2002. Advances in plant health management in the twentieth century. *Ann. Rev. Phytopathol.*, 38: 95–116.

Davanand, B.J., Patil, A.B., Kulkarni, J.H. and Algawadi, A.R., 2002. Effect of plant growth promoting rhizobacteria on growth and yield of pigeonpea (*Cajanus cajan* L.) cultivars. *Karnataka J. Agril. Sci.*, 15: 653–666.

de Weert, S., Vermeiren, H., Mulders, I. H. M., Kuiper, I., Hendrickx, N., Bloemberg, G.V., Venderleyden, J., De Mot, R. and Lugtenberg, B.J.J., 2002. Flagella-driven, chemotaxis towards exudates components is an important trait for tomato root colonization by *Pseudomonas fluorescens. Mol. Plant–Microbe Interact.*, 15: 1173–1180.

de Weger, L.A., van der Bij, A.J., Dekkers, L.C., Simons, M., Wijffelman, C.A. and Lugtenberg, B.J., 1995. Colonization of the rhizosphere of crop plants by plant-beneficial *pseudomonads. FEMS Microbiol. Ecol.*, 17: 221–228.

de Weger, L.A., van der Vlugt, C.I., Wijfjes, A.H.M., Bakker, P.A.H.M., Schippers, B. and Lugtenberg, B.J.J., 1987. Flagella of a plant-growth-stimulating *Pseudomonas fluorescens* strain are required for colonization of potato roots. *J. Bacteriol.*, 169: 2769–2773.

Dommergues, Y.R., 1978. The interaction of higher plants and soil microorganisms. II. Effect of byproducts of plant growth and activity of fungi and actinomycetes. *Soil Science*, 52: 395–413.

Fravel, D.R., Rhodes, D.J. and Larkin, R.P., 1999. Production and commercialization of biocontrol products. In: *Integrated Pest and Disease Management in Greenhouse Crops*, (Eds.) R. Albajes, M.L. Gullino, J.C. van Lenteren, and Y. Elad. Kluwer Academic Publishers, Dordrecht, pp. 365–376 .

Gamalero, E., Trotta, A., Massa, N., Coppeta, A., Martinotti, M.G. and Berta, G., 2004. Impact of two fluorescent pseudomonads and an arbuscular mycorhizal fungus on tomato plant growth, root architecture and P acquisition. *Mycorrhiza*, 14: 185–192.

Garland, J.L., 1996. Patterns of potential C source utilization by rhizosphere communities. *Soil Biol. Biochem.* 28: 223–230.

Germida, J.J., Siciliano, S.D., de Freitas, J.R., and Seib, A.M., 1998. Diversity of root-associated bacteria associated with field–grown canola (*Brassica napus* L.) and wheat (*Triticum aestivum*). *FEMS Microbiol. Ecol.*, 26: 43–50.

Geol, A., Singhu, S. and Daderwal, K., 2002. stimulation of nodulation and plant growth of chickpea (*Cicer arietinum* L.) by *Pseodomonas* spp. antagonistic to fungal pathogens. *Biol. and Fertility of Soil*, 36: 391–396.

Glick, B.R., 1995. The enhancement of plant growth by free-living bacteria. *Can. J. Microbiol.*, 41: 109–117.

Hass, D. and Keel, C., 2003. Regulation of antibiotic production in root colonized *Pseudomonas* spp. and relevance for biological control of plant disease. *Ann. Rev. Phytopathol.*, 41: 117–153.

Hongyou, Z., Hailei, W., Xili, L., Ye, W., Liqun, Z. and Wenhua, T., 2005. improving bicontrol activity of Pseudomonas fluorescens through chromosomal integration of 2, 4-diacetyl-phloroglucinol biosynthesis genes. *Chinese Sci. Bull.*, 50: 775–781.

Howie, W.J., Cook, R.J. and Weller, D.M., 1987. effects of soil matric potential and cell motility on wheat root colonization by fluorescent pseudomonads suppressive to take all. *Phytopathol.*, 77: 286–292.

Hwang, J., Chilton, W.S. and Benson, D.M., 2002. Pyrrolnitrin production by *Burkholderia cepacia* and biocontrol of Rhizoctonia stem rot of poinsettia. *Biological Control*, 25(1): 56–63

Ishimaru, C.A., Klos, E.J. and Brubaker, R.R., 1988. Multiple antibiotic production by *Erwinia herbicola. Phytopathology*, 78: 746–750

Johnson, K.B. and Stockwell, V.O., 1998. Management of fire blight: A case study in microbial ecology. *Annu. Rev. Phytopath.*, 36: 227–248.

Kloepper, J.W., 1993. Plant growth-promoting rhizobacteria as biological control agents. In: *Soil Microbial Ecology: Applications in Agricultural and Environmental Management*, (Eds.) F.B. Metting, Jr., ed. Marcel Dekker Inc., New York, USA, pp. 255–274.

Kloepper, J.W. and Schroth, M.N., 1978. Plant growth-promoting rhizobacteria on radishes. In: *Proc 4th Int. Conf. Plant Pathol. Bact.*, Angers, France, pp 879–882.

Kumar, B.S.D., Berggren, I. and Martensson, A.M., 2001. Potential for improving pea production by co-inoculation with *Pseudomonas* and *Rhizobium*. *Plant and Soil*, 229: 25–34.

Kunc, F. and Macura, J., 1988. Mechanisms of adaptation and selection of microorganisms in the soil. In: *Soil Microbial Associations*, (Ed.) V. Vancura and F. Kunc. Elsevier, Amsterdam, pp. 281–299.

Lugtenberg, B.J.J., Kravchenko, L.V. and Simons, M., 1999. Tomato seed and root exudates sugars: Composition, utilization by *Pseudomonas* biocontrol strains and role in rhizosphere colonization. *Environ. Microbiol.*, 1: 439–446.

Lynch, J.M. and Whipps, J.M., 1991. Substrate flow in the rhizosphere. In: *The Rhizosphere and Plant Growth*, (Eds.) D.L. Keister and B. Cregan. Beltsville Sympos. in Agric. Res. 14. Kluwer, Dordrecht, The Netherlands, pp. 15–24.

Mansouri, H., Petit, A., Oger, P., and Dessaux, Y., 2002. Engineered rhizosphere: the trophic bias generated by opine–producing plants is independent of the opine type, the soil origin, and the plant species. *Appl. Environ. Microbiol.*, 68: 2562–2566.

Markovich, N.A. and Kononova, G.L., 2003. Lytic enzymesof *Trichoderma* and their role in plant defense from fungal diseases: A review. *Appl. Biochem. Microbiol.*, 39: 341–351.

Mathre, D.E., Cook, R.J. and Callan, N.W., 1999. From discovery to use. Traversing the world of commercializing biocontrol agents for plant disease control. *Plant Dis.*, 83: 972–983.

Nehl, D.B., Allen, S.J. and Brown, J.F., 1997. Deleterious rhizosphere bacteria: An integrating perspectives. *Appl. Soil Ecol.*, 5: 1– 20.

Nelson, L.M., 2004. Plant growth promoting rhizobacteria (PGPR): Prospects for new inoculants. Online. Crop Management doi: 10.1094/CM–2004–0301–05–RV.

O'Toole, G.A. and Kotler, R., 1998. Flagellar twitching motility are necessary for *Pseudomonas aeruginosa* biofilm development. *Mol. Microbiol.*, 30: 295–304.

Paulsen, I.T., Press, C.M. and Ravel, J., 2005. Complete genome sequence of the plant commensal *Pseudomonas fluorescens* Pf–5. *Nature Biotechnol.*, 23: 873–878.

Raaijmakers, J., Weller, D.M. and Thomashow, L.S., 2002. Frequency of antibiotic producing *Pseudomonas* spp. in natural environments. *Appl. and Environ. Microbiol.*, 63: 881–887.

Rainey, P.B., 1999. Adaptation of *Pseudomonas fluorescens* to the plant rhizosphere. *Environ. Microbiol.*, 1: 243–257.

Rondon, M.R., Goodman, R.M. and Handelsman, J., 1999. The earth's bounty: Assessing and accessing soil microbial diversity. *Trends Biotechnol.*, 17: 403–409.

Scher, F.M., Kloepper, J.W., Singleton, C., Zaleski, I. and Laliberte, M., 1988. colonization of soybean roots by Pseudomonas and Serratia species: relationship to bacterial motility, chemotaxis and generation time. *Phytopathol.*, p. 1055–1059.

Simons, M., Permentier, H.P., de Weger, L.A., Wijffelman, C.A. and Lutgenberg, B.J.J., 1997. Amino acid synthesis is necessary for tomato root colonization by *Pseudomonas fluorescens* strain WCS365. *Mol. Plant- Microbe Interaction*, 10: 102–106.

Singh, S. and Kapoor, K.K., 1998. Inoculation with phosphate-solubilizing microorganisms and a vesicular-arbuscular mycirrhizal fungus improves dry matter yield and nutrient uptake by wheat grown in a sandy soil. *Biol. and Fertility of Soil*, 28: 139–144.

Smalla, K., Wieland, G., Buchner, A., Zock, A., Parzy, J., Kaiser, S., Roskot, N., Heuer, H., and Berg, G., 2001. Bulk and rhizosphere soil bacterial communities studied by denaturing gradient gel electrophoresis: Plant-dependent enrichment and seasonal shifts revealed. *Appl. Environ. Microbiol.*, 67: 4742–4751.

Somers, E., Vanderleyden, J. and Srinivasan, M., 2004. Rhizosphere bacterial signaling: A love parade beneath out feet. *Crit. Rev. Microbiol.*, 30: 205–240.

Sorensen, J., 1997. The rhizosphere as a habitat for soil microorganisms. In: *Modern Soil Ecology*, (Eds.) J.D. van Elsas, J.T. Trevors and E.M.H. Welington. Marcel Dekker, Inc., New York, pp. 21–46.

Steddom, K., Menge, J.A., Crowley, D., and Borneman, J., 2002. Effect of repetitive applications of the biocontrol bacterium *Pseudomonas putida* 06909–rif/nal on citrus soil microbial communities. *Phytopathol.*, 92: 857–862.

Swanson, R.V., Bourret, R.B. and Simon, M.I., 1993. intermolecular complementation of the kinase activity of Che A. *Mol. Microbiol.*, 8: 435–441.

Tunlid, A., and White, D., 1992. Biochemical analysis of biomass, community structure, nutritional status, and metabolic activity of microbial communities in soil. In: *Soil Biochemistry, Vol. 7*, (Eds.) G. Stotzky and J.-M. Bollag. Marcel Dekker Inc., New York, USA, pp. 229–262

Van Wees, S.C., de Swart, E.A., van Pelt, J.A., Loon, L.C. and Pieterse, C.M.J., 2000. Enhancement of induced disease resistance by simultaneous activation of salicylate- and jasmonate dependent defense pathways in Arabidopsis thaliana. *Proc. Natl. Acad. Sci. USA*, 97: 8711–8716.

Weller, D.M., Raaijmakers, J.M., McSpadden Gardener, B.B. and Thomashow, L.S., 2002. Microbial populations responsible for specific soil suppressiveness to plant pathogens. *Ann. Rev. Phytopathol.*, 40: 309–348.

Zaidi, A., Khan, S. M. and Aamil, M., 2001. Bioassociative effect of rhizospheric microorganisms on growth, yield and nutrient uptake ogf green gram. *J. Plant Nutrition*, 27: 601–612.

Plant Protection, Diversity and Conservation (2011) *Pages* **387–394**
Editors: **D. Prasad & Satya Kumar**
Published by: **BIOTECH BOOKS, NEW DELHI**

Chapter 30

Damage Potential and Management of Thrips Infesting Crops

R. Varatharajan*, O. Dwijamani Singh and T. James Keisa
Department of Life Sciences, Manipur University,
Imphal – 795 003, Manipuir

Thrips are unique among the insects by having long, fine fringes along the margin of wings, protrusible bladder-like structure at the end of tarsus (Physapoda), in the possession of asymmetrical mouthparts with vestigial right mandible, and having the pre-pupal stage in between the larval and pupal stages (Ananthakrishnan, 1979; Bhatti, 1992). Their body length ranges from 500 to 1200µ for the members of the suborder Terebrantia and 0.9 to1.6 mm for Tubuliferans. What is remarkable among thrips is that they compensate their small size with their rapid growth and fecundity. They exhibit remarkable diversity in terms of feeding habit and habitat. With the lacerating/piercing and sucking type of mouthparts, thrips feed on the content of pollen grains, fungal spores and mycelia, besides sucking the cell sap. As a result of feeding, thrips cause damage to plants. When the mouthcone is short as in most Terebrantia, feeding is expected to be superficial with the result only epidermal and upper layer of mesophyll cells get affected. On the other hand, species with longer mouthcone, especially those of Tubulifera, have the deep penetrative mode of feeding. The maxillary stylets have a tongue and groove system, which form a tube through which the liquid contents are sucked up. The degree of thrips infestation depends on the age of the plant, micro-climatic condition including protection level in a concealed niche, water stress and inter or intra specific competition (Lewis, 1973). Owing to such features, thrips have received better attention in recent years as potential pests of crops, gall makers and a few acting as vectors of plant diseases (Ananthakrishnan, 1984; Mound, 2005).

* E-mail: rvrajanramya@rediffmail.com; rvrajanramya@gmail.com

Symptoms and Damaging Pattern

Usually the extent of feeding damage depends on the population build of thrips. As for instance, persistent feeding by cardamom thrips on the tender panicles and flowers leads to stunted and shriveled appearance of the panicle, discolouration of green pods to brown resulting in adverse effects on the visual appearance and aroma of cardamom capsules (Varadarasan, 1995; 2003). In cases of cereals like paddy and wheat, the apices of thrips fed foliage become brown or red and at times of severe infestation the seedling eventually dies. Thrips injury to grains and grasses is referred as "silver top" or white ear symptom with reduced grain weight (Joshi *et al.*, 1985; Ananthakrishnan, 1993). Studies on onion thrips indicate that each individual can destroy plant cells present in an area of 4.93mm^3 in 24 hours (Kendall and Bjostad, 1990). Similarly comparison of leaf area between healthy and thrips infested foliage of *Capsicum annum* revealed a significant reduction in leaf area due to thrips feeding. Such reduction in photosynthetically viable leaf surface ultimately gets reflected in crop yield (Varatharajan and James Keisa, 2000). Likewise, several flower dwelling terebrantians feed on floral parts, as a consequence fruit formation is distorted and surface is scarred with silvery patches. In severe condition, cicaterisation of fruits with thickened black patches, reduction in fruit size and weight have been noticed on fruits such as apple, apricot, cherries, grapes, orange, peaches, prunus etc (Bournier, 1983; Childers and Achor, 1991).

Pest Diversity and Crop Loss

The world record indicates occurrence of nearly 6000 species (Mound, 2007), of which 700 species of thrips are known from India (Sen, 1998). World wide nearly 100 species are economically important pests but with respect to Indian scenario, about a dozen species becomes significant from the view point of crop loss and they are as follows:

1. Cardamom thrips – *Sciothrips cardamomi* Ramakrishna
2. Cashew nut thrips – *Selenothrips rubrocinctus* Giard
3. Castor/Jatropha thrips – *Retithrips syriacus* (Mayet)
4. Cereal thrips – *Anaphothrips sudanensis* Trybom
5. Chilli thrips – *Scirtothrips dorsalis* Hood
6. Groundnut thrips – *Caliothrips indicus* (Bagnall)
7. Flower thrips – *Frankliniella schultzei* (Trybom)
8. Legume thrips – *Megalurothrips distalis* Karny
9. Melon thrips – *Thrips palmi* Karny
10. Onion thrips – *Thrips tabaci* Lindeman
11. Paddy thrips – *Stenchaetothrips biformis* (Bagnall)
12. Tea thrips – *Scirtothrips bispinosus* Bagnall

Thrips that are mentioned above belong to the suborder Terebrantia and family Thripidae. Although the suborder Tubulifera comprises some pest species, the degree of damage caused by them is insignificant in comparison to terebrantians. A few terebrantians have been studied extensively for their crop loss and damage. To quote a few examples, 45-47 per cent yield reduction in cardamom due to *S. cardamomi* (Varadarasan, 1995), 11-17 per cent loss in tea by *S. bispinosus* (Selvasundaram *et al.*, 2004), 30-50 per cent loss in chilli by *S. dorsalis* (Kandasamy *et al.*, 1990), 46-87 per cent crop loss in onion due to *T. tabaci* (Srinivas and Lawande, 2004), 1.6 to 8.1 per cent loss in paddy (Rao *et al.*, 1997) and 40 per cent yield loss in green gram due to infestation by *T. palmi* (Sreekanth *et al.*, 2002). Thus, one can come across diverse range of yield reduction in different crop.

Vector Potential

Thrips are efficient vectors of certain plant diseases caused by bacterial, fungal and viral pathogens. They disseminate either by biological or mechanical means. Tomato-spotted wilt viruses, bud necrosis virus of groundnut and banana, coffee and wheat leaf rust spores, are some of the selective examples to highlight the role of thrips in transmitting diseases (Ananthakrishnan, 1984). Further, species like *Frankliniella schultzei, Scirtothrips dorsalis Thrips palmi* and *Thrips tabaci* ravage a number of crops by transmitting the tospo virus. Thrips disseminate pathogenic fungal spores and leaf rust diseases to wheat, coffee and banana plants. Certain species like *Thrips tabaci, T. palmi* and *Frankliniella schultzei* are efficient vectors of plant viruses, causing appreciable damage to tomato and groundnut crops (Ghanekar *et al.*, 1979). This indicates that thrips exhibit diversity not only in their feeding pattern and habitat of occurrence but also in causing disease to the plants (Ananthakrishnan, 1971). The vector status of thrips is attributed to their dynamic nature of infesting diverse plant hosts, swift movement, feeding and breeding propensity in a concealed niche and possession of potential pathogen in/on their body (Ulman *et al.*, 2007).

Factors Enhancing Pest Density

Pest species generally reproduce prolifically. Some terebrantians reproduce both by sexual and parthenogenetic means, of which the latter method results in production of off springs either by arrhenotoky or thelytoky or at times by amphitoky, which may further be influenced by high temperature as in the case of *T. tabaci* leading to production of males (Lewis, 1973) or by *wolbachia* (bacteria) on certain species of *Aeolothrips* resulting in female progeny by thelytoky (Warren, 1997). In addition, host plant induced morphological forms and reproductive morphs are also evident. In certain cases the pest emerges into strategic "r-pest" with the perennial and monoculture plant host as like thrips attacking the tea plant. Such varied strategies of reproduction enable them to maintain their density as a consequence they pose serious threat to the farmers in combating the pest outbreaks.

On one hand, monoculture habitat provides continuous support to the pest to stabilize its density; the mixed cropping also favours the pest to maintain their population in one or other plant host as like *Thrips tabaci* maintaining its density on plants like cotton, bhendi, onion and tomato in a mixed crop field. Presence of weeds and other alternate hosts in the cropping area provide temporary shelter to thrips (Ananthakrishnan, 1984). Further, repeated application of certain pesticides enhances the chances of resurgence among the sucking pests (Jayaraj and Regupathy, 1987). All these factors promote the pest density in the crop field.

Management Strategies

As thrips occupy diverse niches like flower, leaf, plant gall and soil litter, different strategies need to adopt to control them. The immature stages of thrips usually become victim for pathogens and natural enemies, while incidence of adult thrips is managed by chemical application and trapping mechanisms. Although a number of methods are adopted for pest control, role of climatic factors cannot be ruled out. Earlier studies have unambiguously indicated that low temperature below 15°C inhibits the growth and development of thrips and their population build up enhance with increase of temperature. On the contrary, rainfall shows always negative correlation with thrips density (Varatharajan, 1984). On the basis of the above factors, one can arrive at the fact that thrips populations

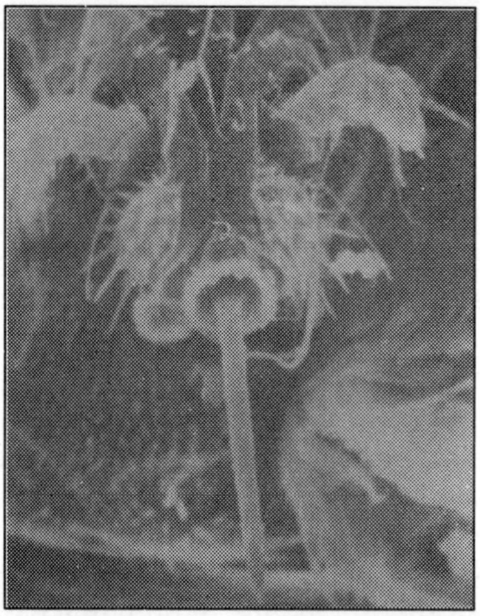

EM Picture Depicting Thrips Mouth Parts

Thrips infested chilli leaves

Normal foliage

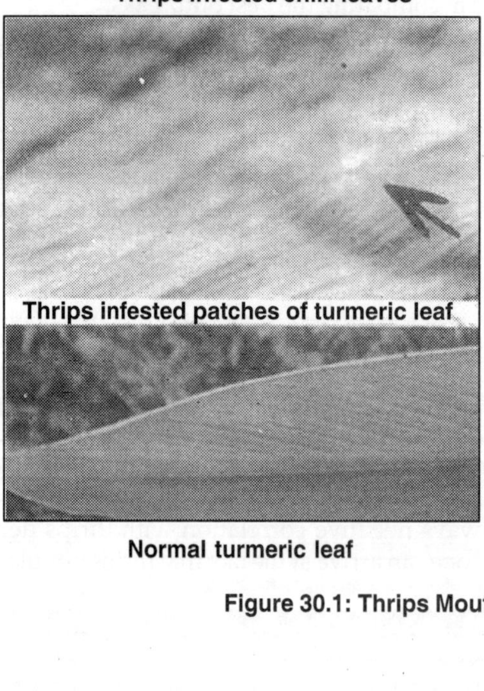

Thrips infested patches of turmeric leaf

Normal turmeric leaf

Thrips infested tea leaf

Figure 30.1: Thrips Mouthparts and Infested Leaves

are naturally regulated by a) climatic factors; b) natural enemies and diseases; and c) availability of plant hosts (Ananthakrishnan *et al.*, 1981)

Predators

The flower and gall thrips are naturally controlled by an array of predatory species like anthocorid bugs, coccinellids, neuropteran and syrphid larvae and predatory mites. To quote a few examples, *Montandoniola moraguesi* is an established predator of many gall thrips. Similarly, *Orius maxidentex* Ghauri, *Orius minutes* L., *Carayonocoris indicus* Muralee, *Micromus timidus* Hagen are common predators of flower thrips (Muraleedharan and Ananthakrishnan, 1978). In addition, predatory thrips such as *Androthrips ramachandrai* Karny and *Androthrips flavipes* Schmutz are potential bio-control agents for gall thrips (Boyd and Held, 2006). *Franklinothrips vespiformis* (Crawford) is an efficient predator of pests like *Thrips palmi* Karny [melon thrips] (Arakai and Okajima, 1998) and *Scirtothrips bispinosus* (Bagnall) [tea thrips] (Personal communication with Radhakrishnan, B. UPASI Tea Research Foundation, Coimbatore). Therefore, field release of such predators can minimize the pest species.

Conservation of Predators

Under the lab condition, thrips can be mass multiplied as prey for the predators. These predators can be later field released by inundative methods. Thrips can be reared in plastic petri-dishes (55 x 15mm) covered with double layered parafilm membrane containing water and 10 per cent honey solution in between two layered transparent membrane (Murai and Ishii, 1982). Fresh pollen of different flowers could be given everyday to feed them. Thrips lay eggs in between the two membranes. The emerging larvae can also be taken out and provided to the predators as a fresh prey; thereby one can maintain a continuous culture of predators which can be field released as and when necessary. This is one of the cost effective method for rearing and conserving predators such as anthocorid bug, predatory mites and coccinellids etc.

Parasitoids

Certain hymenopteran parasitoids belonging to the genera *Ceranisus, Tetrastichus* and *Thripobius* are parasitic on pests like *Thrips tabaci, Rhipiphorothrips cruentatus, Panchaetothrips indicus* etc. The nematode parasite, *Anguillulina aptini* and the microsporadian protozoan parasite, *Mrazekia* sp are also sporadically infect thrips attacking the crops. However, unlike predators, parasitoids could not bring down the pest density to a satisfactory level (Varatharajan and James Keisa, 2000). Pathogenic invasion by fungi appears to give encouraging effect under the lab bioassay. Although a dozen species of fungi has been reported to show appreciable efficacy against thrips, the mycoinsecticides *Hirsutella thompsonii* and *Pochia lecanii* @ 10^8 conidia/ml exhibit good control over thrips in general under field condition (Sithanantham *et al.*, 2007).

Traps

The soil pupating thrips can be periodically trapped with the help of either sticky traps or yellow pan water traps from the cropping areas. Diraviam and Uthamasamy (1992) found that yellow sticky traps are effective not only to monitor thrips density, but also to trap adult thrips. Inter-cropping the groundnut with pearl millet significantly reduced the incidence of thrips (Sreekanth *et al.*, 2002).

Field Sanitation

Weeds are the main reservoirs of pest species. Periodical removal of weeds, leaf sheaths, old panicles etc as practiced by cardamom estates will reduce thrips infestation (Varadarasan, 2003). In the case of tea fields periodical plucking of young tea leaves will minimize thrips incidence (Muraleedharan, 1991).

Botanicals

Various neem products with the active ingredient of azadiractin have been found effective against thrips and this has been observed with rice thrips (Samiayyan and Chandrasekaran, 1995), chilli thrips (James Keisa and Varatharajan, 1995) and onion thrips (Sattar Shah *et al.*, 2005). Field application of volatile organic compounds like jasmonic acid in a cotton field reduced the thrips density to a significant level (Shimoda *et al.*, 1997). Experience indicates that applying suitable combination of different management strategies in an eco-friendly manner will positively reduce the pest outbreak irrespective of the crop concerned.

Acknowledgements

The authors thank the Head, Department of Life Sciences, Manipur University for providing necessary facilities to carry out this work.

References

Ananthakrishnan, T.N., 1971. Thrips (Thysanoptera) in Agriculture, Horticulture and Forestry diagnosis, bionomics and control. *Journal of Scientific and Industrial Research*, 30: 113–146.

Ananthakrishnan T.N., 1979. Biosystematics of Thysanoptera. *Annual Review of Entomology.*, 24: 159–183.

Ananthakrishnan, T.N., 1984. *Bioecology of Thrips*. Indira Publishing House, Michigan, USA., 233pp.

Ananthakrishnan, T.N., 1992. *Dimensions of Insect–Plant Interactions*. Oxford and IBH. Publishing Co. Pvt. Ltd., New Delhi, 184 pp.

Ananthakrishnan, T.N., 1993. Bionomics of Thrips. *Annual Review of Entomology*, 38: 71–92.

Ananthakrishnan, T.N., Varatharajan, R. and Gopinathan, K., 1981. Seasonal periodicity of thrips infesting some Compositae in relation to pollination. *Proceedings of Indian National Science Academy*, B47 (6): 811–815.

Arakai, N. and Okajima, S., 1998. Notes on the biology and morphology of predatory thrips, *Frankliniothrips vespiformis* (Crawford) (Thysanoptera: Aeolothripidae): First record from Japan. *Entomological Science*, 1(3): 359–363.

Bhatti, J.S., 1992. The Order Tubulifera (Insecta): Its characters and classification into families. *Zoology (Journal of Pure and Applied Zoology)*, 3 (2): 127–162.

Diraviam, J. and Uthamasamy, S., 1992. A new sampling technique involving yellow sticky trap for monitoring of thrips infesting different crops. *Journal of Entomological Research*, 16: 78–81.

Bournier, A., 1983. Les thrips, biologie, importance agronomique. *Institut National de la Researche Agronomique*, Paris.

Boyd. W. David and Held W. David, 2006. *Androthrips ramachandrai* (Thysanoptera: Phlaeothripidae): an introduced thrips in the United States. *Florida Entomologist*, 89(4): 455–458.

Childers, C.C. and Achor, D. S., 1991. Feeding and oviposition injury to flowers and developing floral buds of "Navel orange" by *Frankliniella bispinosa* (Thysanoptera: Thripidae) in Florida. *Annals of Entomological Society of America*, 84: 272–282.

Ghanekar, A.M., Reddy, D.V.R., Tizuka, N., Amin, P.W. and Gibbons, R.W., 1979. Bud necrosis of groundnut in India caused by tomato spotted wilt virus. *Annals of Applied Biology*, 93: 173–179.

James Keisa, T. and Varatharajan, R., 1995. Efficacy of two neem products in the field control of Scirtothrips dorsalis Hood on *Capsicum annum*. *Indian Journal of Plant Protection*, 23: 166–168.

Jayaraj, S. and Regupathy, A., 1987. Studies on the resurgence of sucking pests of crops in Tamil Nadu. In: *Resurgence of Sucking Pests,* (Ed.) S. Jayaraj. Proceedings of National Symposium, Tamil Nadu Agricultural University, Coimbatore, India, p. 225–240.

Joshi, R.C., Heinriches, E.A. and Uthamaswamy, S., 1985. A bibliography on rice thrips, *Stenchaetothrips biformis* (Bagnall) (Thysanoptera: Thripidae). *FAO Asia Pacific Plant Protection Commonwealth Quaterly Newsletter*, 28: 27–31.

Kandasamy, C., Mohanasundaram, M. and Karuppuchamy, P., 1990. Evaluation of insecticides for the control of thrips, *Scirtothrips dorsalis* Hood on chillies (*Capsicum annum* L.). *Madras Agricultural Journal*, 77: 169–172.

Kendall, D.M. and Bjostad, L.B., 1990. Herbivory by *Thrips tabaci* Lindeman Induces greater ethylene production in intact onions than mechanical damage alone. *Journal of Chemical Ecology*, 16: 981–991.

Lewis, T., 1973. *Thrips: Their Biology, Ecology and Economic Importance*. Academic Press, 349 pp.

Mound, L.A., 2005. Thysanoptera: Diversity and interaction. *Annual Review of Entomology*, 50: 247–269.

Mound, L.A., 2007. Thysanoptera (Thrips) of the World: A check list. http://www.ento.csiro.au/thysanoptera/worldthrips.html

Murai, T. and Ishii, T., 1982. Simple rearing method for flower thrips (Thysanoptera: Thripidae) on pollens. *Japanese Journal of Applied Entomology and Zoology*, 20: 149–154.

Muraleedharan, N and Ananthakrishnan, T.N., 1978. Bioecology of four species of Anthocoridae predacious on thrips. *Records of Zoological survey of India*, Occasional Paper, 11: 1–32.

Muraleedharan, N., 1991. *Pest Management in Tea.* UPASI Tea Research Institute, Valparai, 130pp.

Rao, J., Anand Prakash, Ghosh, S. K. and Sinha, R.K., 1997. Damage to grain due to cereal thrips, *Haplothrips ganglbaueri* Schmutz in paddy fields. *Oryza*, 34: 250–251.

Samiayyan, K. and Chandrasekaran, B., 1995. Efficacy of botanicals against rice thrips in nursery. *Madras Agricultural Journal*, 82: 497–498.

Sattar Shah, Md. A., Singh, H.C. and Varatharajan, R., 2005. Effect of neemazal on onion thrips, *Thrips tabaci* Lindmann. (Thripidae: Thysanoptera). *Annals of Plant Protection Science*, 13: 470–471.

Selvasundaram, R., Sasidhar, R., Sanjay, R. and Muraleedharan, N., 2004. Seasonal abundance of thrips and crop loss in tea. *Journal of Plantation Crops*, 32: 49–52.

Sen, S., 1998. Faunal diversity in India: Thysanoptera. In: *Faunal Diversity in India*, (Eds.) J.R.B. Alfred A.K. Das and A.K Sanyal. ENVIS Centre, Zoological Survey of India, Kolkata, p. 243– 249.

Shimoda, T., Takabayashi, J., Ashihara, W. and Takafuji, A., 1997. A response of predatory insect, *Scolothrips takahashii* towards herbivore induced plant volatiles under laboratory and field conditions. *Journal of Chemical Ecology*, 23: 2033–2048.

Sithananatham, S., Varatharajan, R., Chandish R. Ballal and Ganga Visalakshy, P.N., 2007. Research status and scope for biological control of sucking pests in India: Case study of thrips. *Journal of Biological Control*, 21 (Special Issue): 1–20.

Sreekanth, M., Sreeramulu, M., Prasada Rao, R.D.V.J., Sarath Babu, B. and Ramesh Babu, T., 2002. Effect of sowing date on *Thrips palmi* Karny population and pea nut bud necrosis virus incidence in green gram (*Vigna radiate* L. Wilezek). *Indian Journal of Plant Protection*, 30: 16–21.

Srinivas, P.S. and Lawande, K.E., 2004. Impact of planting dates on *Thrips tabaci* Lindeman infestation and yield loss in onion (*Allium cepa* L.). *Pest Management in Horticultural Ecosystems*, 10: 11–18.

Ullman, D., Moyer, J., Goldbach, R. and Moritz, G., 2007. VIII International Symposium on Thysanoptera and Tospoviruses September 11–15, 2005 Asilomar, Pacific Grove, California. 49pp. *Journal of Insect Science*, 7: 28, available online: *insectscience.org/7.28*.

Varadarasan, S., 2003. Relevance of integrated pest management in cardamom cultivation. *Spice India*, Cochin, May, p. 11–22.

Varatharajan, R., 1984. Bioecological studies on some *Anthophilous* thrips from Southern India. *Ph.D. Thesis*, Madras University.

Varatharajan, R. and James Keisa, T., 2000. Bioecology and management of thrips. In: *IPM System in Agriculture*, (Eds.) Rajeev K. Upadhyay, K.G. Mukerji and O.P. Dubey. Aditya Books Pvt. Ltd., New Delhi, 219–234 pp.

Warren, J.H., 1997. Biology of *Wolbachia*. *Annual Review of Entomology*, 42: 587–609.

Plant Protection, Diversity and Conservation (2011) Pages 395–416
Editors: D. Prasad & Satya Kumar
Published by: BIOTECH BOOKS, NEW DELHI

Chapter 31

Integrated Disease Management of Onion and Garlic

R.K. Mishra[1]*, *D.K. Srivastava*[2], *R.K. Jaiswal*[3] *and R.P. Gupta*[4]

[1]*National Horticultural Research and Development Foundation, Regional Research Station, Salaru, Karnal – 132 001, Haryana*

[2]*Council of Science and Technology, Lucknow – 226 001, U.P.*

[3]*Division of Nematology, IARI, New Delhi – 1100 12*

[4]*Director, National Horticultural Research and Development Foundation, Nasik*

Onion (*Allium cepa* L.) and Garlic (*Allium sativum* L) is the most important *Allium* species cultivated in India and used as vegetable, salad and spice in the daily diet by large population. In India, onion and garlic crops are grown almost all over the country, especially in the states of Haryana, Punjab, Madhya Pradesh, Gujarat, Rajasthan, Uttarakhand, Jammu and Kashmir, Bihar, Uttar Pradesh and Maharastra, A.P., Karnataka. M.P. is the leading state in garlic production while in onion, Maharastra is the leading state accounting for more than 22.83 per cent of area and 28.42 per cent of production with an average yield of 12.37 t/ha. In India per hectare yields are highest in Gujarat (22.65 t/ha) followed by Punjab (22.63 t/ha). The area under garlic during 2006-07 is 0.14 million ha and production is 0.75 million tones with average productivity of 5.38 t/ha. In India per hectare yield are higher in Punjab (16.67 t/ha) followed by H.P. (13.14 t/ha) and Haryana (12.38 t/ha). The crop is attacked by many diseases, which vary from region to region, season to season and variety to variety. Various biotic factors like fungi; bacteria, viruses, phytoplasmas and nematodes are associated with garlic at different stages of growth and cause considerable damage/losses in yield as well as quality of garlic. There are many diseases affecting the garlic production through out the country. These diseases can also affect at production, harvesting, processing and marketing stages, which lower the quality, reduce the yield there by increase the cost of production and export potential also. The diseases alter the cropping pattern and affect local and export markets. In the year 1993 in Maharastra state 60-80 per

* E-mail: rajpathologist@yahoo.com

cent losses were reported due to Purple blotch. Consistent use of fungicides and other chemicals to manage the diseases in crop plants not only poses a serious threat to the environment and mankind but also slowly build up resistance in the pathogens. Most of the new generation pesticides are systemic in their mode of action leads to certain level of toxicity in the plant system and thus resulting health hazards. Further, it disturbs complete microbial diversity of whole ecosystem. All these factors have led to new dimension in research for biological control and integrated disease management. The effect of disease can be lessened through various means including cultural, chemical and biological. In recent IDM approach has been found to be one of the most important tools to minimize the incidence of diseases of onion and garlic. Important diseases affecting the onion and garlic and their integrated management are reviewed here.

Integrated Disease Management: Basic Concept

Plant disease control means absolute check of a disease, which is neither possible nor feasible and replaced by more appropriate term management. No pathogen is completely eradicated from natural ecosystem, but its population can be reduced below economic injury level. Disease management does not employ against the only pathogen but also considers the other three components of disease triangle *i.e.* host, environment and time factor which gives the concept of integrated disease management (IDM). The main objective of IDM is to maintain the loss below an economic injury level and minimize the recurrence of disease by interrupting disease cycle, survival period and inoculums source. Pathogen management involves reduction in inoculum, eradication and prevention of inoculum while management of host involves improving plant vigour, induced resistance through nutrition and genetic manipulation and protection by chemical means. The time factor could be managed by adjustment of planting date so that favorable time of disease should not coincide without any adverse effect on yield. The principles of IDM are based on avoidance, exclusion, eradication, protection, host resistance and therapy. Avoiding disease by planting at times when or in areas where inoculum is ineffective due to environmental conditions. It could be also achieved by selecting geographic area for disease free seed production, using disease escaping varieties, selection of disease free seeds and modification cultural practices. Exclusion means preventing the inoculum from entering or establishing in the field or area where it does not exit. Eradication is applied when pathogen has already been entered in area or crop in spite of the above-mentioned precautions. At this conditions, biological control methods, crop rotation, rouging, sanitation, soil treatments, heat and chemical treatments of diseased plants can be applied. In present scenario integrated disease management (IDM) strategies is the only sustainable approach. IDM should be effective against more one disease, long lasting, economically viable and eco-friendly. There is an urgent need to retinalize the use of pesticides and develop IDM packages for onion and garlic diseases that are capable to managing plant pathogens more effectively and have minimum impact on humans, wild life and environment.

Integrated Disease Management: Key Strategies

The success of integrated disease management approach in onion and garlic is only possible when we apply all the following practices and precautions every year on community basis in large area.

- ☆ Avoid indiscriminate use of fungicides and do not mix incompatible pesticides, phytohormones and micro-nutrients at a time.
- ☆ Soil as a reservoir of harmful as well beneficial microorganisms, therefore soil health should be properly managed by timely tillage, summer ploughing, green manuring, optimum C: N ratio, balanced macro and micro-nutrients, aeration etc.

- ☆ Summer ploughing followed by irrigation in summer and then subsequent ploughing is very effective.
- ☆ Seed health must be maintained up to prescribed standard and ensure that seed should be free from internal and external pathogen, contamination of any infected crop debris, sclerotial bodies etc.
- ☆ Field sanitation as essential practices where removal and burning of infected crop debris, alternate and collateral weed hosts should be carried out periodically.
- ☆ Crop rotation is very important for soil borne diseases where non-host crop preferably cereals should be selected for particular pathogen.
- ☆ Avoid off season vegetables and intensive cropping because it prolongs the perpetuation period of a pathogen and do not break the life cycle of the pathogen.
- ☆ Knowledge of correct diagnosis of the diseases and disease cycle of the pathogen is very important to apply different methods of critical phase of life-cycle for maximum efficacy.
- ☆ Application of biocontrol agent's particularly resident antagonist accompanied with green manuring, FYM or any organic matter should be maximized.
- ☆ Always prefer to grow tolerant varieties. However, it is difficult to get resistance against diseases with all desirable traits in hybrid
- ☆ Proper postharvest management practices should be followed to avoid any rotting of onion and garlic during storage and transportation.

Knowledge of the following aspects of disease development is essential for effective and economic management:

- ☆ Cause of the diseases.
- ☆ Mode of perennation and dissemination of the infectious causes.
- ☆ Host-parasite relationship and means of secondary spread.
- ☆ Effect of environment on pathogenesis in the plant and spread of the disease in the plant population.

Important Diseases

Pre-harvest Diseases

Damping-off

Pathogen

- ☆ *Pythium* sp.
- ☆ *Fusarium* sp.
- ☆ *Rhizoctonia* sp.

Symptoms

Damping-off of onion seedlings occurs in two stages:

1. Pre emergence
2. Post emergence

Figure 31.1: Damping-off

In pre emergence stage, the younger seedlings are killed before they reach the soil surface. They may, infect, be killed even before the hypocotyls has broken the seed coat. The radicle and plumule, when come out of the seed, undergo complete rotting. Since this happen below the soil surface and the disease often not seen, the failure in emergence of seedlings is attributed to the poor quality of seeds. The post emergence damping off is characterized by the toppling over of infected of infected seedlings anytime after they emerge from the soil. It usually occurs at or before the ground level and infected tissues soft and water soaked. As the disease advances, the stèm becomes constricted at the base and plant collapses. It is observed that most of the loss is due to pre-emergence damping-off. It is more common during Kharif season when temperature and humidity are very high.

Management

- ☆ Proper drainage is essential.
- ☆ Sowing of clean uninfected seed on raised beds.
- ☆ Crop rotation.
- ☆ Soil solarization of nursery beds with transparent polythene for 30 days before sowing gave good control of damping-off.

- ☆ Seed treatment with Thiram, Captan @ 2.5 g/kg seed or Carbendazim @ 0.1 per cent or *Trichoderma viride* @ 4-5 g/kg is effective.
- ☆ Drenching the nursery by Thiram @2.5 per cent or Carbendazim @ 0.1 per cent or *Trichoderma viride* @ 4-5g/litre.

Purple Blotch

Pathogen

- ☆ *Alternaria porri.*

Symptoms

This is a most important disease prevalent in all the onion and garlic growing areas in the country. The disease is reported to occur in hot and humid climate. First it appears on the leaves as small, whitish, sunken lesions. These spots later enlarge and eventually encircle the leaf. Later, darkened zones appear on the surface of the leaves, retaining the characteristic purple colour. The leaves and stems fall over gradually. Infection can cause a semi-watery rot of neck of the bulbs, which turn yellow red in colour, bulb tissues eventually become papery. Infected plant debris is the main source of inoculums.

Figure 31.2: Purple Blotch

Figure 31.3: Purple Blotch

Management

- ☆ Use of healthy seeds.
- ☆ Crop rotation of 2-3 years with non-related crops should be followed.
- ☆ Spraying of Dithane M-45 (2.5gram/litre water), Kavach (2.0gram/litre water) should be sprayed at 15 days intervals just after 45 days after transplanting.
- ☆ Spraying of *Trichoderma viride* (5gram/litre) and *Baeveria bassiana* (2-3g/litre) at weekly intervals would be effective against this disease.
- ☆ Seed treatment with Captan @ 2.5 g/kg seed, Thiram @2.5 g/kg or *Trichoderma viride* @ 4-5 g/kg seed before nursery sowing.
- ☆ Summer ploughing is reduced the incidence.
- ☆ Grow resistant/tolerant cultivars like, G-1, G-50, and G-323

Stemphylium Blight

Pathogen: *Stemphylium vesicarium.*

Symptoms

This is a serious problem in Northern parts of the country especially in bulb crops. The disease is also prevalent in some parts of the southern parts of the country. Symptoms appears as small, yellow to orange flecks or streaks on the leaf. These soon develop into elongated, spindle shaped to ovate elongate, diffused spots, often reaching the leaf lips. They usually turn grey at the centre, brown to dark olive brown. The spots frequently coalesce into extended patches blight of the leaves and gradually the entire foliage. The pathogen survives on the dead host tissue and serves as the source of inoculums.

Management

- ✰ Spraying of Mancozeb (2.5gram/litre water) along with Dhanuvit is effective to control the disease. Spraying is started just after appearance of disease and repeated fortnightly.
- ✰ Spraying of Sixer (carbendazim + D. M-45) @ 2.5gram/litre of water with Dhanuvit (Sticker) at 15 days intervals.
- ✰ Crop rotation with other vegetables and cereals to reduced the disease severity.

Figure 31.4: Stemphylium Blight

Cercospora Leaf Spot

Pathogen

- *Cercospora duddiae*

Symptoms

The disease appears on the leaves as small ash colored and irregular shaped on leaf lamina. The spots coalesce gradually and result in blighting of the foliage. High temperature and prolonged wet conditions favors the disease development.

Management

- Spraying of Dithane M-45 (2.4gram/litre water) and Copper oxychloride (3.0gram/litre water) at 15 days intervals give good results.
- Spraying of Kavach (Chlorothalonil) @ 2.0 gram/litre water and sixer (Carbendazim + D. M-45) @ 2.5 gram/litre water at 15 days intervals.
- Deep summer ploughing.

Powdery Mildew

Pathogen

- *Leveillula taurica*

Symptoms

Whitish, circular to oblong lesions of variable size occurs on abaxial surface of leaves. Older leaves infected first. The lesions are covered with white powdery mass of fine hypae and conidia.

Management

- Use of sulpher fungicides @ 0.2 per cent at fortnightly intervals.
- Proper drainage is essential.

Rust

Pathogen

- *Puccinia allii* or *Puccinia porrii*

Symptoms

This fungus was considered to be minor importance in onion and garlic production. Initial symptoms occur on the foliage and stem as small white flecks that develop into orange spots (spores) or pustules. The bulbs become shrunken and deformed. Heavily infected plants may turn yellow and die. Disease incidence is highest in stressed plants.

Management

- Use of healthy seeds
- Spraying of hexaconazole @ 0.1 per cent and propiconazole (Tilt) @ 0.1 per cent at 15 days intervals.
- Rotate with non-allium crops.

Anthracnose Disease (Twister)

Pathogen

☆ *Colletotrichum gloeosporioides*

Perfect Stage

☆ *Glomerella cingulata.*

Symptoms

Curling and twisting of leaves, chlorosis and abnormal elongation of neck portion are major symptoms. Bulb development is already affected. White oval sunken lesions are developed on the leaves. In advance stage roots become sparse and plant may die. Bulbs are slender and may rot before harvest or during storage.

Management

☆ Use of good quality of seeds.

☆ Seed treatment with Thiram or Capaton (2.0-2.5gram/kg seed) before nursery sowing.

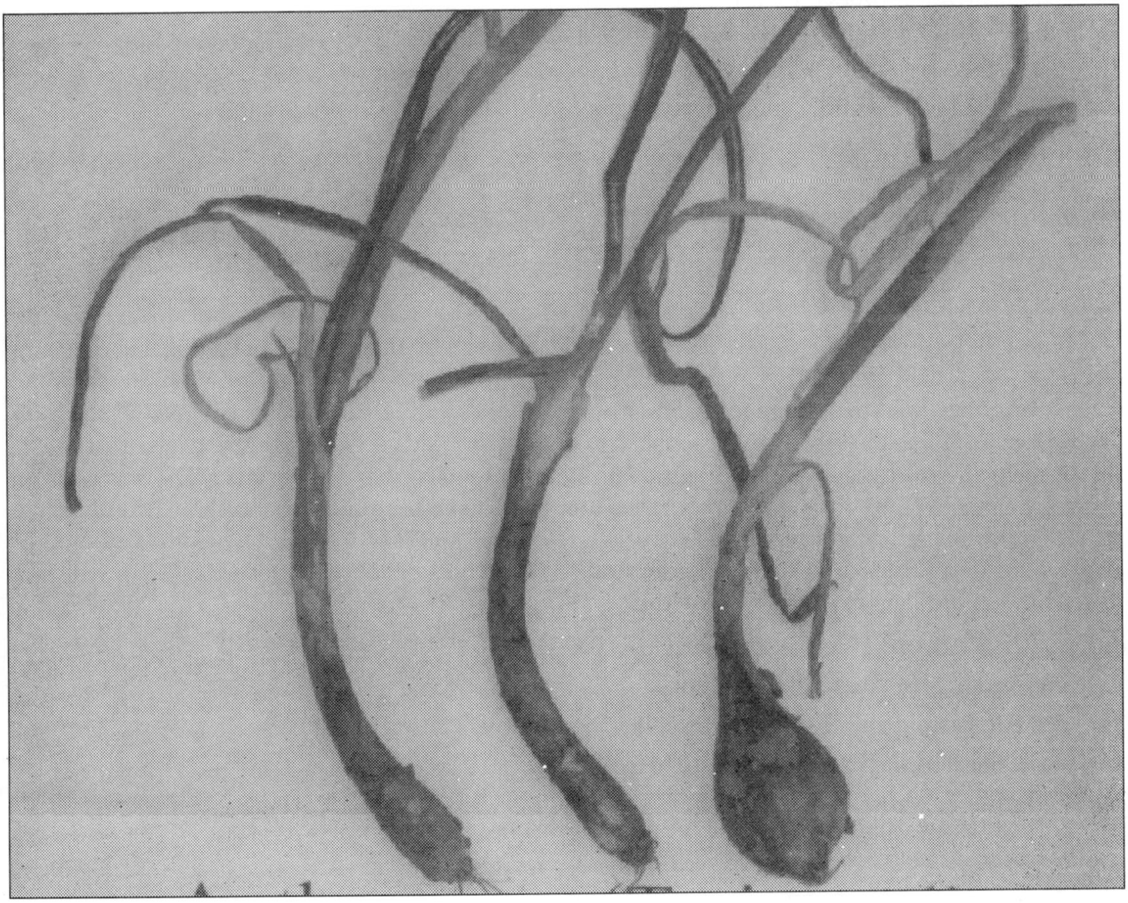

Figure 31.5: Anthracnose Disease (Twister)

- ☆ Seed treatment with *Trichoderma viride* (5.0gram/kg seed).
- ☆ Soil treatment with Benomyl or Benlate (@ 2.0gm/m^2) and spraying of D.M-45 (2.5g/lit water) were effective.
- ☆ Sanitation and destruction of infected plant debris helps in reducing the disease incidence.

Garlic Mosaic

Pathogen

- ☆ Virus

Symptoms

Garlic plants infected with mosaic virus show typical symptoms of chlorotic mottling and strips on the emerging leaf followed by pale yellow broken stripes resulting typical mosaic pattern on matured leaves. Yellowish dots on leaves, whitish leaf margin, or twisting of leaves are also recorded on some cultivars. Generally symptoms are more pronounced in young leaves.

Management

- ☆ Spraying of Monocrotophas @ 0.05 per cent, Endosulphan @0.25 per cent or methyl dematon @ 0.075 per cent at 10-15 days intervals
- ☆ Cloves treatment with Bavistin @ 0.1 per cent before planting
- ☆ Use of HNPV and some neem based botanicals at weekly intervals

Onion Yellow Dwarf

Pathogen

- ☆ Virus (Poty virus)

Symptoms

Onion Yellow Dwarf Virus (OYDV) is an important pathogen of garlic and onion also, causing severe losses in garlic clones. It is an aphid-borne poty virus. It produces symptoms of mild chlorotic strips to bright yellow strips depending on virus isolate and cultivars. Reduction in growth and bulb size also occurs. Infection by other viruses such as Leek yellow stripe, Garlic common latent virus and Shallot latent virus also occurs and may aggravate the symptoms further. However, OYDV is recognized as a major element of the virus disease complex. Enzyme Linked Immunosorbant Assay (ELISA) is the main diagnostic method for large-scale routine detection of OYDV in garlic. The other method of OYDV detection is based on reverse transcriptase polymerase chain reaction (RT-PCR), which is 10-100 times more sensitive than ELISA.

Management

- ☆ Removal of virus infected plants.
- ☆ Alternate spray of systemic insecticides with neem based botanicals.
- ☆ Collect healthy seed from disease free plants.

Iris Yellow Spot Virus (IYSV)

Pathogen

- ☆ Virus (Tospovirus)

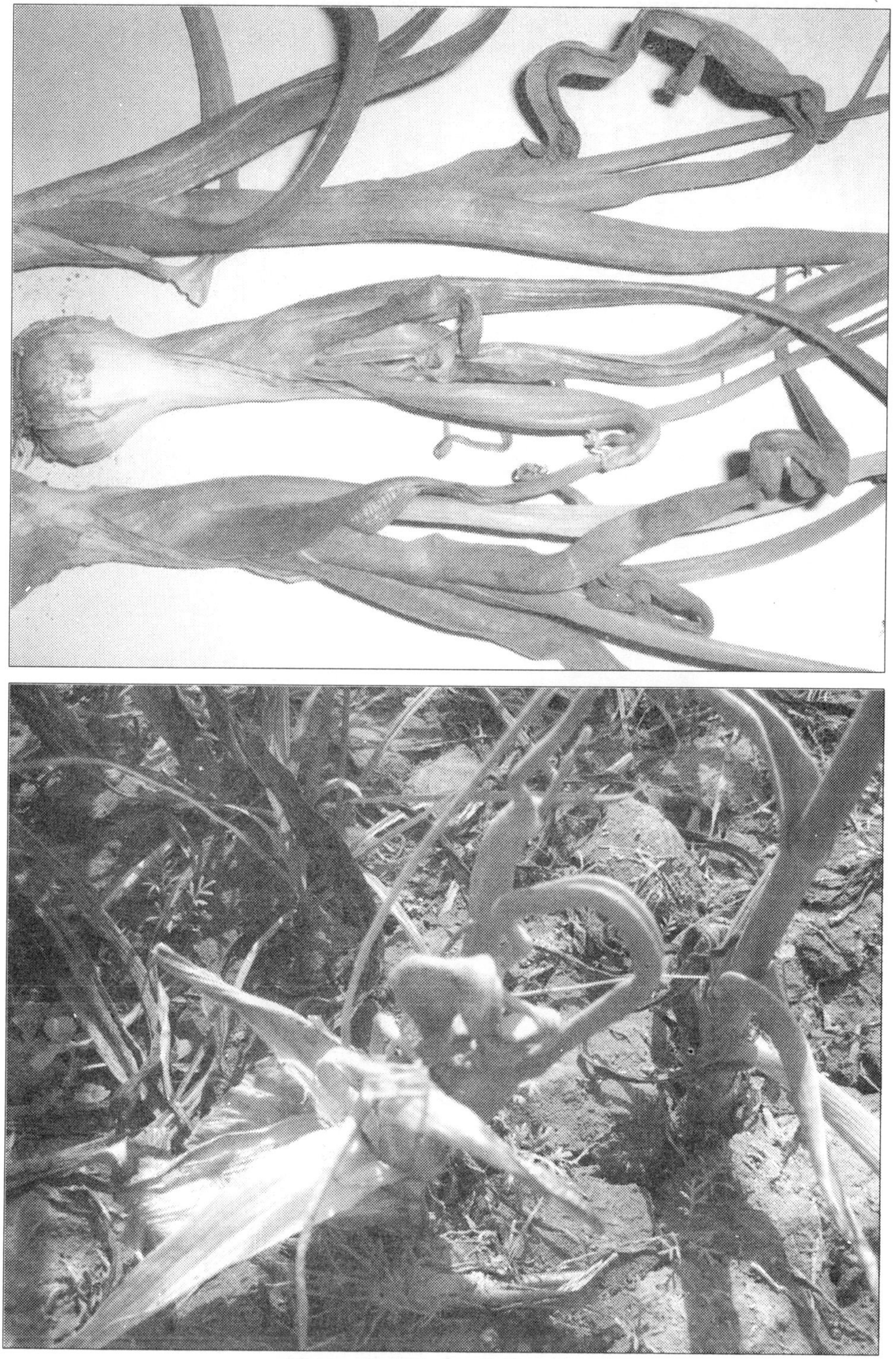

Figure 31.6: Onion Yellow Dwarf

Figure 31.7: Iris Yellow Spot Virus (IYSV) on Onion Bulb Crop and Seed Crop

Symptoms

The disease symptoms vary among onion bulb and seed crops, but often appear as straw-colored, diamond shaped lesions on leaves and scapes with twisting or banding flower bearing stalks were observed in onion plants. Some lesions have distinct green centre with yellow or tan borders, other lesions appears as concentric rings of alternating green and yellow/tan tissue. Infected plants can be scattered or generalized through out a field. Large necrotic region may develop on scapes and cause a collapse of the escape. Diseased plants may be scattered or wide spread across a field, but the highest incidence of disease is often found on the field edges.

Management

- ☆ Iris yellow spot virus (IYSV) is a top virus, similar to tomato spotted wilt virus, which is currently through to be vectored solely by onion thrips (*Thrips tabaci*), so alternate spray of softer insecticides formulations (spinosad, neem extract) and organic mulches (straw) is effective against IYSV.
- ☆ An integrated approach is essential for management of IYSV.

Postharvest Diseases

White Rot

Pathogen

- ☆ *Selerotium cepivorum*

Symptoms

Pathogen is soil inhabiting and invades the roots and the basal part of the bulb scales. The first symptoms of the disease is yellowing and dying back of leaf tips. The roots are generally destroyed and there is semi watery decay of the scale with abundance of superficial white truly mycelium. Brown to black sclerotia is developed on surface or within tissue. The organism is most active when the temperature is cool. In Northern climates it usually attacks in the spring.

Management

- ☆ Removal of infected plants during season reduces the sclerotial population and also avoids incorporation of the same to the soil.
- ☆ Solarization of soil at high temperature *i.e.* 35°C for 18hours or 45°C for 6 hours reduces the incidence.
- ☆ Iprodione (Rovral) @ 0.25 per cent was effective and Benomyl @ 0.1 per cent thiophemate methyl @ 0.1 per cent have reduced disease incidence.
- ☆ The organisms, which have given promising result, are *Trichoderma viride, Gliocladium Zeae, Penicillium nigricans, Bacillus subtilus* and *Trichoderma harzianum* were highly effective against postharvest diseases.

Basal or Bottom Rot

Pathogen

- ☆ *Fusarium oxysporum* f. sp. *cepae*

Symptoms

In garlic, pre-emergence decay of cloves, stem plate and postharvest decay of cloves in stored

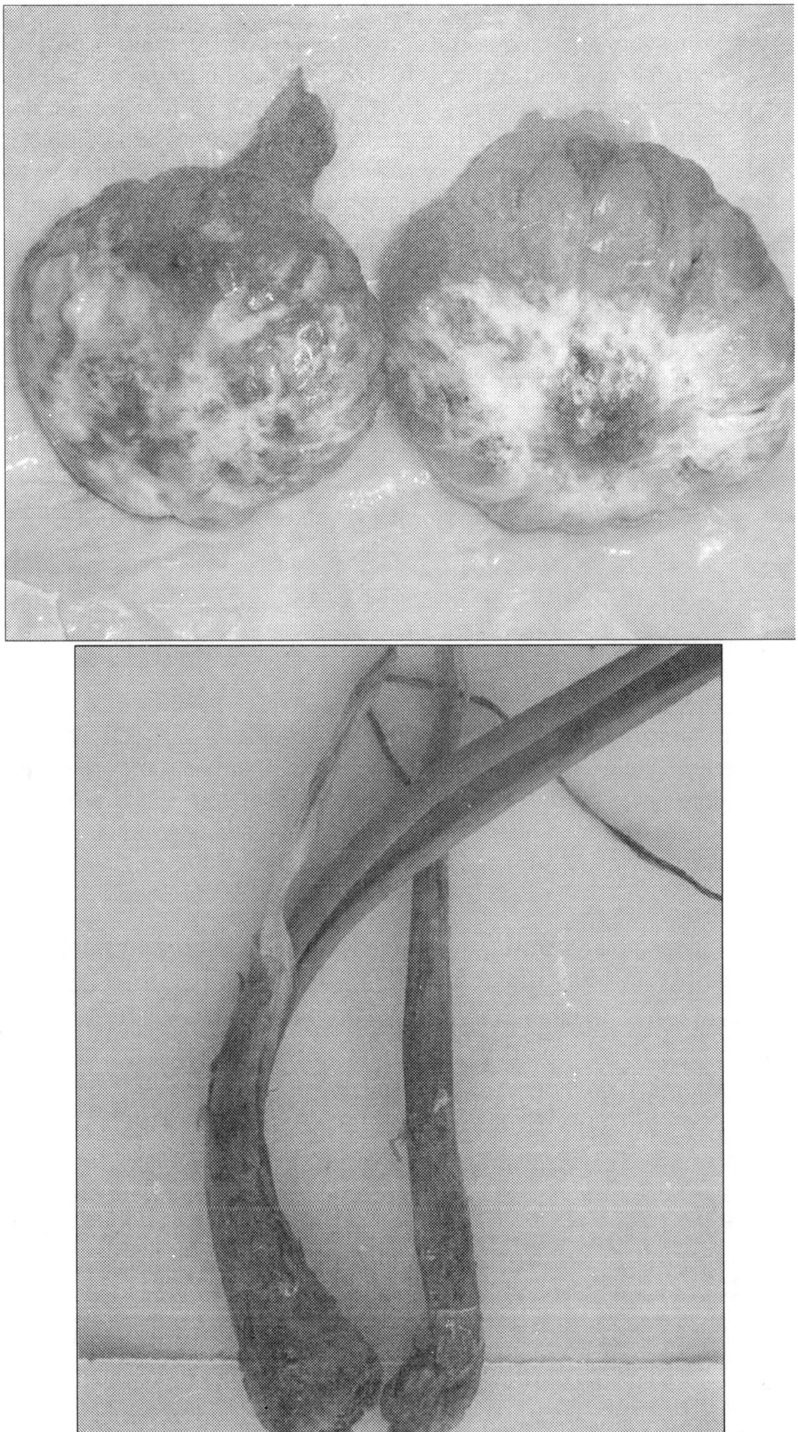

Figure 31.8: White Rot

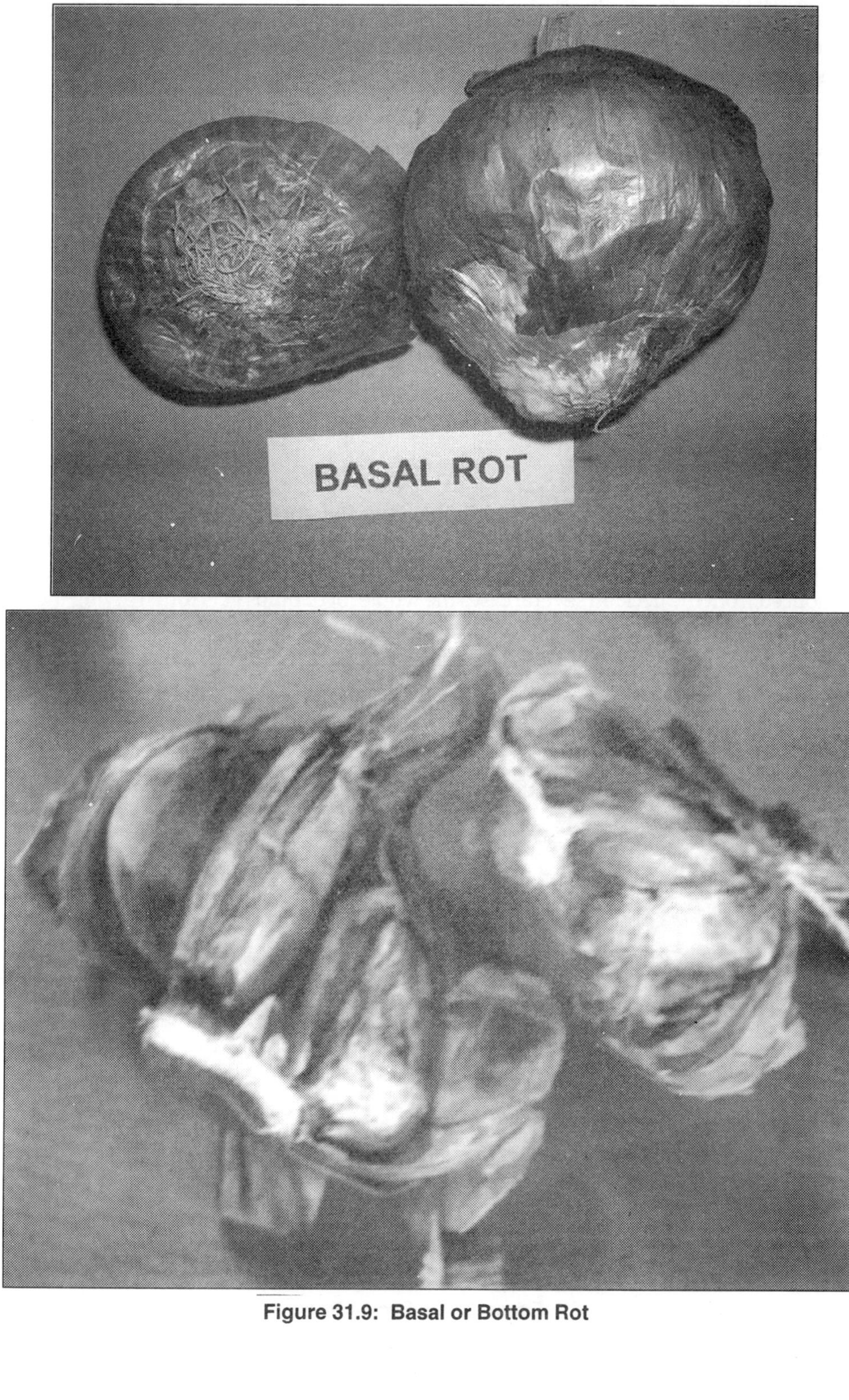

Figure 31.9: Basal or Bottom Rot

bulbs are main symptoms. Infected garlic show reduced emergence, yellowing and/or browning (necrosis) of leaves beginning at tips. On stem and bulb early in the season with some discolouration on bulb sheath at harvest. This will eventually wither and die.

Management

- ☆ Use of *Trichoderma viride* @4-5 kg/ha before transplanting is effecting against this disease.
- ☆ Proper drainage is essential during entire cropping period.
- ☆ Proper drying and curing
- ☆ Use of *Trichoderma viride, Pseudomonas* spp., *Baeveria bassiana* before transplanting is effecting against this disease.
- ☆ Deep ploughing and avoiding injury during cultural practices.

Brown Rot

Pathogen

Pseudomonas aeruginosa

Symptoms

Dark brown discolouration in bulb scale is the characteristics feature of this disease. Browining of inner scale along with rotting is the main symptoms of disease. The rotting starts from the inner scale and spread to outer scales. Apparently, the bulb seems to be healthy, but when pressed, the white oozing is noticed from the neck. In several cases the whole lot of bulbs gets rotten giving the bad odour in storage.

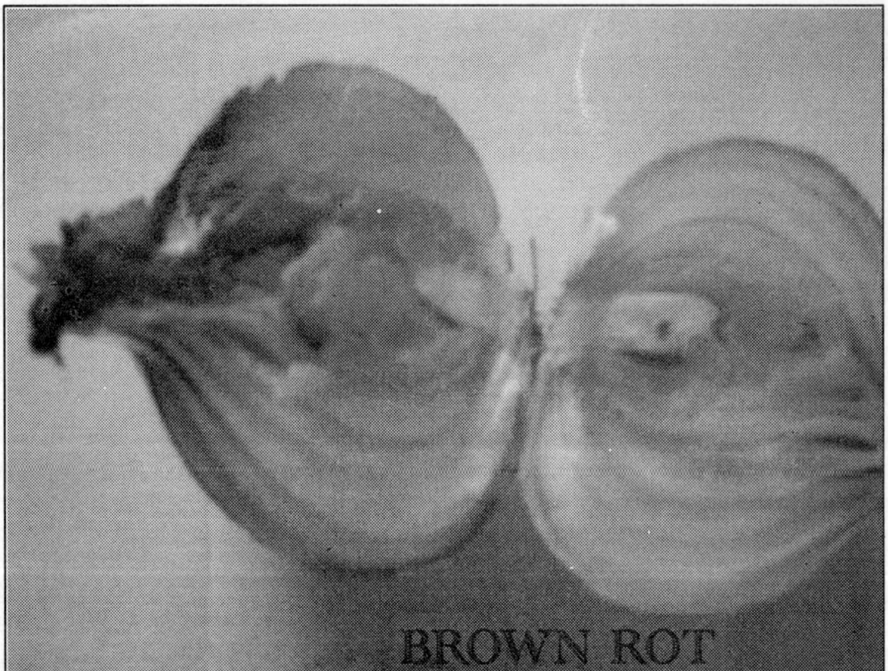

Figure 31.10: Brown Rot

Management

- ☆ Proper curing is required before storage.
- ☆ Use of maleic hydrazide (20 ml/litre) before one month of harvest.
- ☆ Neck cutting is about 2.5-3.0 cm.long above the bulb is reduced the bacterial infection.
- ☆ Light irrigation is required during entire cropping period.

Soft Rot

Pathogen

- ☆ *Erwinia caratovora*

Symptoms

Severe discolorations with soft rotting of onion bulbs are the main symptoms of this disease. Severe infection occurs at high temperature. The affected fleshy scale tissues are water soaked and pale yellow to light brown and become soft as the rot progresses. The whole bulb may break down and a watery liquid may ooze from the having foul odour as the disease advanced and, if squeezed.

Management

- ☆ Proper curing is required after harvest.
- ☆ Proper drainage is required during cropping period.
- ☆ Reduced the doses of nitrogenous fertilizers.

Botrytis Rot

Pathogen

- ☆ *Botrytis allii*

Symptoms

Infection usually takes place through neck tissue and occasionally else where. The first sign is the softening of the effected tissue, which takes a sunken cooked appearance. Later stage, decaying starts, which is separated from healthy tissue by a definite margin. A dense grayish mycelial mat after develops upon the decaying tissue of the scope bearing short conidiophores with conidia. The disease progresses rapidly down the scales of neck tissue. Black sclerotial bodies are developed later. When diseased bulbs are cut open, water soaked brown tissues are seen near the neck region.

Management

- ☆ Seed treatment with benomyl @1 g/kg seed or benomyl + thiram @ 1 g/kg seed reduced the disease incidence.
- ☆ Pre harvest spray of benomyl @0.1 per cent reduced the fungal infection.
- ☆ Proper curing is required before storage
- ☆ Neck cutting is 2.5-3.0 cm above the bulb to reduce the infection.

Pink Root

Pathogen

- ☆ *Pyrenochaeta terrestris*

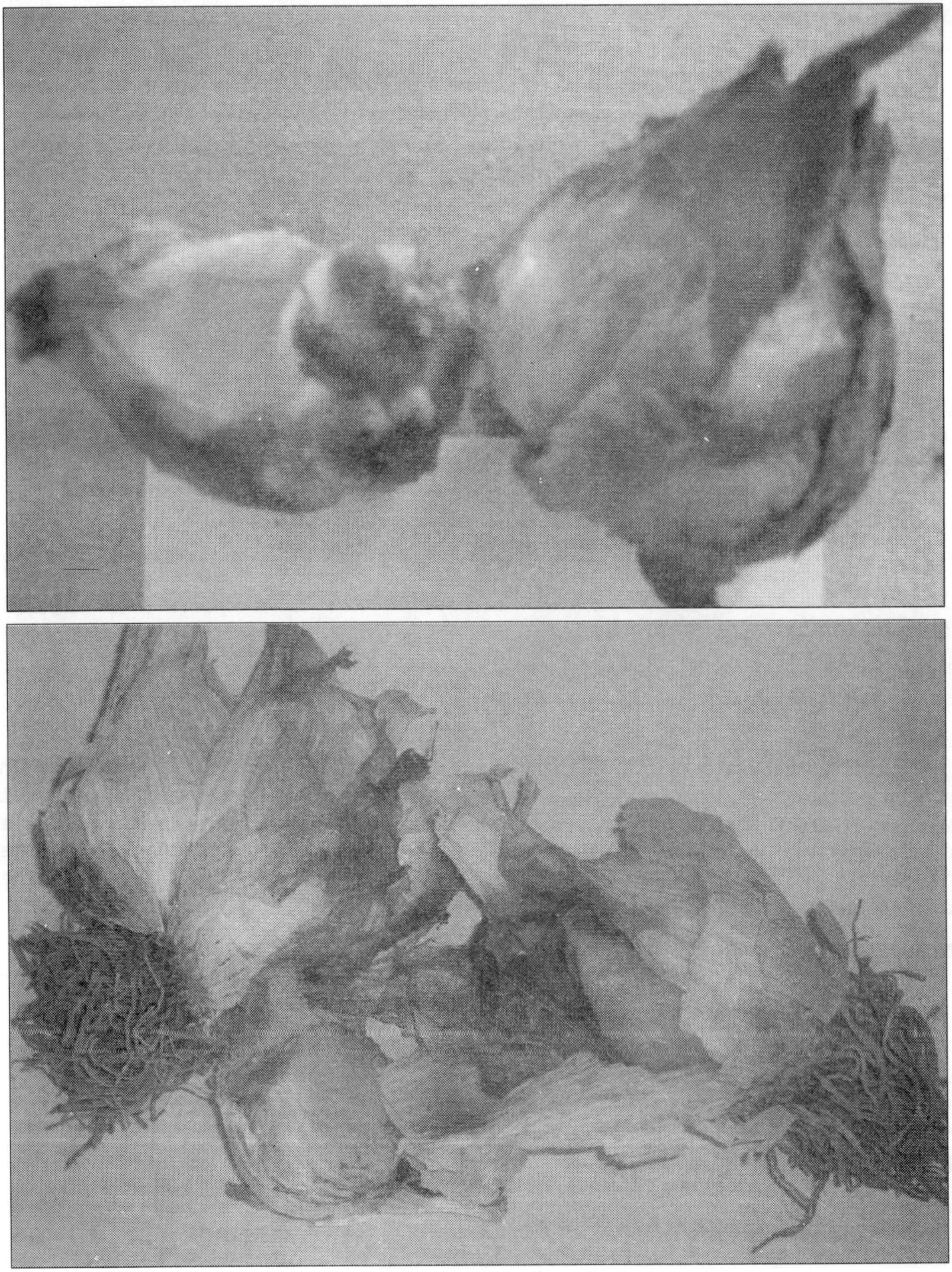

Figure 31.11: Soft Rot

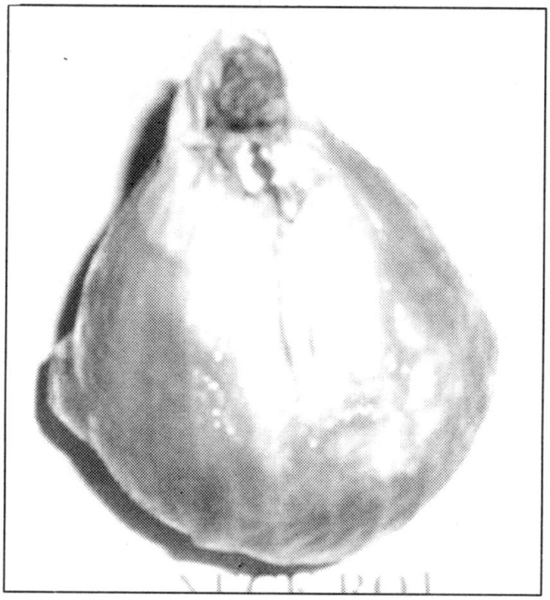

Figure 31.12: Botrytis Rot

Symptoms

The characteristic symptom is the pink colouration of the roots. The affected roots initially turn yellow but later on become soft and ultimately take distinctly deeper pink colour. The new roots, which grow from the infected plants, get immediately infected and become functionless. The affected plants are not killed but development is retarded as leaf number and size are reduced. Bulbing starts earlier, but the size is reduced at maturity. The disease is confined to the roots only.

Management

- ✰ 5-6 years crop rotation gives good control of pink root.
- ✰ Soil solarization also helps in reducing the disease.
- ✰ Use of dichloropropene @ 450 litres/ha increases the yield.

Black Mould

Pathogen

- ✰ *Aspergillus niger*

Symptoms

Fungus infects the neck of bulbs and occasionally penetrates from side and balsa end of bulb, where damage to the dry skins has exposed the bulb scales. In advance stages of the symptoms, the entire surface of bulbs turns black and all scales are infected. It reduces the market value of the bulb.

Management

- ✰ Proper drying and curing of the bulbs after harvest.
- ✰ Bulbs are dusted with calcium carbonate and godowns will fumigate with nitrogen trichloride @ 430mg/m^2.

Figure 31.13: Black Mould

Blue Mould

Pathogen

- ✰ *Penicillium corymbiferum.*

Symptoms

The first symptoms of the disease are pale yellowish blemishes, watery soft spots, or occasionally a purplish red stain on the scales. A green to blue green mould may develop on the surface of lesions when bulbs are cut longitudinally, one or more of the fleshy scales may appear water soaked and exhibit a light tan or gray colour.

Management

- ✰ Proper drying and curing after harvest and before storage.
- ✰ Injury should be avoided during postharvest handling.
- ✰ Mercurial dip of bulb/cloves was practiced before drying.
- ✰ Bulbs stored at temperature of 5°C or less with RH as low as possible may present infection.

Garlic Bulb Canker

Pathogen

- ✰ *Embellissia allii.*

Symptoms

This is new disease recorded in garlic bulbs (variety: G-313) produced in Himachal Pradesh. The initial symptoms of the disease are small blackish dark on the outer scales of the bulbs, which later enlarges and cover whole bulbs with black powdery mass. Under advance conditions, decay of the cloves has been observed. It was recorded by NHRDF for the first time in India.

Figure 31.14: Garlic Bulb Canker

Management

- ☆ Spray of Carbendazim (1.0gram/litre) should be given before 15-20 days before harvest of the crops.
- ☆ Proper curing of bulbs should be taken before storage.
- ☆ Seeds (Cloves) will be treated with bavistine (1.0 g/kg seed) before planting.
- ☆ The exotic material of garlic not is used for planting.

Future Strategies

- ☆ A series of cultural and chemical control methods for garlic pests and diseases with targets varying with crop growth stage.
- ☆ Monitoring of new diseases and appropriate management strategies should be developed.
- ☆ Exploitation of disease tolerance observed in the screened improved lines.
- ☆ Large-scale use of fungal and bacterial bioagents in a consortium to achieve better disease management.
- ☆ Studies on the use of Plant growth promoting Rhizobacteria (PGPR) in improving plant growth and offering protection from diseases.
- ☆ Develop techniques for quick diagnosis of viral diseases.
- ☆ Conversion of onion and garlic waste in to an antimicrobial agent.
- ☆ Develop IPM modules for garlic diseases.
- ☆ Since many diseases of garlic reported in other parts of the world and not present in our country, strict embargo on import of seed, planting material and the produce for domestic consumption should be followed.
- ☆ Quarantine and legislation should be mandatory in these crops.

Plant Protection, Diversity and Conservation (2011) *Pages* **417–447**
Editors: **D. Prasad & Satya Kumar**
Published by: **BIOTECH BOOKS, NEW DELHI**

Chapter 32

Microbial Toxins and their Application as Plant Protectants

K. Majumder and D.P. Ray
West Bengal Agricultural Services (Research)
Department of Agriculture, Govt. of West Bengal

The introduction of DDT in 1945 followed by the use of various other synthetic organochlorine pesticides has played a key role in the increase of agricultural productivity, protection of crops and forests and in the control of insect vectors of human diseases. However the unscrupulous use of chemical insecticides has led to wide spread contamination of water, food, non-target fauna and flora etc besides the presence of high concentration in the food chain and development of insect pest populations resistant to the chemical insecticides. Increased public concern about the potential adverse environmental effects associated with the heavy use of chemical insecticides has prompted the examination of alternative methods for insect pest control. One such alternative is the use of microbial pesticides employing microorganisms or their by-products.

Microbial pesticides comprise of microorganisms such as bacterium, fungus, virus or protozoan as the combat weapon. These can control different kinds of pests, although a specific organism may often be needed for a target pest(s). For example, a set of fungi may control weeds, and another the insects. Their toxicity to non-target animals and humans is extremely low. Their ecological advantages are that they tend to be highly selective, infecting or killing a very narrow range of target pests.

These organisms produce a mixture of structurally related compounds, of which one or more are in relative abundance. They are classified as primary or secondary metabolites. The formers are concentrated in vegetative storage organs and are generally required for the organism's physiological development. The relatively less abundant secondary metabolites, on the other hand, have apparently no function in the 'physiological development of microbes and are often derived from primary metabolites as a result of chemical adaptation to environment stress. These serve as chemical defense against pest infestation. Some of the microbes are storehouse of extractable secondary metabolites in quantities sufficient to be economically viable as raw material for biopesticides as well as for pharmaceutical and other applications. Advances in the chemical and biochemical sciences since 1950s made possible the isolation and biological characterization of the main active ingredients of all currently used pest control agents. Screening of fermentation broths for antibacterial activity has resulted in the discovery of thousands of antibiotics, a few of which have commercial utility. Rapid advances in test systems, isolation techniques, structure elucidation and recognition of structural diversity of natural products of microbial origin recently encouraged broader testing for diverse utility enabling alternative strategies to coup up with the increasing problems associated with chemical pesticides. As mentioned these products can affect several different organisms, they can be categorized according to their main use as follows: 1) insect control agents, 2) disease control agents including fungicides and bactericides; 3) weed control agents and nematicides and rodenticides.

Insect Control Agents

A large number of insecticides, either commercial or structural prototypes, obtained from microorganisms are described below.

Avermectins

Avermectin (Avid®, Agrimek®) was the most successful natural product discovered in *vivo* screen involving fermentation broths of the naturally occurring soil actinomycetes *Streptomyces avermitilis* MS and D in mice against nematode *Nematospiroides dubius* Dujardin in a dual mice-nematode system. It is a complex of eight closely related novel macrocyclic lactones, subsequently named avermectins (Campbell *et al.*, 1983). A mixture of avermectin B1a and avermectin B1b (Figure 32.1) (Springer *et al.*, 1981) was introduced as an insecticide/acaricide by Merck, Sharp and Dohme Agvet, and is now marketed by Novartis. The target for avermectin B1a involves γ-aminobutyric acid (GABA) receptor in the peripheral nervous system. The compound stimulated the release of GABA from the nerve endings and enhanced the binding of GABA to receptor sites on the postsynaptic member of inhibitory motor neurons of nematodes and on the post junction membrane of muscle cell of insects and other arthropods. The enhanced GABA binding resulted in an increased flow of chloride ions into the cells, with consequent hyper polarization and elimination of signal transduction resulting in inhibition of neurotransmission (Dybas, 1983; Turner and Schaeffer, 1989;). Avermectins are insecticides and acaricides with contact and stomach action.

Milbemectins

Milbemectins (Figure 32.2) were isolated from the soil actinomycete *Streptomyces hygroscopicus* (Jensen) Waksman and Henrici subsp. *aureolacrimosus* (Takiguchi *et al.*, 1980) and have the same mode of action as avermectins but with a narrow spectrum of activity. A mixture of milbemycin A3 and A4 has been developed as an acaricide, applicable at 5–6 g ai ha^{-1}. It stimulated Cl^- uptake by the leg muscles of the American cockroach at as low as 10^{-7} mol dm^{-1} concentration, suggesting that its mode of action involved opening the chloride channel in the plasma membrane of the nervous and neuromuscular systems (Tanaka and Matsumara, 1985). However, irreversible agonistic activity to GABA appears not to be directly mediated through binding to a GABA receptor, since bicuculline, a typical antagonist of GABA, has no effect on the activity of milbemycin. The mixture (milbemectins) has been used for the control of Tetranychidae on tea and eggplants since 1990 (Figure 32.2).

X-Y = CH = CH
R = CH_3
R = H

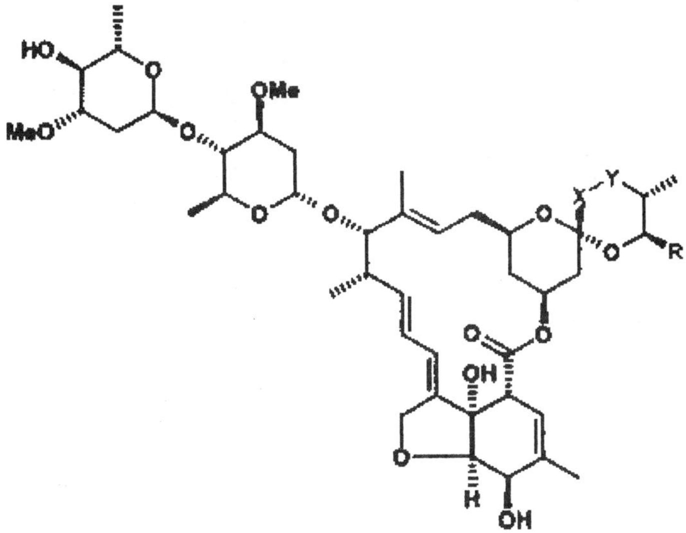

Figure 32.1

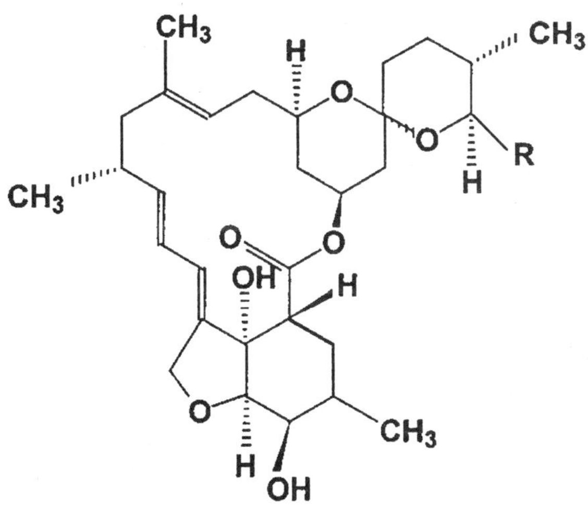

Milbemycin A_3: R = CH_3; Milbemycin A_4: R = CH_2CH_3

Figure 32.2

Nemadectin and Moxidectin

Over 20 macrocycles closely related to the milbemycins have been reported from a culture of *Streptomyces cyaneogriseus* . The major difference is the presence of an alkene group at C_{25}. One of them LL-F 28249a, named as nemadectin (Figure 32.3), is being developed as an anthelmintic agent. Another member of this series is named moxidectin (Figure 32.4) that differs from the nemadectin (Figure 32.3)

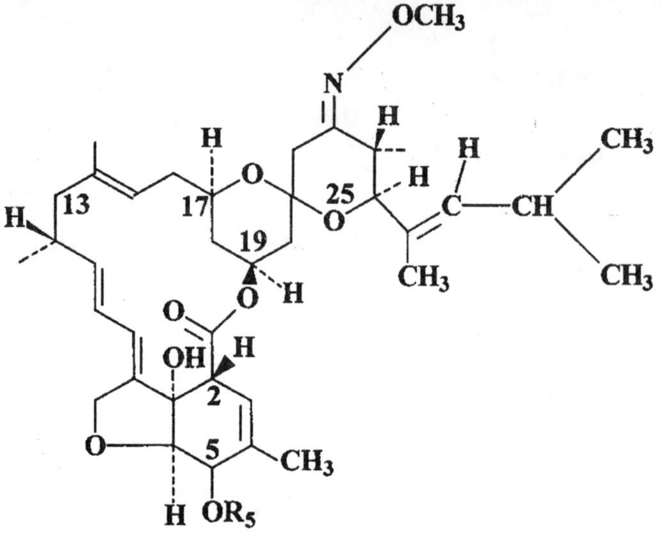

Figure 32.3: Nemadectin

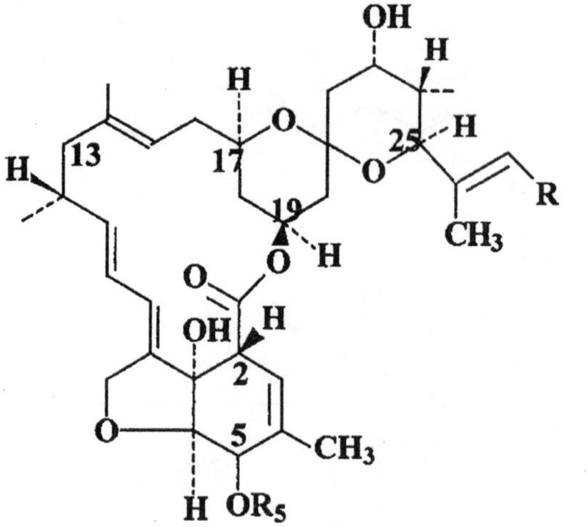

Figure 32.4: Moxidectin

by replacement of the C_{23} hydroxyl with the methyl ether of the oxime is being marketed as an acaricide. Ramsay *et al.*, 1987, have isolated milbemycin type compounds identical to the LL-F 28249 from *Streptomyces thermoarchaensis.* Several milbemycin analogues from various *Streptomyces* strains, distinguished by acyl substituents on the C_4 methyl group, and a hydroxyl at C_{22} have also been reported.

Spinosyns

The spinosyns, a new class of highly active natural insecticides were discovered in 1989 from a culture of soil actimomycete *Saccharopolyspora spinosa* (Boeck *et al.*, 1990). It is producer of very effective insecticidal compounds and the commercial product is a mixture of spinosyn A and spinosyn D (also called A83543A and A83543D, Figure 32.5) hence the name spinosad (approximately 85 per cent A and 15 per cent D). The name "spinosad" originated from a contraction of the spinosyns "A" and "D". Each spinosyn has a 12-membered macrocyclic ring that is part of an unusual tetracyclic ring system to which two different sugars, the amino-sugar forosamine and the neutral sugar 2N, 3N, 4N-tri-*O*-methylrhamnose are attached. This unique structure sets the spinosyns apart from other macrocyclic compounds (Occlowitz 1992). Spinosyn A is dextrorotary and is biologically active especially against lepidopteran insects like tobacco bud worm (*Helicoverpa virescens*), the cotton boll worm (*Helicoverpa zea*) (Thompson, 1997), American boll worm (*Helicoverpa armigera*), army worm (*Spodoptera exegua, S. littoralis, Sfurgiperda*) etc.

Spinosad (Figure 32.5) has a novel mode of action, with effects on target insects consistent with activation of the nicotinic acetylcholine receptor, but at a different site than that from nicotine or imidacloprid. It also affects GABA receptor, but their role in the overall activity is unclear. There is currently no known cross resistance to other insecticide classes. The compound causes rapid death of target phytophagous insects such as caterpillars, leaf miners, thrips and foliage feeding beetles.

Dioxapyrrolomyein

Carter *et al.*, 1987, reported the isolation of dioxapyrrolomycin (Figure 32.6) from a strain of *Streptomyces fumanus* and described its insecticidal activity. Dioxapyrrolomycin, halogen containing nitropyrroles called pyrrolomycins (Nakamura, 1987) are moderately active towards different insects and mites (Hunt and Treacy, 1998), with LC_{50}S of 10, 32, 40 and > 100 ppm for *Tetranychus urticae, H. virescens, S. eridania* and *Empoasca abrupta*, respectively (Addor *et al.*, 1992). Inspite of its undesirable toxicity to vertebrates, it has served as a lead structure for some effective analogues (Hunt and Treacy, 1998). The mode of action appears to be uncoupling of oxidative phosphorylation.

Bacillus thuringiensis

Bacillus thuringiensis is a gram positive, spore forming bacterium. There are several subspecies with a diverse spectrum of activity; for example, *B. kurstaki* attacks Lepidoptera; *B. israelensis* the aquatic Diptera; *B. tenebrionis* Coleoptera etc. Several other subspecies, infecting a limited range of insects adds further to their usefulness as biocontrol agents (Knowles, 1994).

The toxins produced by bacteria lyse the midgut cells, allowing it to penetrate the hemocoel and replicate. The four types of toxins include: α-exotoxin, β-exotoxin, γ-exotoxin, and δ-endotoxin (Figure 32.7). The α-exotoxin is toxic to vertebrates and is produced by the subspecies *B. thuringiensis* and *B. aizawai*. It is effective against a broad range of insects including Lepidoptera, Diptera, and Coleoptera. The β-exotoxin is a specific inhibitor of DNA-dependant RNA polymerase, thus interfering with RNA transcription and mitosis during molting and metamorphosis. It also can affect mammals by producing lesions in liver, adrenal glands, and kidney. The δ-endotoxin has been used for biocontrol purposes and is produced by *B. thuringiensis* or BT. During sporulation, *Bt* forms crystalline parasporal inclusion bodies which contain δ-endotoxin, either free or as part of a large protein. After uptake by feeding, parasporal bodies dissolve in insect gut and δ-endotoxin is liberated. It then docks into epithelial cells

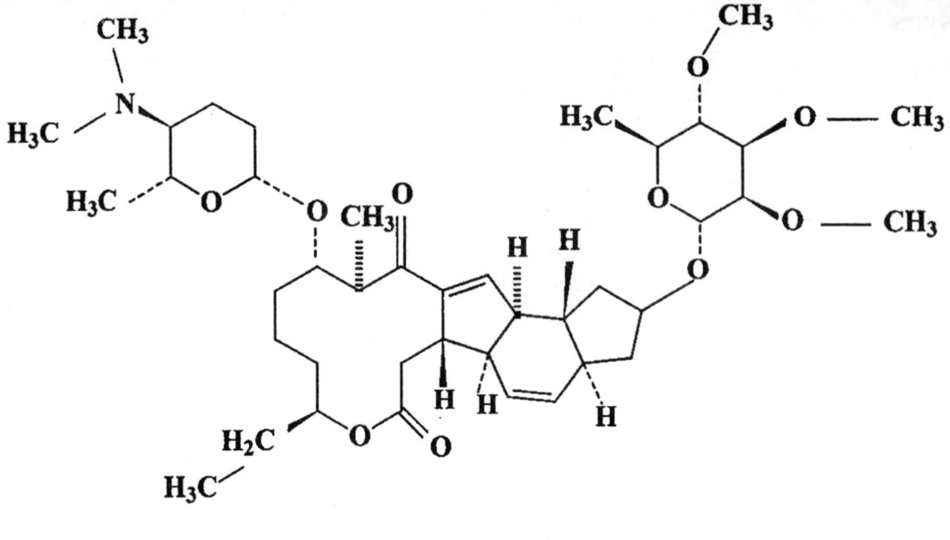

Spinosyn A

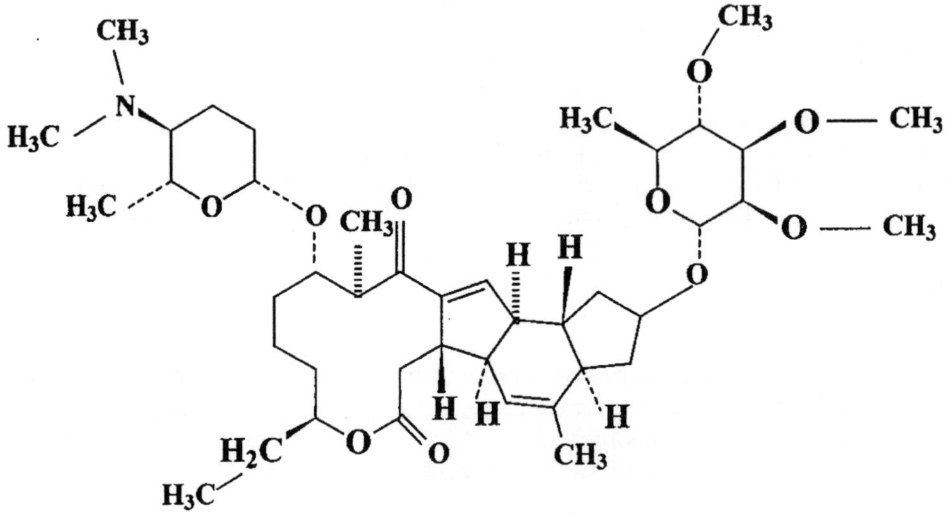

Spinosyn D

Figure 32.5

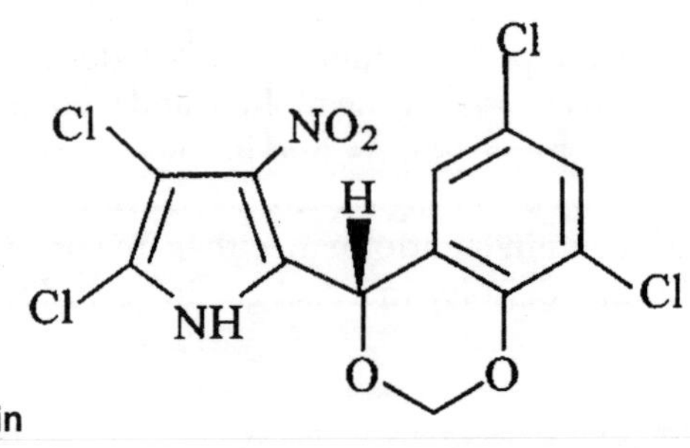

Figure 32.6: Dioxapyrrolomycin

and causes them to swell and burst which leads to death of the insect (Special issue, 1994). Due to its highly insect specific mode of action, δ-endotoxin is not toxic to other living organisms. Its first use as an insecticide was reported in 1938, and commercialization started in 1957. The β-exotoxin, "heat-stable exotoxin", is water soluble and highly toxic to larvae of several species of flies, while α-exotoxin "heat-unstable" is toxic per se to mice and to the diamond moth, *Plutella xylostella* (maculipennis). The γ-exotoxin has been defined chemically as an adenine nucleotide and ATP analogue and given the name "thuringiensin". It is composed of adenine, ribose, glucose, and phosphorylated allaric acid.

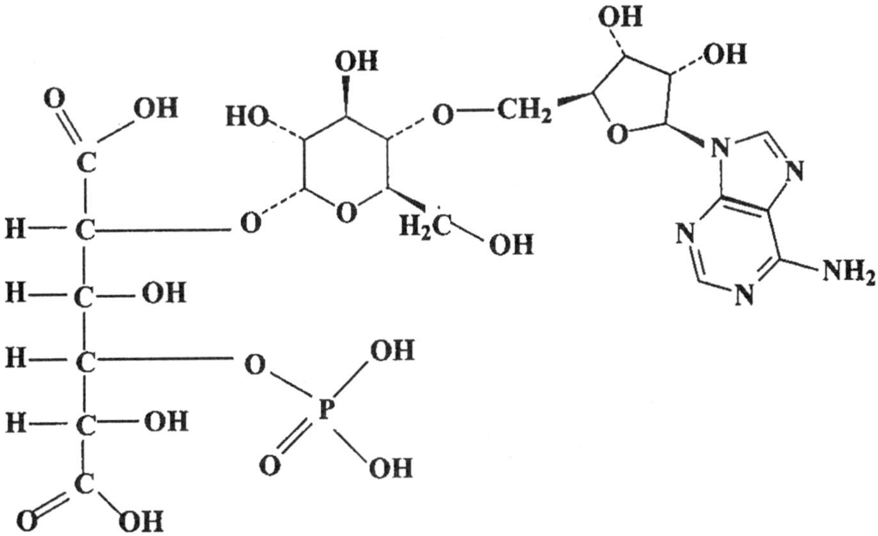

Figure 32.7: δ-Endotoxin

Many regulatory authorities opted to prevent its use in agriculture, because it has a terratogenic effect in insects and is mutagenic.

Photorhabdus luminescens

Photorhabdus luminescens, a bacterium produces a family of insecticidal toxins that, in nature, infects and kills insects with the help of a tiny worm or nematode. The toxins are active against a wide range of insects and are at least as potent as the insect-killing poisons produced by *B. thuringiensis* or *Bt*. *Photorhabdus* is expected to become an important alternative to *Bt*, or be deployed in concert with it to prolong the effective life of both biological insecticides by delaying the evolution of resistant strains of insect pests. In nature, *Photorhabdus* bacteria live inside the guts of nematodes that invade insects. Once inside an insect host, the bacteria are released from the nematode, kill the insect, and set up rounds of bacterial and nematode reproduction that turns the insect into a "protein soup," food for large numbers of nematodes. Moreover, the insect corpses left behind glow in the dark as the microbe produces luminescent proteins in addition to potent insecticides. Previous studies have shown that, in concentrated doses, the toxin can be used as a spray or fed directly to insects. The greatest potential application, however, lies in transferring the toxin-producing genes from the bacteria to crop plants.

Tetranactin

The macrotetrolide tetranactin (Figure 32.8) isolated from a culture of *Streptomyces aureus* 8-3466 has excellent acaricidal activity against *Tetranychus cinnabarinus* and other mites with low toxicity to mammals (Oishi *et al.*, 1970). It acts by uncoupling oxidative phosphoralization. The recommended use is for the control of spider mites on fruit trees. To reduce any risk of development of resistance, the polynactins, consisting of a mixture of dinactin, trinactin and major component tetranactin, are usually sold as combination products with other acaricides, under the trade name "Mitecidin" and "Mitedown". These are considered to be non-toxic to mammals with an acute oral LD_{50} to mouse of > 15000 mg kg^{-1}. Schmidt and Werner (1986) reported its total synthesis. Tetranactin had no major impact in the crop protection market, possibly because it is predominantly an adulticide and has no effect on eggs.

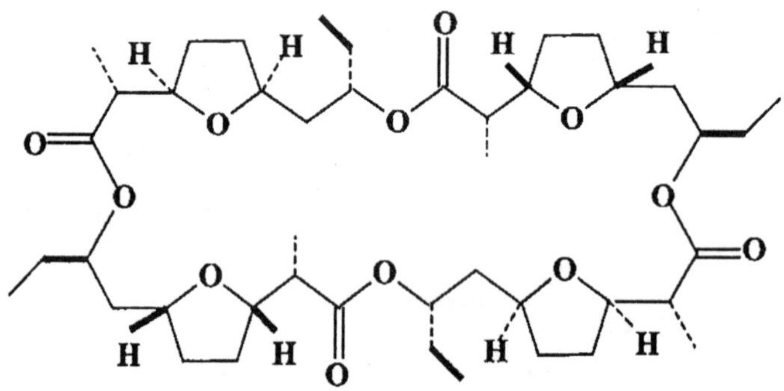

Figure 32.8: Macrotetrolide Tetranactin

Nikkomycins

The fungicidal and acaricidal nikkomycins were isolated in 1972 from culture of *Streptomyces tendae* (Dahn *et al.*, 1976) with nikkomycin X (Figure 32.9) and Z (Figure 32.10) being the main members of this group. Like the related polyoxins, the nikkomycins inhibit chitin synthase (Brillinger, 1979) and are active against spider mites even under field conditions.

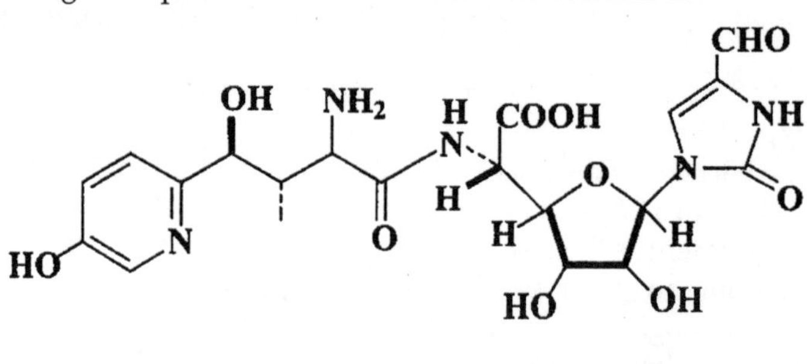

Figure 32.9: Nikkomycin X

Thiangazole

Thiangazole (Figure 32.11) has been isolated from a culture of the myxobacterium *Polyangium* sp. (Jansen *et al.*, 1992) with remarkable insecticidal activity against *Heliothis virescens* and *Lucilia sericata.*

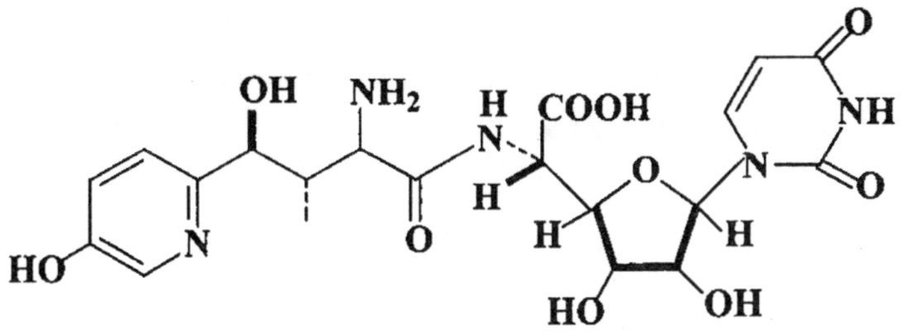

Figure 32.10: Nikkomycin Z

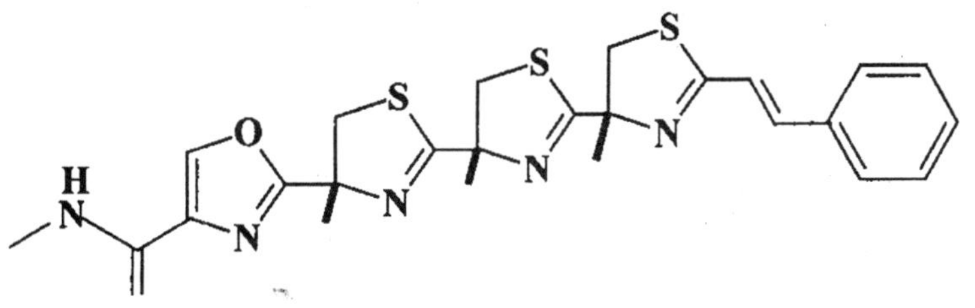

Figure 32.11: Thiangazole

Cyclophosphates

Cyclophosphates (Figure 32.12) isolated from culture filtrate of *Streptomyces antibioticus* have been shown to be very efficient inhibitors of acetylcholine esterase, with compound Tü-1718-P (R = n-propyl) being as effective on the enzyme and on insects as the commercial insecticide carbofuran (Neumann and Peter, 1987). However, this discovery came at a time, when organic phosphates are in no more demand.

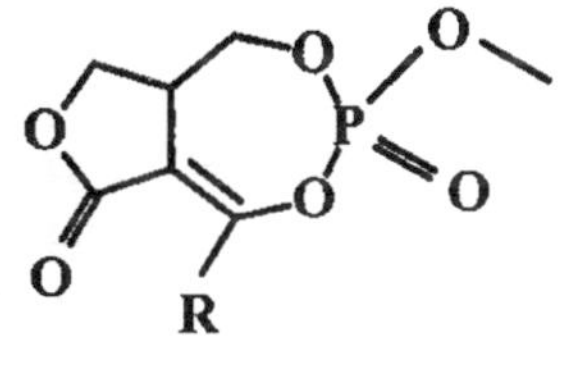

Figure 32.12: Cyclophosphate

Diabroticin

The polar insecticidal diabroticin A (Figure 32.13) produced by *Bacillus subtilis* and *B. cereus* is highly active against the southern com rootworm *Diabrotica undecimpunctata,* with an LD_{50} of 20û4 ppm when incorporated in the diet (Stonard *et al.,* 1994).

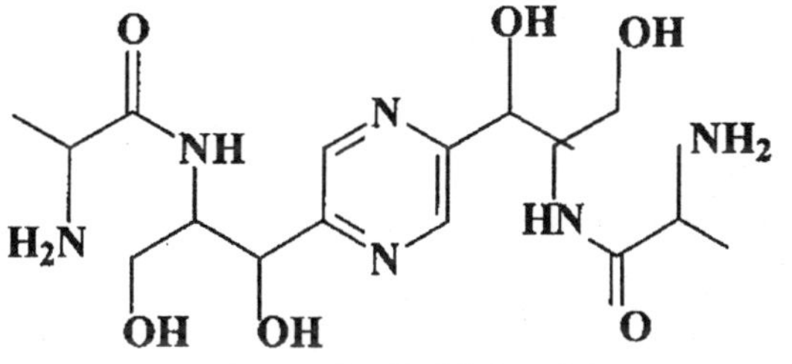

Figure 32.13: Diabroticin

Destruxins

The first systematic study of toxin production by fungal entomopathogens *in vitro* was conducted on *Metarhizium anisopliae* and led to the discovery of destruxins A and B (Kodaira, 1961). Their basic structure backbone consists of five amino acids (β-alanine, alanine, valine, isoleucine, and proline) and a ahydroxy acid. To date 28 structurally different, related destruxins (Figure 32.14) have been identified (Strasser *et al.*, 2000). Recently, destruxins A4, A1 and AS and homodestruxin B were isolated from *Aschersonia* sp. (Krasnoff and Gibson, 1996). Relationships between the chemical structure of destruxins and their biological activity have also been investigated. Destruxins A and E appear to be the most toxic than destruxin D (Dumas *et al.*, 1994). Early experiment with destruxin A and B injected into *Bombyx mori* fifth instars caused larval paralysis, but larvae also recovered within 24 hours. Destruxin cause depolarization of muscle membranes, inhibition of ATPase, and inhibit acidification of cellular compartments. Inhibition of ATPase may account for most of the effect of destruxins.

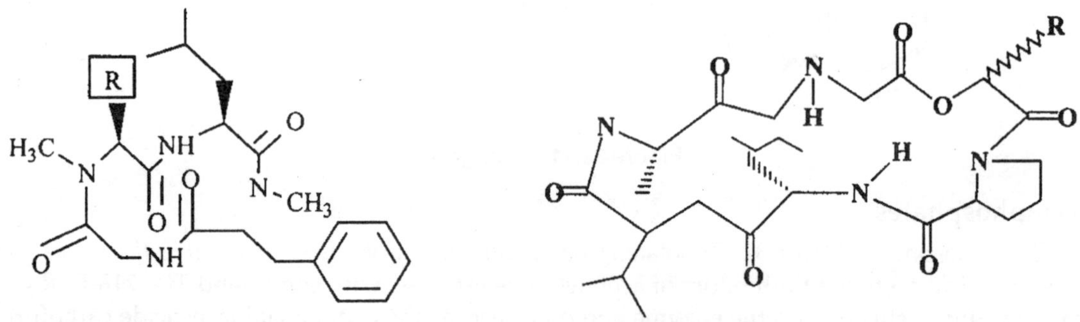

R= CH_3
CH_2-Obu
CH_2-OH
CH_2-CO_2-t-Butyl
CH_2-COOH

R= CH_3
CH_2-$CH(CH_3)$-CH_2OH
CH_2-$CH(CH_3)$-COOH
CH_2-CH=CH_2
CH_2-$CH(CH_3)_2$
CH_2-C_6H_5,
CH_2-C=CH

Figure 32.14: Destruxins

Bassianolide

A cyclodepsipeptide composed of four moles of each of L-N-methylleucine and D-α-hydroxyisovaleric acid has been isolated as the toxic principle from the mycelia of two entomopathogenic fungi *Beauveria bassiana* and *Verticillium lecanii*. (Suzuki *et al.*, 1977). Bassianolide (Figure 32.15) was lethal to silkworm larvae at 8 ppm in artificial diet. It was synthesized along with the related cyclodepsipeptides enniantin C (Figure 32.15) and decabassianolide (Figure 32.15). Enniatin A1 (Figure 32.15) has been isolated from cultures of *Fusarium avenaceum* present on balsam fir foliage and implicated as the possible causative agent in aiding the collapse of spruce budworm infestations.

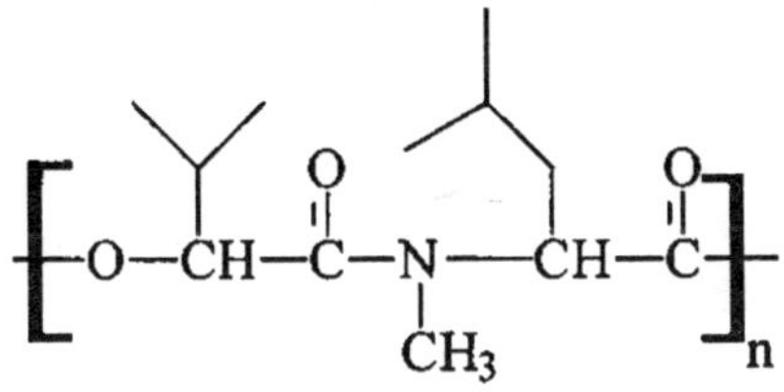

Enniatin C
Bassianolide $n = 4$
Decabassianolide $n = 5$

Figure 32.15: Bassianolide

Isarolides

From the fungus *Isaria fetina,* isarins B, C and D (Figure 32.16) were isolated. Jaspamide, an unusual cyclodepsipeptide, which incorporates the two rare amino acids ptyrosine and 2-bromoabrine in the unnatural β-configuration, was isolated from bright orange sponge *Jaspis* sp.

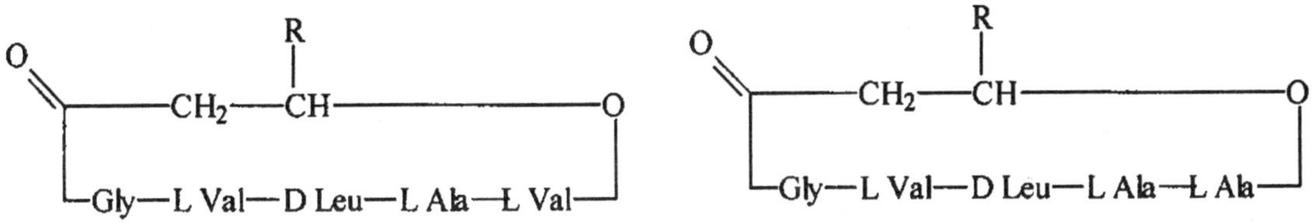

Isarin ($R= n\text{-}C_9H_{19}$)
Isarin B ($R= n\text{-}C_6H_{13}$)
Isarin C ($R= n\text{-}C_6H_{13}$)
Isarin D ($R= n\text{-}C_4H_9$)

Figure 32.16: Isarolides

Valinomycin

Valinomycin (Figure 32.17), isolated from cultures of *Streptomyces fulvissimus* shown to be effective against southern armyworm and Mexican bean beetle larvae. Valinomycin has been recently identified as a mosquito larvicide.

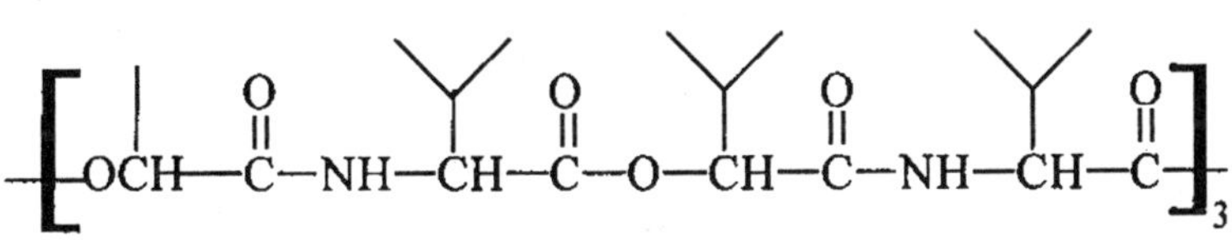

Figure 32.17: Valinomycin

Beauverolides

Two very similar cyclotetradepsipeptides, beauverolides H and I were isolated from mycelium of *B. bassiana* from South Africa (Elsworth and Grove, 1974, 1976). This is moderately effective against mosquito and fly.

Beauvericin

It is a hexadepsipeptide isolated from *Beauvericin* spp. and *Paecilomyces* spp and comprises a cyclic repeating sequence of three molecules of N-methyl phenylalanine alternating with three molecules of 2-hydroxyisovaleric acid (Figure 32.18) (Steinrauf, 1985). It is structurally and functionally similar to member damaging enniatins A, B and C. It is a specific cholesterol acyltransferase inhibitor of certain cell lines and is somewhat toxic to mosquito larvae, brine shrimp, bacteria, housefly adults and cockroach adults *in vitro*.

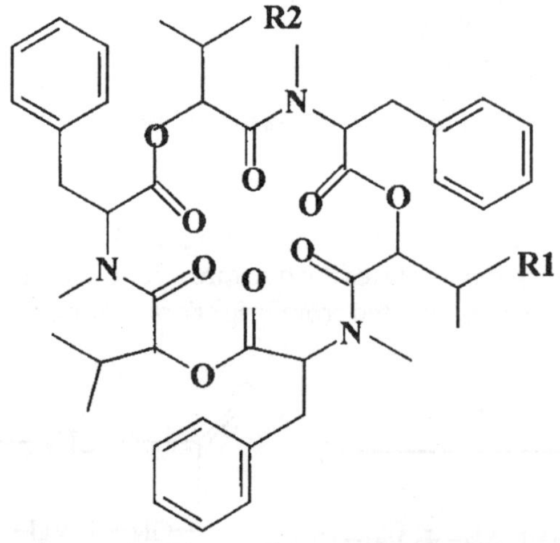

Figure 32.18: Beauvericin

Fungicides and Bactericides

With the exception of kasugamycin which is mainly used as a fungicide, the natural products and analogues which have so far been developed for controlling bacterial plant pathogens are based on the same modes of action as some of the antibiotics used in medicine. For the control of bacterial plant diseases, antibiotics like streptomycin and oxytetracyclin are being used in some countries. There is a concern however, that their application in the environment might cause natural resistance, rendering these antibiotics useless for medical treatment. Compared with the economic losses entailed by bacterial plant diseases, those caused by fungi are much larger. The search for novel bactericides suitable for crop protection also turned out to be exceedingly difficult in contrast, to the useful antifungal natural products and analogues was very successful.

Blasticidin S

From culture filtrate of *Streptomyces griseochromogenes* (Takeuchi *et al.*, 1958; Swaminathan, *et al.*, 1981) blasticidin S (Figure 32.19) was discovered in Japan in 1958 to control rice blast *Pyricularia oryzae* at 10–30 g ai ha^{-1} (Yamaguchi, 1996), however, due to toxic and phytotoxic side effects, it has lost ground in the market to other better performing products like kasugamycin (Figure 32.20).

Kasugamycin

Umezawa *et al.* (1965) discovered kasugamycin (Figure 32.20) as a bactericidal and fungicidal metabolite of *Streptomyces kasugaensis*. Simultaneously, its structure (Suhara *et al.*, 1966) and total synthesis (Suhara *et al.*, 1968) was reported. It acts as an inhibitor of protein biosynthesis in

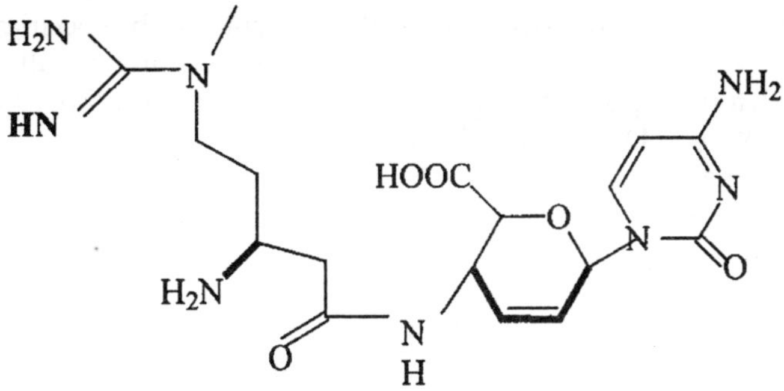

Figure 32.19: Blasticidins

microorganisms but not in mammals (Umezawa, 1966), and has excellent toxicological properties with systemic in action. Hokko Chern. Ind. developed a fermentative production process to market it for the control of rice blast *Pyricuiaria oryzae* and bacterial *Pseudomonas* diseases in several crops.

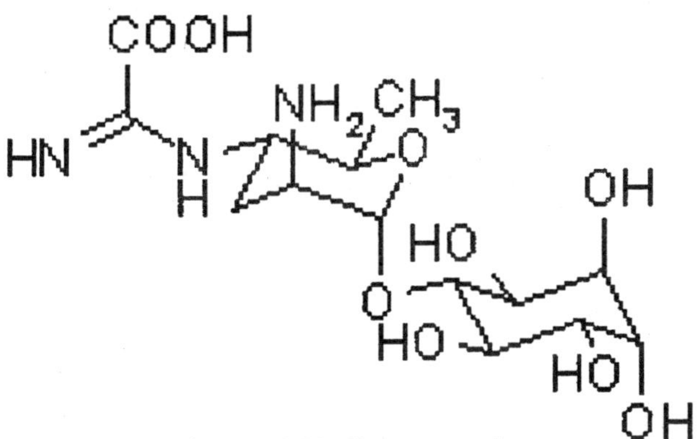

Figure 32.20: Kasugamycin

Polyoxins

Suzuki *et al.* (1965) isolated polyoxin B (Figure 32.21) and D (Figure 32.22) as metabolites of *Streptomyces cacaoi* var *asoensis*. Polyoxins are a new class of natural fungicides (Isono *et al.*, 1965; Isono

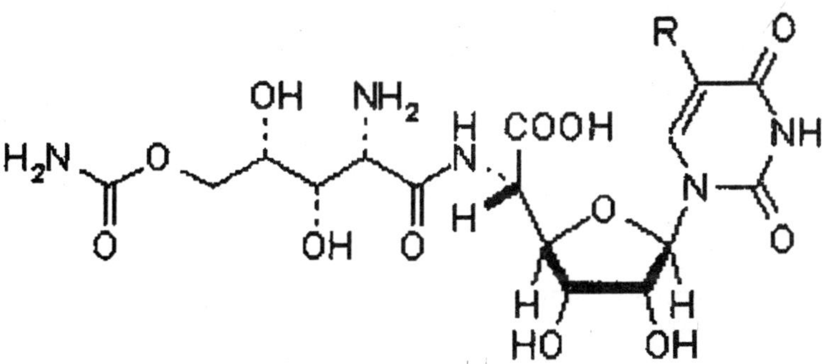

Figure 32.21: R = CH_2OH polyoxin B
Figure 32.22: R = COOH polyoxin D

and Suzuki, 1968a;)which interfere with the fungal cell wall synthesis by specifically inhibiting chitin synthase (Endo and Misato, 1969). Both polyoxin B and D are commercially produced through fermentation. Polyoxin B found application against a number of fungal pathogens in fruits, vegetables and ornamentals, and polyoxin D is marketed as Zn-salt by several companies to control rice sheath blight caused by *Rhizoctonia solani.*

Validamycin

The validamycin family was discovered by Takeda researchers in 1968 in a greenhouse assay while screening streptomycete extracts for activity against rice sheath blight caused by *Rhizoctonia solani* (Shibata, *et al.*, 1970). Validamycin A (Figure 32.23) was commercialized by developing a fermentation process with *Streptomyces hygroscopicus var. limoneus*. The same group of antibiotics was discovered independently in China and given the name jinggangmycins. The structure of validamycin A (Suami, 1980) and the total synthesis were subsequently reported (Ogawa, 1983). Validamycin A was found to be a prodrug which is converted within the fungal cell to validoxylamine A (Figure 32.23b), an extremely strong inhibitor of trehalase (Kameda, 1987). This mode of action gives validamycin A (Figure 32.23) a favourable biological selectivity, because vertebrates do not depend on hydrolysis of the disaccharide trehalose.

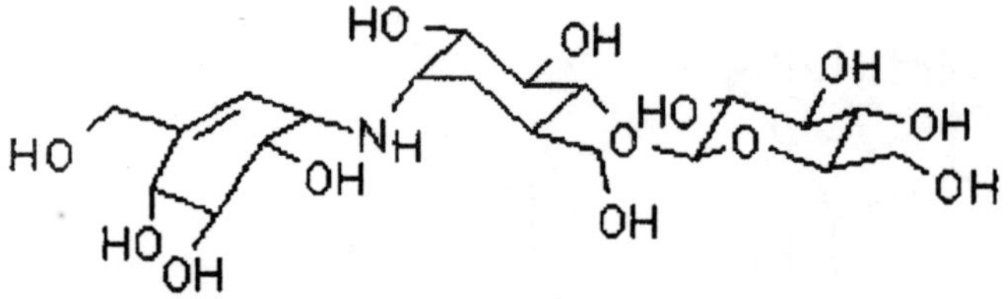

Figure 32.23: Validamycin A

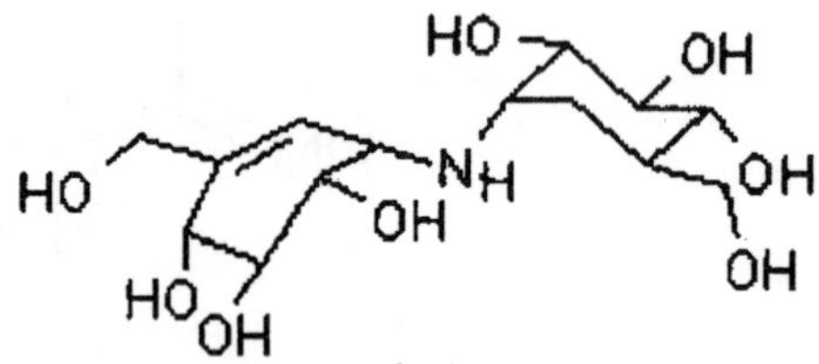

Figure 32.23b: Validoxylamine A 2.5

Pyrrolnitrin

Arima *et al.* (1964, 1965) described the antifungal activity of pyrrolnitrin (Figure 32.24) isolated from *Pseudomonas pyrrocinia.* It was subsequently synthesized (Nakano *et al.*, 1966; Gosteli, 1972). Although pyrrolnitrin showed excellent activity *in vitro* and greenhouse against the fungal plant pathogens *Botrytis cinerea* and *Pyricularia oryzae,* its performance in the field was disappointing, as it rapidly decomposed when exposed to sunlight. Replacement of the chi oro substituent in the 3-position of the pyrrole by a cyano group led to a remarkable improvement in its stability. Thus, the half-life of fenpiclonil in simulated sunlight was 48 has compared to 1/2 h for its 3-chloro analogue (Nyfeler and Ackermann, 1992). The biological activity was subsequently optimized by appropriate substitution on the phenyl ring (Nyfeler and Ackermann, 1992). Two commercial products namely fenpiclonil (Nevill, 1988) and fludioxonil (Gehmann *et al.*, 1990), emerged as excellent seed treatment agents against fungal pathogens like *Fusarium graminearum* in maize and *Gerlachia nivalis* in wheat. Studies on the mode of action of these products and of pyrrolnitrin concluded, that these inhibited a protein kinase PKIII that is involved in the osmosensing signal transmission pathway. It was suggested that this inhibition might lead to an increased concentration of a non-phosphorylated regulatory protein, causing a deregulation of a osmosensing mitogen-activated protein kinase cascade (Pillonel and Meyer, 1997; Pillonel, 1997)

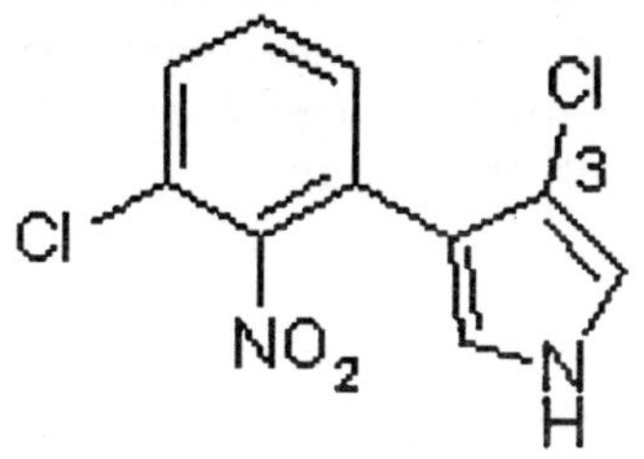

Figure 32.24: Pyrrolnitrin

Strobilurins

The antifungal antibiotic strobilurin A (Figure 32.25) initially named as mucidin, was isolated from a culture of the basidiomycete *Oudemansiella mucida* (Musilek, 1965; Musilek *et al.*, 1969). Anke *et al.* (1977) isolated strobilurin A (Figure 32.25) and B (Figure 32.26) from the basidiomycete *Strobilurus tenacellus*. On the basis of spectral data it became apparent that mucidin and strobilurin A were identical (Anke *et al.*, 1984). Such antifungal compounds are reported to be produced by several basidiomycetes (Anke *et al.*, 1989; Weber *et al.*, 1990). The strobilurins were also found as metabolites of an ascomycete, *Bolinia lutea* (Fredenhagen *et al.*, 1990a; 1990b). The antifungal activity, especially of strobilurin B, *in vitro* and in the greenhouse against plant pathogens like *Venturia inaequalis, Cercospora arachidicola, Plasmopara viticola* and *Phytophthora infestans* was excellent. *Becker et al.* (1981) showed that the strobilurins strongly inhibited mitochondrial respiration. The substructure responsible for the biological activity is the [beta]-methoxy acrylic ester, in short [beta]-MAE. This acronym is often used to describe compounds having this or a closely related toxophore. In field tests the natural strobilurins were a failure, due to their inherent photoinstability. The half-life of strobilurin A in simulated sunlight was only 12 sec (Godwin, 1992). The challenge to design photostable analogues of strobilurins with equal or even improved antifungal activity led to the development of kresoxim methyl (Figure 32.27), azoxystrobin (26) and SSF 126 (Figure 32.29), which are stable, highly active, and same mode of action as the natural strobilurins. The analogue kresoxim methyl with a half-life in

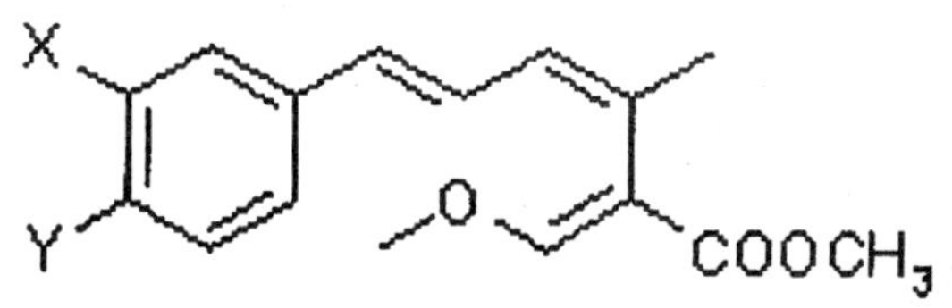

Figure 32.25: X = H, Y = H strobilurin A
Figure 32.26: X = OCH_3, Y = Cl strobilurin B

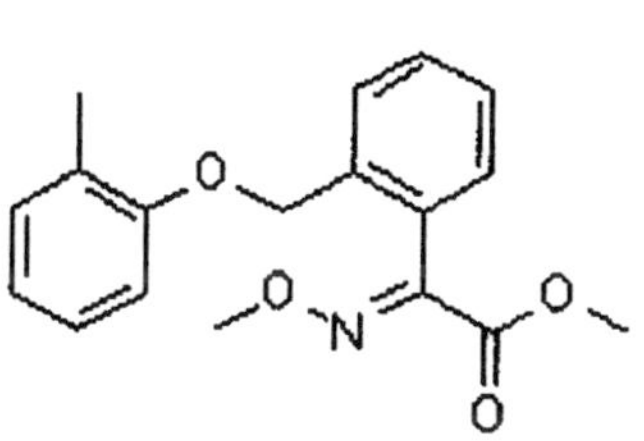

Figure 32.27: Kresoxim Methyl

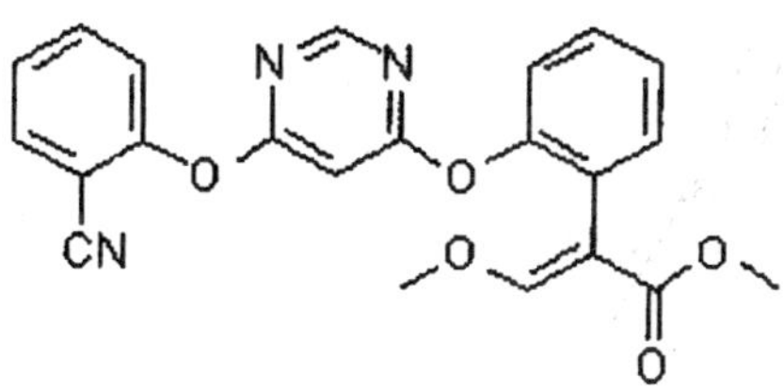

Figure 32.28: Azoxystrobin

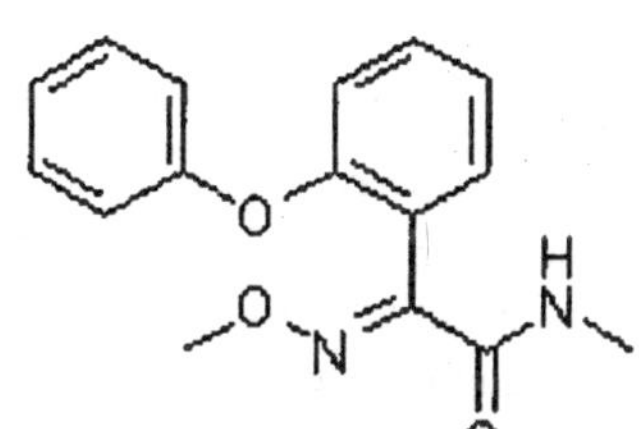

Figure 32.29: SSF 126

simulated sunlight of more than 24 h (Gold, 1994) was marketed in 1996 as a broad spectrum fungicide followed by azoxystrobin for use in cereal, fruit and vegetable market. Strobilurin analogues have also been reported to exhibit insecticidal activity (Benoit, 1992).

Mildiomycin

Antifungal mildiomycin (Figure 23.30) was reported from a culture of *Streptoverticillium rimofaciens* (Benoit *et al.*, 1978; Harada *et al.*, 1978; Harada *et al.*, 1981). It is highly active against several powdery mildews on various crops (Harada *et al.*, 1978; Harada *et al.*, 1981; Misato, 1984), through inhibition of the fungal protein biosynthesis (Feduchi *et al.*, 1985). Its low toxicity to vertebrates makes it an environmentally sound crop protection agent (Harada *et al.*, 1978), but it has never appeared in the market (Kishimoto, 1997).

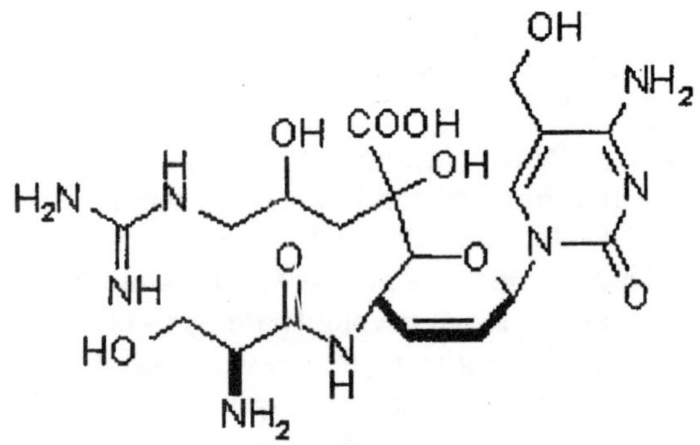

Figure 32.30: Mildiomycin

Soraphen A

Soraphen A (Figure 32.31) was isolated from myxobacteria *Sorangium cellulosum* strain So ce26, which showed broad antifungal *in vitro* activity (Bedorf *et al.*, 1993; Gerth *et al.*, 1994). The total synthesis of soraphen A has also been achieved (Abel *et al.*, 1994). Its high potency has been demonstrated in greenhouse tests (Reichenbach *et al.*, 1988; Bedorf *et al.*, 1990; Boehlendorf *et al.*, 1997) and a number of derivatives have been prepared. Most of the derivatives are less active, but some ether derivatives at the 11-OH group like (Figure 32.32) showed improved activity. However, this gain did not make up for the increased production costs, and further development was therefore focused on the natural product soraphen A, both as a broad spectrum seed dressing agent and a fungicide for foliar application. Mechanistic studies showed soraphen A to be the first antifungal agent inhibiting acetyl-CoAcarboxylase (ACC) (Vahlensieck *et al.*, 1994). The corresponding enzyme in plants is known to be the target of the aryloxyphenoxy-propionate and the cyclohexanedione herbicides (Burton, 1997) Soraphen A however, has no effect on the ACC of plants, and did not show phytotoxic effects in the field. Its efficacy under practical conditions was outstanding. As a seed treatment agent at 0.3 g ai/kg seeds it completely controlled mildew *Erysiphe graminis* in barley and snow mold *Gerlachia nivalis* in rye. Full control of apple scab *Venturia inaequalis* on apples and grey mold *Botrytis cinerea* on grapes was achieved with 10-25 g ai ha^{-1}. Apart from fungal ACC, soraphen A (Figure 32.31) also inhibits the mammalian enzyme. Due to difficulty of production and a flaw in its toxicology profile, soraphen A never reached the crop protection market.

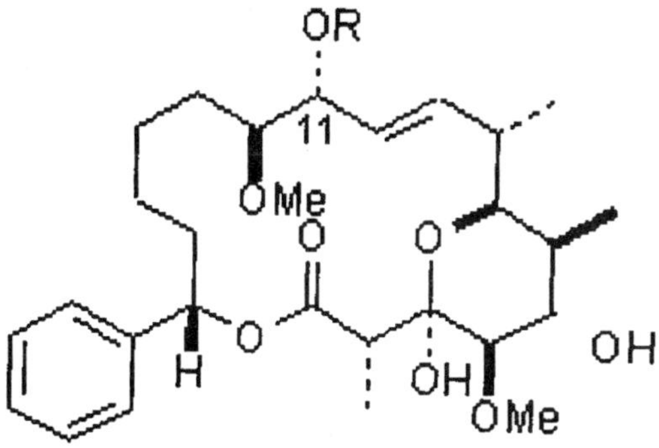

Figure 32.31: R = CH_3: soraphen A; **Figure 32.32: R = $CH_2OCH_2CH_2OCH_3$**

Other Fungicidal Leads

Antibacterial fumaramidmycin (Figure 32.33) was isolated from *Streptomyces kurssanovii* (Maruyama *et al.*, 1975). F-21223 (Figure 32.34) has been reported from a culture of the fungus *Sordaria sp.* due to its antifungal activity against *Pythium ultimum.* Based on the structure of coniothyriomycin (Figure 32.35), which has been reported from fungus Coniothyrium *sp.* (Krohn *et al.*, 1992) a number of analogues have been prepared for activity against oomycetes (Loubinoux, *et al.*, 1991).Although improvements in activity were achieved, as in the case of compound 33 with an EC_{50} of 60 ppm against *Phytophthora infestans* and *Plasmopara viticola* and of Figure 32.36 with an EC_{50} of 6 ppm against *Pythium ultimum*, a competitive level of activity was not reached. Peptides Figure 32.37–32.39 named as rhizocticines (Rapp *et al.*, 1988; Kugler *et al.*, 1990) were able to control grey mold *Botrytis cinerea* on apples and vines. It could be shown after proteolytic digestion, which L-2-amino-5-phosphono-3-(Z)pentenoic acid (Fredenhagen *et al.*, 1995) was the actual active agent. The corresponding 3-(E) compound did not show any activity. A mixture of rhizocticine A, B and D (Figures 32.37–32.39) was tested in the field against grey mold *Botrytis cinerea* on grapes.

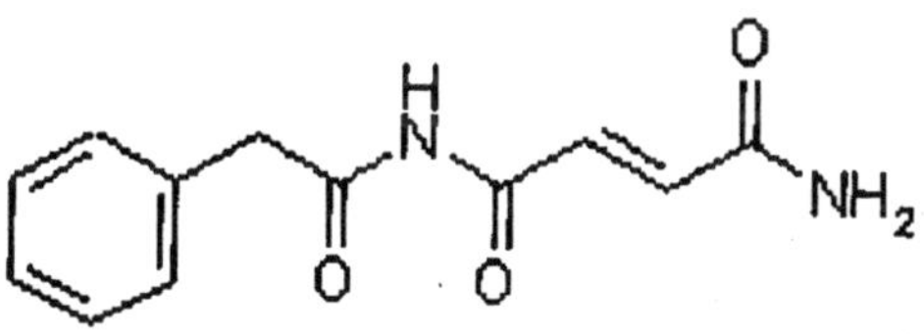

Figure 32.33: Fumaramidmycin

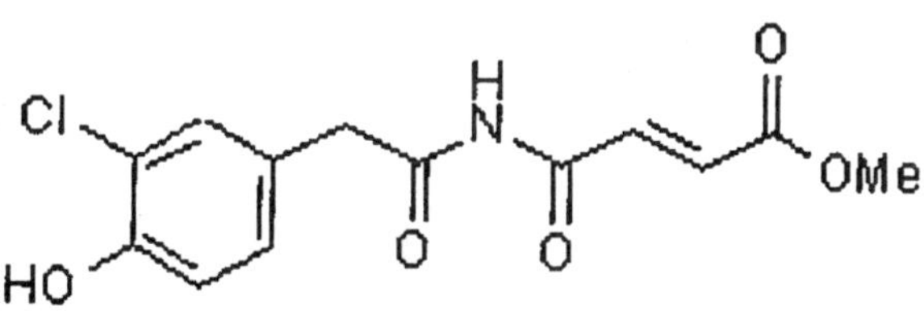

Figure 32.34: Fumarimid F-21223

Gliovirin (32.40), a natural fungicide active against *Pythium ultimum*, was isolated from *Gliocladium virens* (Stipanovic and Howell, 1982). Gliovirin was also detected in a culture of *Aspergillus viridinutans* F-4464 and tested in the greenhouse against oomycete pathogens. Its good activity against *Pythium ultimum* reported earlier was confirmed when gliovirin was incorporated into soil at 10 ppm. In other tests and especially against other oomycetes the performance of gliovirin was insufficient.

Bioherbicides

Bialanaphos

Bialanaphos (phosphinothricylalanylalanine, Figure 32.41) produced by *Streptomyces hygroscopicus* (JENSEN) Waksman and Henrici and *Streptomyces viridochromeogenes* (Krainsky)

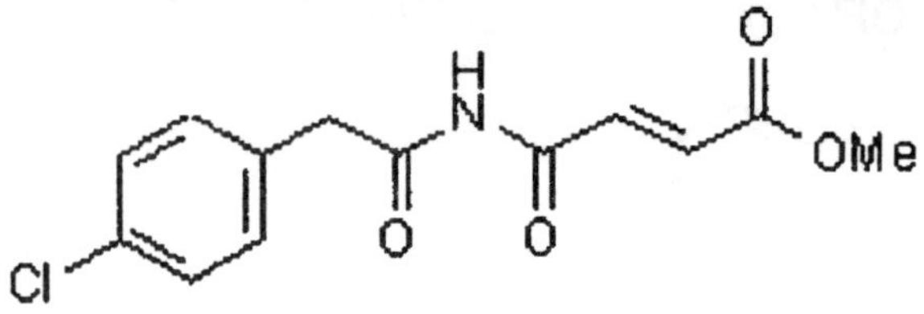

Figure 32.35

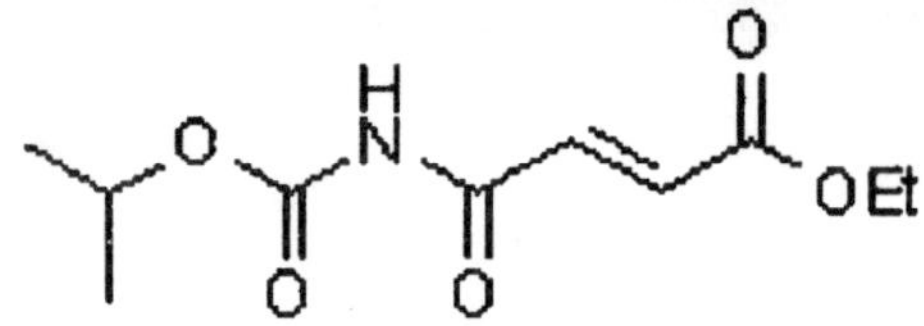

Figure 32.36

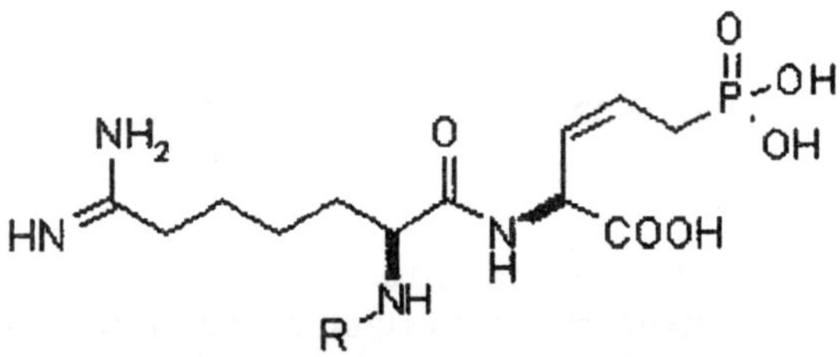

Figure 32.37: R = H rhizocticine A
Figure 32.38: R = L-val rhizocticine B
Figure 32.39: R = L-leu rhizocticine D

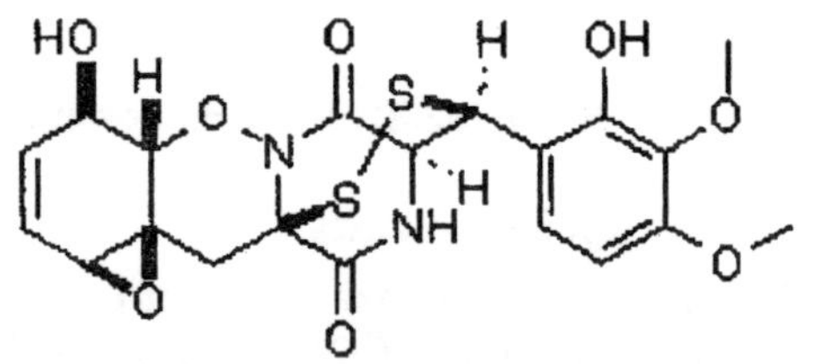

Figure 32.40: Gliovirin

Waksman and Henrici (Bayer, 1972; Ogawa *et al.*, 1973) is an alanylalanine amide of the biologically acid phosphinothricin and a potent irreversible inhibitor of glutamine synthase, causing ammonia accumulation and inhibition of photophosphorylation in photosynthesis (Omura *et al.*, 1984) It is used post emergence in vines, apples, brassicas, cucurbits, and many other crops. Bilanaphos has no *in vitro* activity on the enzyme but is converted to phosphinothricin (Figure 32.42) within the treated plants for herbicidal activity.

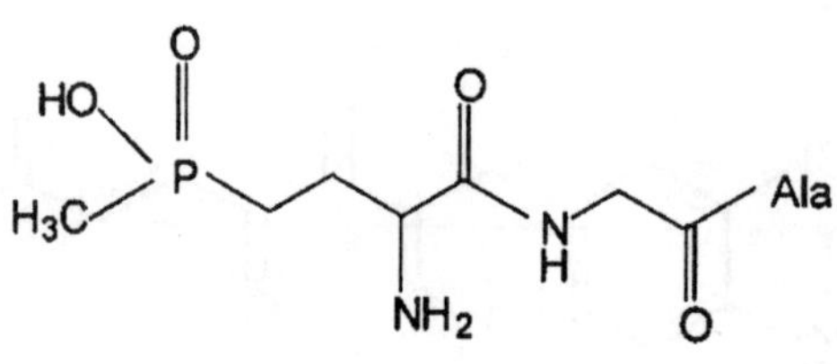

Figure 32.41: Bilanaphos

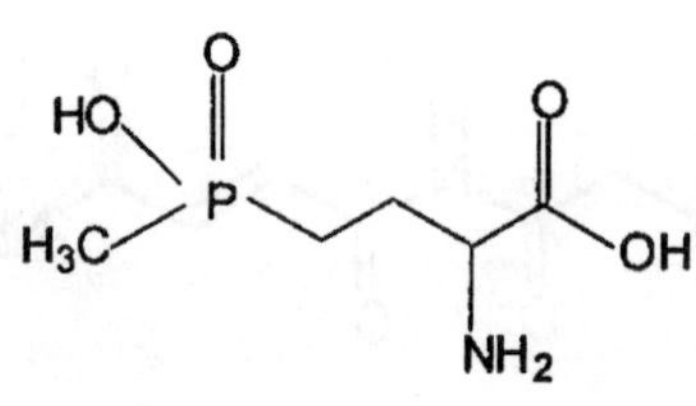

Figure 32.42: Phosphinothricin

Hydantocidin

Hydantocidin (Figure 32.43) a nonselective powerful herbicide isolated from *S. hygroscopicus* is of at least the same potency as the commercial products glyphosate and bilanaphos (Duke *et al.*, 1998; 2000). It is converted within the plant to hydantocidin-5'-phosphate, the actual herbicide which efficiently inhibits adenylosuccinate synthase.

Pyridazocidin

Pyridazocidin (Figure 32.44) is another *Streptomyces sp.* phytotoxin with an interesting mode of action (Gerwick *et al.*, 1997). It is only a weak phytotoxin but the only natural phytotoxin known to act

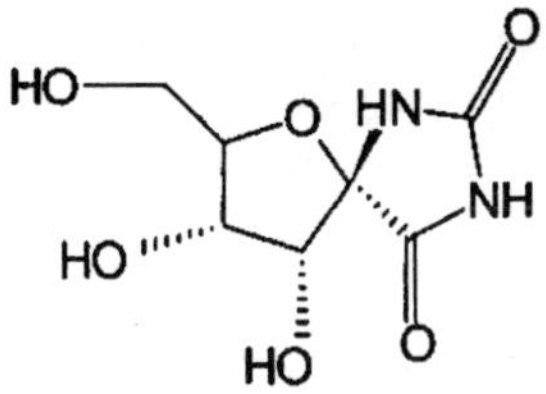

Figure 32.43

by the same mechanism as paraquat, by accepting electrons from photo system I and transferring them to molecular oxygen to generate superoxide radical.

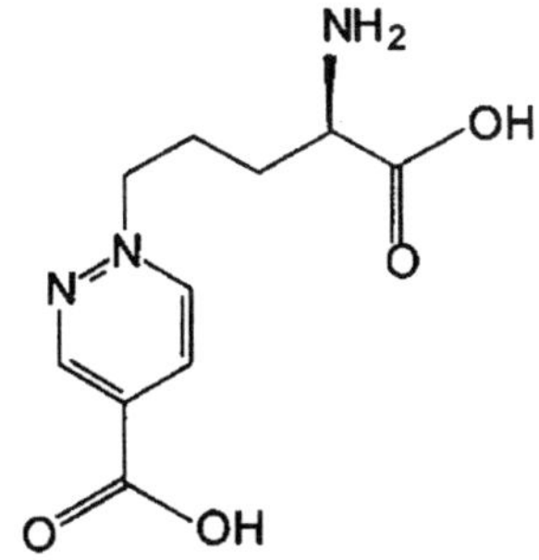

Figure 32.44

Comexistin and Hydrocornexistin

Comexistin (Figure 32.45) and hydrocomexistin are products of non pathogenic fungus *Paecilomyces variotii.* Both of these products are very phytotoxic and have been patented as herbicides (Fields *et al.*, 1996). Comexistin have unique molecular target site, asparatate amino transferases (Amagasa *et al.*, 1994). The two compounds have different selectivity on crops and weeds.

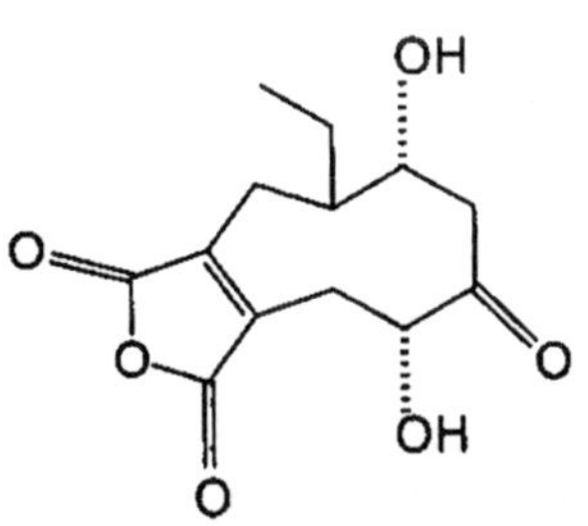

Figure 32.45

Some more interesting phytoxins from microbes are being discovered. For example phomalactone (Figure 32.46) isolated from *Nigrospora sacchari* was recently found to cause rapid electrolyte leakage of plant plasma membranes, leading to plant death (Fukashima *et al.*, 1998).

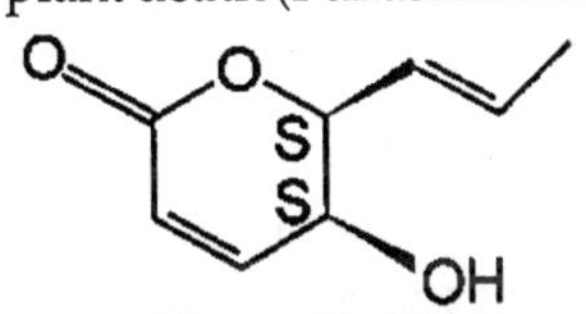

Figure 32.46

Resormycin (Figure 32.47) isolated from *Streptomyces platensis* (Igarashi *et al.*, 1997) and the nigrosporins (Figure 32.48) from fungus *Nigrospora oryzae* (Tanaka *et al.*, 1997) are interesting new phytotoxins with activity against grasses and broadleaf weeds.

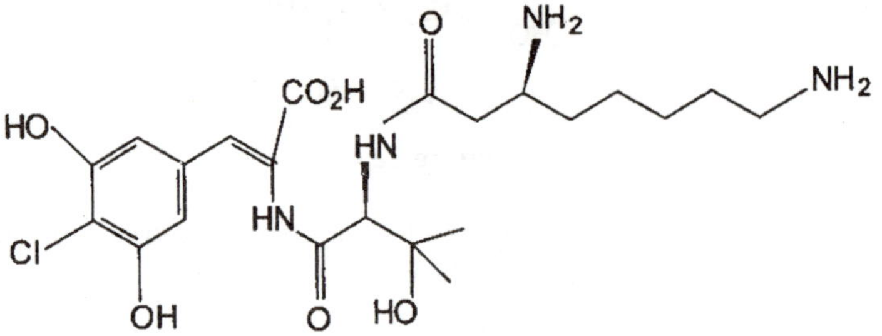

Figure 32.47: Resormycin

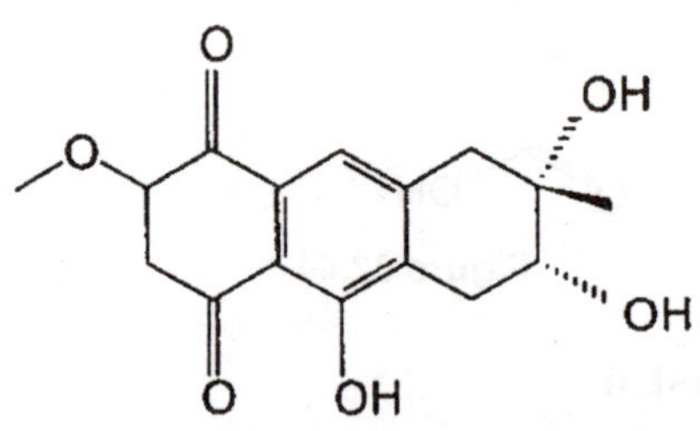

Figure 32.48: Nigrosporins

Vulgamycin

The herbicidal activity of the streptomycete metabolite vulgamycin (Figure 32.49) was detected. The post emergence application at a rate of 250 g ha^{-1} gives excellent control of several weeds without damaging cotton, barley or maize (Babczinski *et al.*, 1991) Phosphonothrixin (Figure 32.50) isolated from a culture of *Saccharothrix* sp., induces chlorosis in a non-selective way when applied to plant leaves, but the mode of action is not yet known. It controls several mono and dicotyledonous weeds at an application rate of 500 g ha^{-1} (Nakamura *et al.*, 1995). Herboxidiene (Figure 32.51) was isolated from a culture of *Streptomyces chromofuscus* A 7847, which fully controlled several weed species, while having no effect on the wheat and soybean (Isaac *et al.*, 1992). It induces apoptosis in the G2 phase of the cell cycle (Horiguchi *et al.*, 1996). Isoxazole-4-carboxylic acid, a metabolite of *Streptomyces* sp., has been reported to exert herbicidal activity in pot tests (Kobinata *et al.*, 1991). Homoalanosine (Figure 32.52) has been isolated from a culture of Streptomyces galilaeus. It acts as an antimetabolite of L-aspartic acid and L-glutamic acid. It gives a full control of weeds in the paddy field without impairing rice crop (Fushimi *et al.*, 1989).

Some Recent Works Done On Microbial Toxins

Total Synthesis of the Macrolide Antibiotic Cytovaricin (Evans *et al.*, 1990)

A convergent asymmetric synthesis of the antineoplastic macrolide antibiotic cytovaricin has been achieved through the synthesis and coupling of the illustrated spiroketal and polyol glycoside subunits. All absolute stereo-chemical relationships within the target structure were ultimately controlled by the use of asymmetric aldol, alkylation, or epxidation methodology. Union of the two subunits was accomplished by Julia-Lythgoe trans olefination, providing direct access to a suitable macrolactonization substrate. A high yielding ring closure (92 per cent) and subsequent three-step reftmctionalization of the macrocyclic product afforded cytovaricin. In supporting studies, the solution

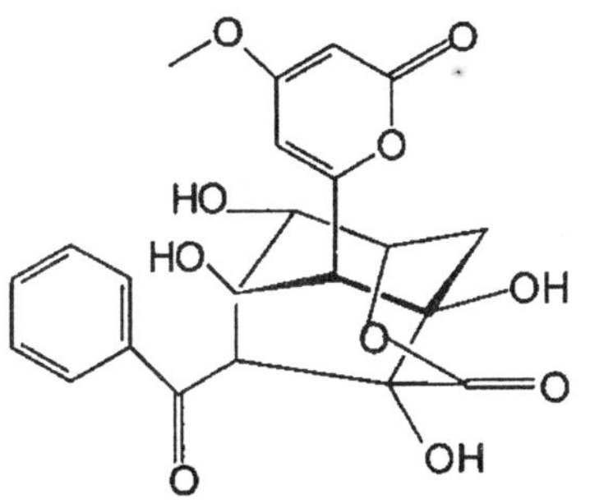

Figure 32.49: Vulgamycin

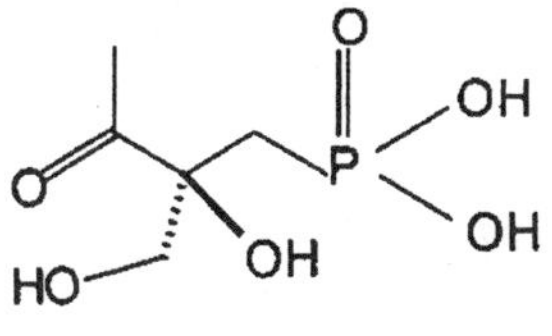

Figure 32.50: Phosphonothrixin

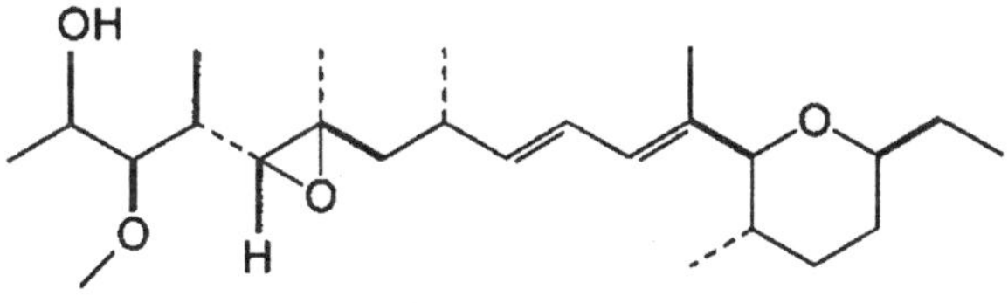

Figure 32.51: Herboxidiene

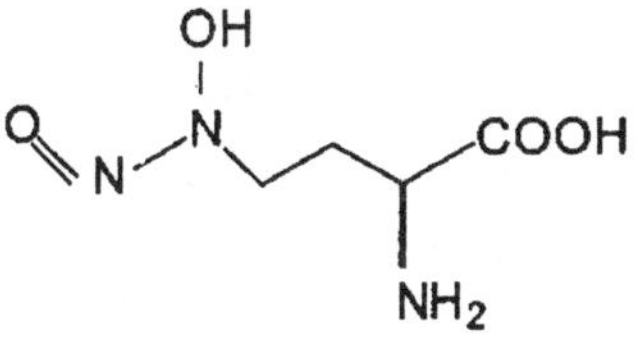

Figure 32.52: Homoalanosine

conformation and chemical reactivity of the natural product were also examined. Three-dimensional overlay of cytovaricin with rutamycin A indicates an unexpected homology between the two structures, in turn suggesting a potential mode of action for cytovaricin.

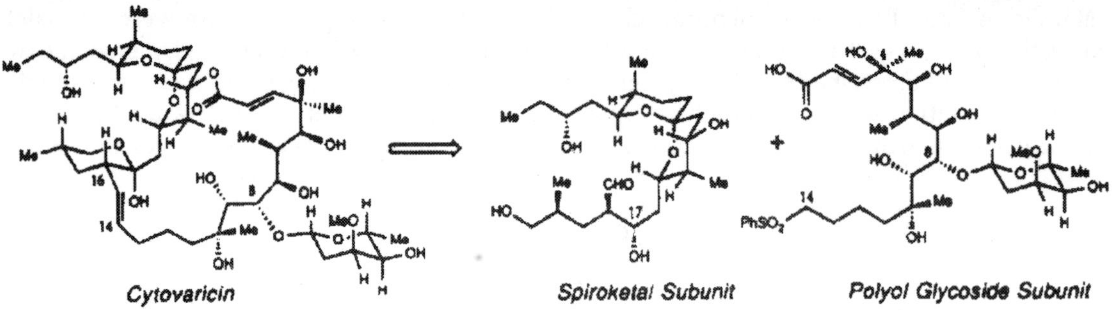

The synthetic scheme is given below:

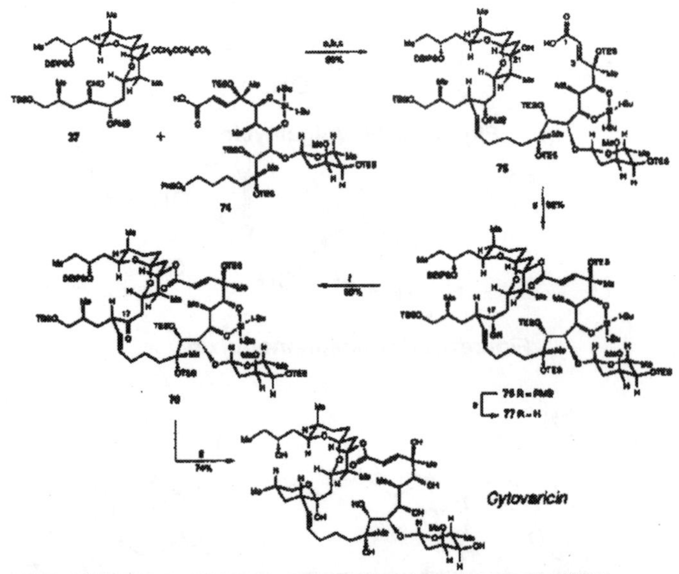

Asymmetric Total Synthesis (+)-Milbemycin D (Crimmins *et al.*, 2000)

The enantioselective total synthesis of the potent antiparasitic agent milbemycin D (1) has been achieved. The spiro ketal fragment is prepared through a novel spiroketalization of a hydroxy pyrone to set the anomeric stereocenter and establish functionality for the stereocontrolled attachment and subsequent extension of the connecting chain between the spiroketal and the hexahydrobenzofuran fragment. The hexahydrobenzofuran fragment is constructed through the exploitation of a sequential electrophilic cyclization-[2,3]-sigmatropic rearrangement to close the oxygen containing ring and incorporate the C5 hydroxyl. A lithium bromide accelerated Wittig olefination joins the spiroketalcontaining subunit and the hexahydrobenzofuran subunit at the C10, 11 double bond in high yield. Subsequent oxidation of the C1 hydroxyl provides access to the seco acid, which smoothly undergoes macrolactonization. The sensitive C2 stereochemistry and the C3,4 double bond are incorporated without epimerization at C2 or migration of the C3,4 double bond.

Scheme 1

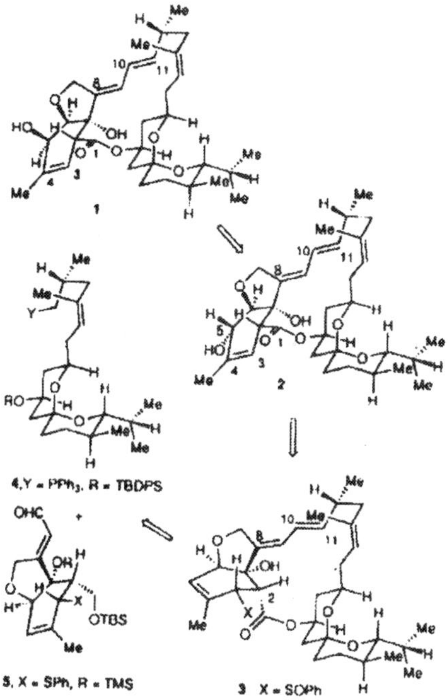

Scheme 2

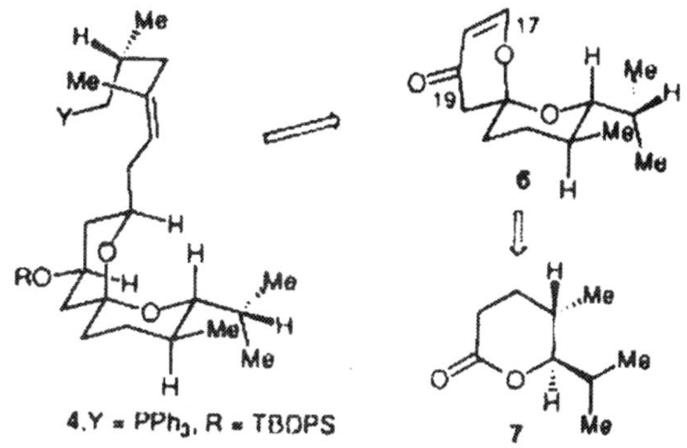

Scheme 3

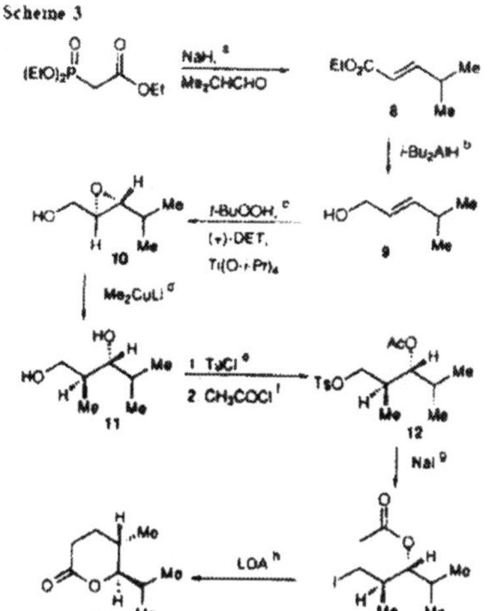

[a] THF, 25 °C, 98%. [b] THF, −78 to 0 °C, 97%. [c] CH_2Cl_2, 4 Å molecular sieves, 20 °C, 74%. [d] 0 °C, Et_2O, then $NaIO_4$, THF, H_2O, 90%. [e] Et_3N, CH_2Cl_2. [f] CH_2Cl_2, Et_3N (75%, 2 steps). [g] Acetone, 60 °C, 88%. [h] THF, HMPA, −78 °C, 94%.

Scheme 4

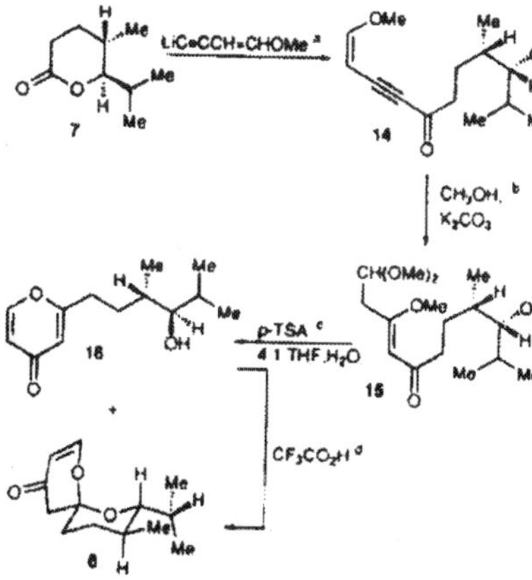

[a] THF, −78 °C, 99%. [b] 88%. [c] 55 °C. [d] C_6H_6, 90%, 2 steps

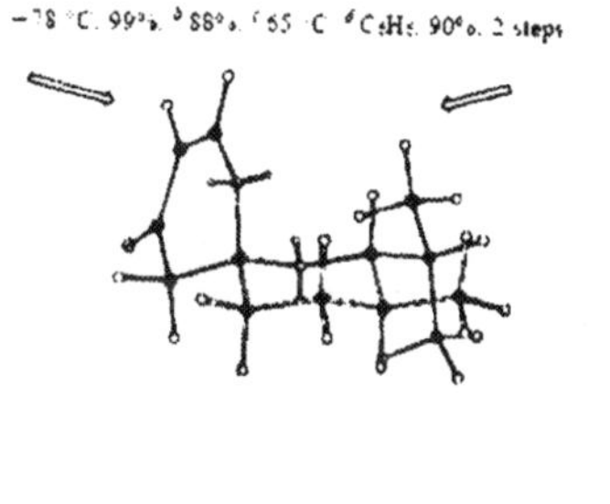

Synthesis of Anthelmintically Active Analogous of the Cyclic Octadepsipeptide PF1022A (Jeschke *et al.*, 2002)

PF1022A is the active ingredient of fungus imperfectus *Mycelia sterilia* having powerful anthelmintic activity with low toxicity in animals.

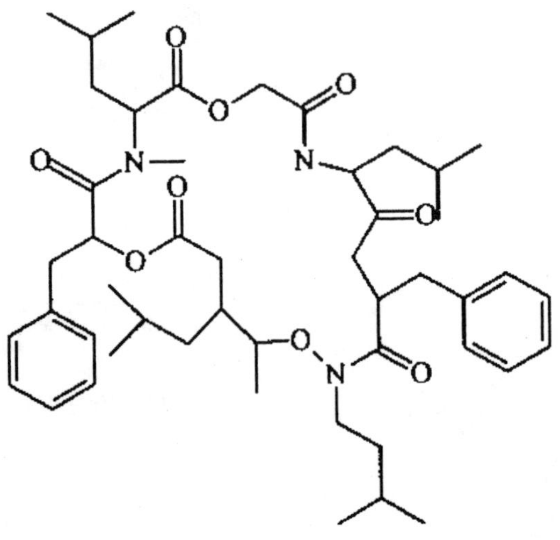

Synthesis of Analogues

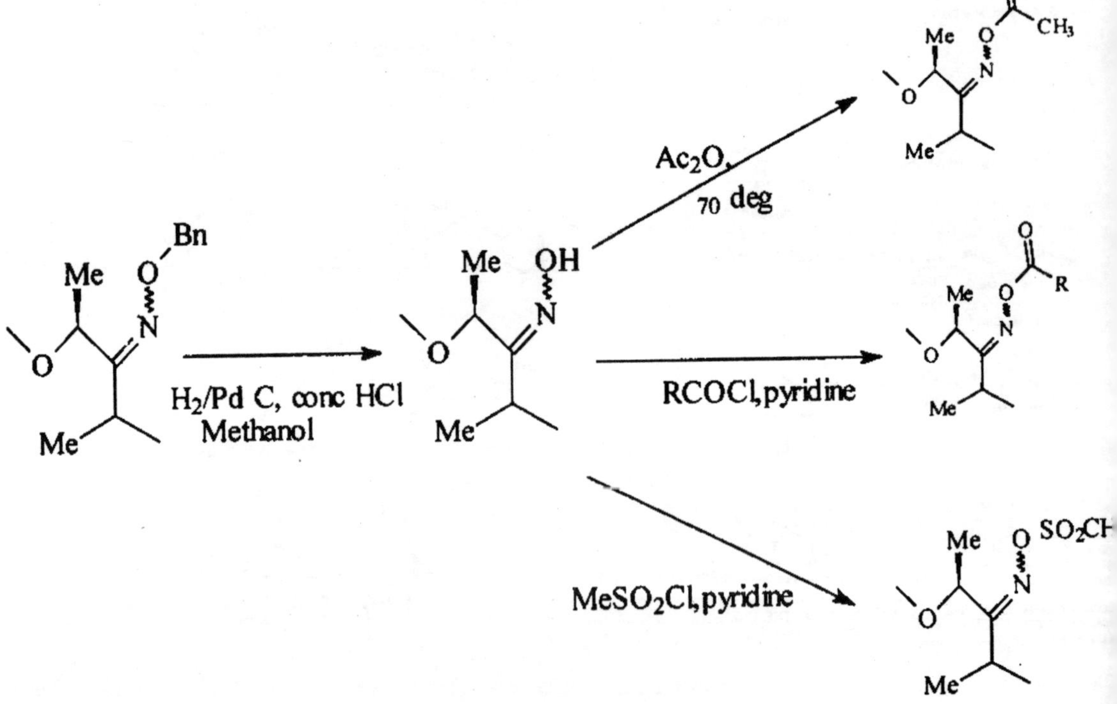

The analogues show very good broad spectrum activity against *Hymenolepsis nana* and *Haemonchus contortus*.

Synthesis of Avamectin (Raymer, 2002)

Major Synthetic Steps:

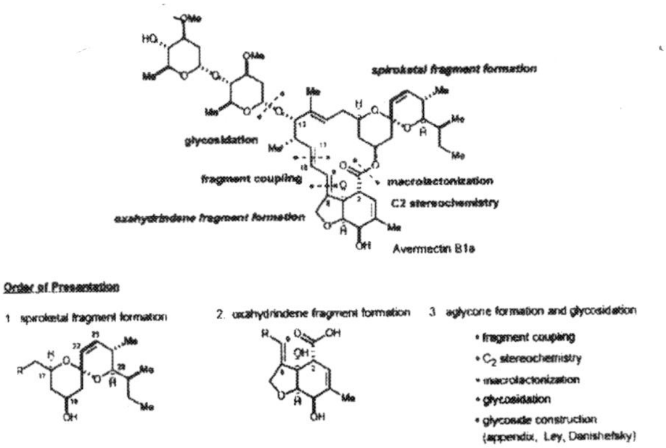

Biosynthesis: Oxahydrindene Formation

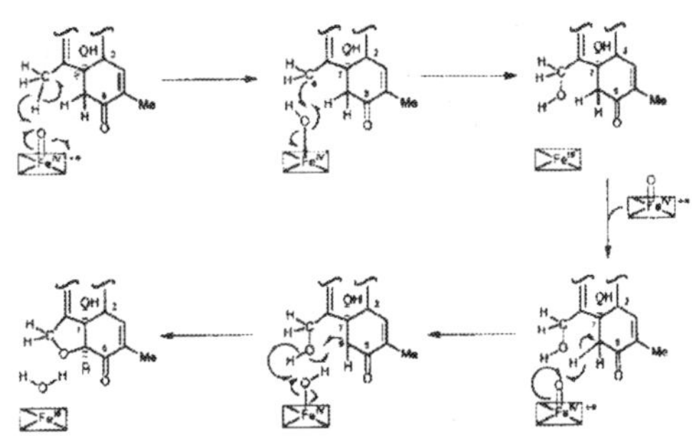

Biosynthesis: C_5 reduction and C_{13} Glycosidation

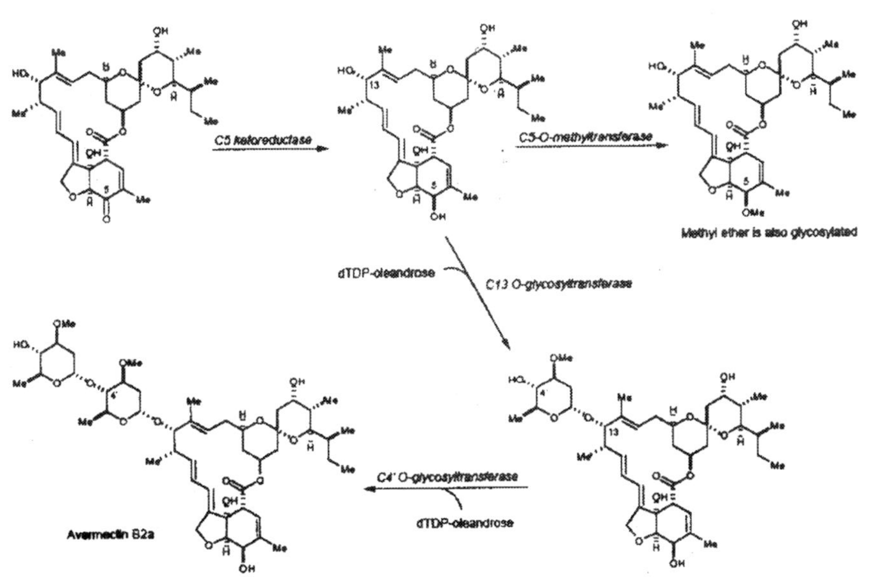

Conclusion

Microbes contain a virtually untapped reservoir of pesticides that can be used directly or as templates for synthetic pesticides. Numerous factors have increased the interest of the pesticide industry and the pesticide market in this source of natural products as pesticides. The potential bioactivity evident from the overview on secondary metabolites of microbial presented above brings out clearly their potential for use in pest control. They can also be exploited as a skeleton for new synthesis strategies.

References

Abel, S., Faber, D., Hilter, O. and Giese, B., 1994. *Angew. Chem.*, 106: 2522.

Addor, R.W., Babcock, T.J., Black, B.C., Brown, D.G., Diehl, R.E. and Furch, J.A., 1992. Insecticidal Pyrroles: Discovery and overview. In: *Synthesis and Chemistry of Agrochemicals,* (Eds.) D.R. Baker, J.G. Fenyes, J.J. Steffens. American Chemical Society, Washington DC, pp. 283–297.

Amagasa, T., Paul, R.N., Heitholt, J.J. and Duke, S.O., 1994. Physiological effects of comexistin on *Lemna pausicostata. Pestic. Biochem. Physiol.*, 49: 37–52.

Anke, T., Oberwinkler, F., Steglich, W. and Schramm, G., 1977. *J. Antibiot.*, 30: 806.

Anke, T., Schramm, G., Schwalge, G., Steffan, B. and Steglich, W., 1984. Antibiotics from basidiomycetes 20. Synthesis of strobilurins A and revision of stereochemistry of natural strobilurins. *Liebigs. Ann. Chem.*, 1616.

Anke, T. and Steglich, W., 1989. In: *Biologically Active Molecules,* (Ed.) U.P. Schlunegger. Springer Verlag, Berlin and Heidelberg, p. 9.

Arima, K., Imanaka, H., Kousaka, M., Fukuta, A. and Tamura, G., 1964. *Agr. Biol. Chem.*, 28: 575.

Arima, K., Imanaka, H., Kousaka, M., Fukuta, A. and Tamura, G., 1965. Studies on the pyrrolnitrin: A new antibiotic, isolation and properties of pyrrolnitrin. *J. Antibiot.*, 18: 201.

Babczinski, P., Dorgerloh, M., Lobberding, A, Santel, H.J., Schmidt, R.R., Schmitt, P. and Wunche, C., 1991. Herbicidal activity and mode of action of vulgamycin. *Pestic. Sci.*, 33: 439.

Bayer, E., Gugel, K.H., Hagele, K., Hageenmaier, H., Jessipow, S., Konig, W.A. and Zahner, H., 1972. Stoffwechselprodukte von Mikroorganismen, Phosphinothricin und phosphinothricinyl-alanyl-alanine. *Helv. Chim. Acta.* 55: 224.

Becker, W.F., Von Jagow, H., Anke, T. and Steglich, W., 1981. *FEBS Lett.*, 132: 329.

Bedorf, N., Boehlendorf, B., Forche, E., Gerth, K., Hoefle, G., Irschik, H., Jansen, R., Kunze, B. and Reichenbach, H., 1990. EP Pat. 358606 A2 (Gesellschaft fuer Biotechnologische Forschung mbH, Ciba–Geigy AG).

Bedorf, N., Schomburg, D., Gerth, H., Reichenbach, H. and Hofle, G., 1993. *Liebigs Ann. Chem.*, 1017.

Benoit, M., Brayer, J-L. and Demout, J.P., 1992. EP Pat. 489660 Al (Roussel–Uelat).

Boeck L.D., Chio, H., Eaton, T.E., Godfrey, O.W., Michel, K.H., Nakatsukasa, W.M. and Yao, R.C.F., 1990. EP Pat. 375316 AI (Eli Lilly Co.) 1990 (priority 30, Oct, 1989).

Boehlendorf, S., Hoefle, G., Kiffe, M., O'Sullivan, AC., Schummer, D. and Sutter, M., 1997. ACS Symp. Ser., 658: 249.

Brillinger, G.U., 1979. Metabolic products of microorganisms 181. Chitin synthase from fungi *Coprinus cinereus*, a test model for substances with insecticidal properties. *Arch. Microbiol.*, 121:71–74.

Burton, J.D., 1997. In: *Herbicide Activity: Toxicology, Biochemistry and Molecular Biology*, (Ed.) R.M. Roe. IOS Press, Amsterdam, p. 187.

Campbell, W.C., Fisher, M.H., Stapley, E.O., Albers-Schonberg, and Jacob, T.A., 1983. Invermectins: A potent new antiparasitic agent. *Science*, p. 373.

Carter, G.T., Nietsche, J.A, Goodman, J.J., Torrey, M.J., Dunne, T.S., Borders, D.S. and Testa, R.T., 1987. Japanese contribution to the development of allelochemicals. In: *Allelochemicals: Role in Agriculture and Forestry*, (Ed.) P.A Hedin. *J. Antibiotics*, 40: 233.

Crimmins, M.T., Al-awar, R.S., Vallin, I.M., Hollis, W., O'Mahony, R., Lever, J.G. and Bankaitis-Davis, D.M., 2000. Asymmetric Total Synthesis of (+)–Milbemycin D. *J. Am. Chem. Soc.*, 118: 7513–7528.

Dahn, U., Hagenmaier, H., Hohne, H., Konig, W.A, Wolf, G. and Zahner, H., 1976. Nikkomycin ein never hemmstoff der chitinsynthese bei Pilzen. *Arch. Microbiol.*, 107: 143.

Duke, S.O., Dayan, E.E. and Rimando, A.M., 1998. Natural products as tool for weed management. *Proc. Jpn. Weed Sci. Soc.* I–II.

Duke, S.O., Dayan, E.E. and Rimando, A.M., 2000. Natural products and herbicidal discovery. In: *Herbicides and their Mechanism of Action*, (Eds.) Cobb, A.H. and Kirkwood, R.C. Sheffield, UK, Academic Press, pp. 105–133.

Dumas, C., Robert, P., Pais, M., Vey, A. and Quiot, J.M., 1994. Insecticidal and cytotoxic effects of natural and hemisynthetic destruxins. *Comparative Biochemistry and Physiology*, 108C: 195–203.

Dybas, R.A., 1983. Avermectins: Their chemistry and pesticidal activity. In: *IUPAC Pesticide Chemistry*, (Ed.) J. Miyamoto and P.C. Kearney. Pergamon Press, Oxford Vol. 1, p. 83.

Elsworth, J.F. and Grove, J.F., 1974. *S.J. Afican Jour. Sci.*, 70: 379.

Elsworth, J.F. and Grove, J.F., 1976. *J. Chem. Soc. Perkin Trans*, 13: 270–273.

Endo, A. and Misato, T., 1969. *Biochem. Biophys. Res. Commun.*, 37: 718.

Evans, D.A., Kaldor, S.W., Jones, T.K., Clardy, J. and Stout, T.J., 1990. Total Synthesis of the Macrolide Antibiotic Cytovaricin. *J. Am. Chem. Soc.*, 112: 7001–7031.

Feduchi, E., Cosin, M. and Carrasco, L., 1985. Mildiomycin: A nucleoside antibiotic that inhibits protein synthesis. *J. Antibiot.*, 38: 415–419.

Field, S.C., Mireles, I. and Gerwick, B.C., 1996. Hydroxycornexistin: A new phytotoxin from *Paecilomyces variotii*. *J. Nat. Prod.*, 59: 698–700.

Fisher, M.H. and Mrozik, H., 1990. In: *Ivermectin and Abamectin*, (Ed.) W.C. Gampbell. Berlin/Heidelberg, Springer-Verlag.

Fredenhagen, A., Angst, C. and Peter, H.H., 1995. *J. Antibiot.*, 48: 1043.

Fredenhagen, A., Hug, P. and Peter, H.H., 1990. *J. Antibiot.*, 43: 661.

Fredenhagen, A., Kuhn, A., Peter, H.H., Cuomo, V. and Giuliano, U., 1990. *J. Antibiot.*, 43: 655.

Fukashima, T.M., Tanaka, M., Gohbara, M. and Fujimori, T., 1998. Phytotoxicity of three lactones from *Nigrospora sacchari*. *Phytochemistry*, 48: 625–630.

Fushimi, S., Nishikawa, S., Mito, N., Ikemoto, M., Sasaki, M. and Seto, H., 1989. Studies on a new herbicidal antibiotic homoalanosine. *J. Antibiot.*, 42: 1370.

Gehmann, K., Nyfeler, R., Leadbeater, A.J., Nevill, D. and Sozzi, D., 1990. *Brighton Crop Prot. Conf. – Pests and Diseases*, 1: 399.

Gerth, K., Bedorf, N., Irschik, H., Hofle, G. and Reichenbach, H., 1994. *J. Antibiot.*, 47: 23.

Gerwick, B.C., Fields, S.S., Graupner, P.R., Gray, J.A., Chapin, E.L., Cleveland, J.A. and Heim, D.R., 1997. Pyridazocidin: A new microbial phytotoxin with activity in the Mehler reaction. *Weed Sci.*, 45: 654–657.

Godwin, J.R., Anthony, V.M., Clough, J.M. and Godfrey, C.R.A., 1992. *Brighton Crop Protection Conference–Pests and Diseases*, 2: 435.

Gold, R.E., Koehle, H.H., Akers, A. and Sauter, H., 1994. *Brighton Crop Protection Conference–Comparing Glasshouse and Field Pesticide Performance II, Monograph*, 59: 47.

Gosteli, J., 1972. *Helv. Chim. Acta*. 55: 451.

Harada, S. and Kishi, T., 1978. Isolation and characterization of mildiomycin a new nucleoside antibiotic. *J. Antibiot.*, 31: 519–524.

Harada, S., Mizuta, E. and Kishi, T., 1981. *Tetrahedron*, 37: 1317.

Horiguchi, T., Shirasaki, M. and Tanida, S., 1996. *Takeda Kenkyushoho*, 55: 149.

Hunt, D.A. and Treacy, M.F., 1998. Pyrolle insecticides: A new class of agriculturally important insecticides functioning as uncoulpler of oxidative phosphorylation. In: *Insecticide with Novel of Action*, (Eds.) I. Ishaaya and D. Degheele. Springer, New York, pp. 138–151.

Igarashi, M., Kinoshita, N., Ikeda, T., Kameda, M., Hamada, M. and Takeuchi, T., 1997. Resormycin, a novel herbicidal and fungicidal antibiotic produced by a strain of *Streptomyces platensis*. I. Taxonomy, production, isolation and biological properties. *J. Antibiotic.*, 50: 1020–1025.

Isaac, M.B., Ayer, S.W., Elliott, R.C. and Stonard, R.J., 1992. Herboxidine: A potent phytotoxic polyketide from *Streptomyces* sp. A7847. *J. Org. Chem.*, 57: 7220.

Isono, K., Nagatsu, J., Kobinata, K., Sasaki, K. and Suzuki, S., 1965. Studies on polyoxins, antifungal antibiotics. Part I. Isolation and characterization of polyoxins A and B. *Agric. Biol. Chem.*, 29: 848–854.

Isono, K. and Suzuki, S., 1968. Structures of polyoxins D, E, F, G, H, I, J, K and L. *Agr. Biol. Chem.*, 32: 1193.

Iwasa, T., Suetomi, K. and Kusuka, T., 1978. Taxonomic study and fermentation of producing organisms and antimicrobial activity of mildiomycin. *J. Antibiot.*, 31: 511–518.

Jansen, R., Kunze, B., Reichenbach, H., Jurkiewicz, E., Hunsmann, G. and Hofle, G., 1992. Liebigs. *Ann. Chem.*, 357.

Jeschke P., Harder A., Etzel W., Gau W. and Bonse G., 2002. Synthesis of anthelmintically active analogues of the cyclic octadepsipeptide PF 1022A. *Pest Manag. Sci.*, 58: 1205–1215.

Kameda, Y., Asano, N., Yamaguchi, T. and Matsui, K., 1987. Validoxylamines as trehalase inhibitor. *J. Antibiot.*, 40: 563–565.

Kishimoto, K. and Sawada, H., 1997. *Seibutsu Kogaku Kaishi*, 75: 245.

Knowles, B.H., 1994. Mechanism of action of *Bacillus thuringiensis* insecticidal delta-endotoxins. In: *Advances in Insect Physiology*, Volume 24 (Ed. P.D. Vans).

Kobinata, K., Sekido, S., Uramoto, M., Ubukata, M., Osada, H., Yamaguchi, I. and Isono, K., 1991. Isooxazole-4-carboxylic acid as a metabolite of *Streptomyces* sp. and its herbicidal activity. *Agric. Biol. Chem.*, 55: 1415.

Kodaira, Y., 1961. Biochemical studies on the Muscardine fungi in silkworms, *Bombyx mori. J. Fac. Text. Sci. Tecnol.*, Shinshu Univ., No. 29: Ser. E(5): 1–68.

Krasnoff, S. and Gibson, D.M., 1996. New destruxins from the entomopathogenic fungus *Aschersonia* sp. *Jour. Natural Products*, 59: 485–489.

Krohn, K., Franke, C., Jones, P.G., Aust, H.-J., Draeger, S. and Schulz, B., 1992. *Liebigs Ann. Chem.*, 789.

Kugler, M., Loeffler, W., Rapp, C.A. and Kern, G., 1990. *Jung. Arch. Microbiol.*, 153: 276.

Loubinoux, B., Gerardin, P., Kunz, W. and Herzog, J., 1991. *Pestic. Sci.* 33: 263.

Lukacova, V., Polonsky, J., Moretti, C., Pettit, G.R. and Schmidt, J.M., 1982. *J. Nat. Prod.*, 45.

Maruyama, H.B., Suhara, Y., Suzuki-Watanabe, J., Maeshima, Y., Shimizu, N., Ogura-Hamada, M., Fujimoto, H. and Takano, K., 1975. *J. Antibiot.*, 28: 636.

Misato, T., 1984. *Brighton Crop Prot. Conf. Pests and Diseases*, 3: 929.

Mishima, H., Kurabayashi, M., Tamura, C., Sato, S., Kuwano, H. and Saito, A., 1975. *Tetrahedron Lett.*, 711–714.

Musilek, V., 1965. *Czech Pat.* 136492.

Musilek, V., Cerna, J., Sasek, V., Semerdzieva, M., Vondracek, M., 1969. *Folia Microbiol. (Prague)*, 14: 377.

Nakamura, H., Shiomi, K., linuma, H., Naganawa, H., Obata, T., Takeuchi, T., Umezawa, H., Takeuchi, Y. and Litaka, Y., 1987. *J. Antibiot. (Tokyo.* 1(6): 899–903.

Nakamura, K., Kimura, T., Kanno, H. and Takahashi, E., 1995. *J. Antibiot.*, 48: 1134.

Nakano, H., Umio, S., Kariyone, K., Tanaka, K., Kishimpto, T., Noguchi, H., Ueda, J., Nakamura, H. and Morimoto, Y., 1966. *Tetrahedron Lett.*, 737.

Neumann, R. and Peter, H.H., 1987. 43: 1235.

Nevill, D., Nyfeler, R. and Sozzi, D., 1988. *Brighton Crop Proto Conf.: Pests and Diseases*, 1: 65.

Nyfeler, R. and Ackermann, P., 1992. *ACS Symp. Ser.*, 504: 395.

Occlowitz, J. W., Paschel, J., Deeter, R. and Thompson, G.D., 1992. Discovery, isolation and structure elucidation of a family of structurally unique fermentation-derived tetracyclic.

Ogawa, S., Ogawa, T., Nose, T., Toyokuni, T., lwasawa, Y. and Suami, T., 1983. *Chem. Lett.*, 921.

Ogawa, Y., Yoshida, H., Inouye, S. and Niida, T., 1973. Studies on a new antibiotic SF-1293, III. Synthesis of a new phosphorus containing amino acid, a component of antibiotic SF–1293. *Meiji Seika Kenkyu Nempo*, 13: 49.

Oishi, H., Sugawa, T., Okutomi, T., Suzuki, K., Hayashi, T., Sawada, M. and Ando, K., 1970. *J. Antibiotics*, 23: 105.

Omura, S., Murata, M., Hanaki, H., Hinotozawa, K., Oiwa, R. and Tanaka, H., 1984. Phosphinothricin, a new herbicidal antibiotic phosphinothricin. containing fermentation, isolation, biological activity and mechanism of action. *J. Antibiot.*, 37: 829.

Pillonel, C., 1997. *Personal Communication.*

Pillonel, C. and Meyer, T., 1997. *Pestic. Sci.*, 49: 229.

Ramsay, M.V.J., Roberts, S.M., Russell, J.C., Shingler, A.H., Slawin, A.M.J., Southerland, D.R., Tiley, E.P. and Williams, D.J., 1987. *Tetrahedron Letter*, 28: 5353.

Rapp, C., Jung, G., Kugler, M., Loeffler, W. 1988. Liebigs Ann. Chern., 655.

Raymer B (2000) Total Synthesis of Avamectin. *Pure Appl. Chem.*, 72: 1337–1362

Reichenbach, H., Hoefle, G., Augustiniak, H., Bedorf, N., Forche, E., Gerth, K., Irschik, H., Jansen, R. and Kunze, R., 1988. EPP at. 282455 A2 (Gesellschaft flir Biotechnologische Forschung mbH, Ciba–Geigy AG), 1988 (prior. 12. Mar. 1987).

Schmidt, U. and Werner, J., 1986. *J. Chem. Soc. Chem.*, Communication, p. 996.

Shibata, M., Iwasa, O., Wakae, K., Matsuura, H., Yamamoto, M., Asai, K. and Mizuno, D.E., 1970. Pat. 1954110 (Takeda Chemical Industries, Ltd.), 1970 (prior. 29. Oct. 1968. Special issue 1994. OECD Workshop on Ecological Implications of Transgenic Crops containing *Bacillus thuringiensis* Toxin Genes. Academic Press, London. Biocontrol Science and Technology Volume 4. pp. 275–308.

Springer, J.P., Arison, R.H., Hirshfield, J.M. and Hoogsteen, K., 1981. *J. Amer. Chem. Soc.*, 103: 4221.

Steinrauf, L.K., 1985. Beauvericin and the other enniatins. In: *Metal ions in Biological systems–Antibiotics and their Complexes*, (Eds.) Sigel, H. and Sigel, A. Marcel Dekker, New York, pp. 139–171.

Stipanovic, R.D. and Howell, C.R., 1982. *J. Antibiot.*, 35: 1326.

Stonard, R.J., Ayer, S., Kotyk, J.J., Letendre, L.J., McGary, C.l., Nickson, T.E., LeVan, P. and Lavrik, P., 1994. ACS Symp. Ser., 551: 25.

Strasser, H., Vey, A., and Butt, T.M., 2000. Are there any risk in using entomopathogenic fungi for pest control, with particular reference to the bioactive metabolite of *Metarhizium, Tolypocladium* and *Beauveria* species. *Biocontrol Science and Technology* 10: 717–735.

Suami, T., Ogawa, S. and Chida, N., 1980. The revised structure of validamycin. *J. Antibiot.*, 33: 98–99.

Suhara, Y., Maeda, K. and Umezawa, H., 1966. Chemical studies on kasugamycin. V. The structure of kasugamycin. *Tetrahedron Lett.*, 1239–1244.

Suhara, Y., Sasaki, F., Maeda, K., Umezawa, H. and Ohno, M., 1968. Total synthesis of kasugamycin. *J. Amer. Chem. Soc.* 90: 6559.

Suzuki, A., Kanaoka, M., Isogai, A., Murakoshi, S., Ichinoe, M., and Tamura, S., 1977. Bassianolide: A new insecticide cyclodepsipeptide from *Beauveria bassiana* and *Verticillium lecanii. Tetrahedron Lett.*, 25: 2167–2170.

Suzuki, S., Isono, K., Nagatsu, J., Mizutani, T., Kawashima, Y. and Mizuno, T., 1965. A new antibiotic polyoxin-A. *J. Antibiot. Ser. A*, 18: 131.

Swaminathan, V., Smith, J.L., Sundaralingam, M., Coutsogeorgopoulos, C. and Kartha, G., 1981. *Biochim. Biophys. Acta*, 655: 335.

Takeuchi, S., Hirayama, K., Veda, K., Sakai, H. and Yonehara, H., 1958. Blasticidin S, a new antibiotic. *J. Antibiot.* 11: 1–5.

Takiguchi, Y., Mishima, H., Okuda, M., Terao, M., Aoki, A., Fukuda, R., 1980. Milbemycins: A new family of macrolide antibiotics: Fermentation, isolation and physico-chemical properties. *J. Antibiot.*, 33: 1120–1127.

Tanaka, K. and Matsumara, F., 1985. Action of avermectin B la on the leg muscle and nervous system of American cockroach. *Pestic. Biochem. Physio.*, 24: 124.

Tanaka, T., Fukushima, T., Tsujino, Y. and Fusimori, T., 1997. Nigrospora A and b, new phytotoxic and antibacterial metabolites produced by a fungus *Nigrospora oryzae*. *Biosci. Biotech. Biochem.*, 61: 1848–1855.. The Pesticide Manual, 10th Edition, Ed. C. Tomlin, British Crop Protection Council, 1994, p. 824.

Thompson, G.D., Michel, K.H., Yao, R.C., Mynderse, J.S., Mosburg, C.T., Worden, T.V., Chio, E.H., Sparks T.C. and Hutchins, S.H., 1997. The discovery of *Saccharopolyspora spinosa* and a new class insect control products. *Down to Earth*, 52: 1–5.

Turner, M.J. and Schaeffer, J.M., 1990. In: *Ivermectin and Abamectin*, (Ed.) Cambell W.C., Springer Verlag, New York, p. 73.

Umezawa, H., 1966. *Chemotherapy (Tokyo.* 14: 464.

Umezawa, H., Okami, T., Hashimoto, T., Suhara, Y., Hamada, M. and Takeuchi, T., 1965. A new antibiotic, kasugamycin. *J. Antibiot. Ser.*, A18: 101–103.

Vahlensieck, H.F., Pridzun, L., Reichenbach, H. and Hinnen, A., 1994. *Curr. Genet.*, 25: 95.

Weber, W., Anke, T., Steffan, B. and Steglich, W., 1990. Antibiotic from basidomycetes 32. Strobilurins E, a new cystatic and antifungal E-beta methoxyacrylate from *Crepidotus fulvotomentosus*. *J. Antibiot.*, 43: 207.

Yamaguchi, I., 1996. Pesticides of microbial origin and application of molecular biology. In: *Crop Protection Agents from Nature: Natural Products and Analogues*, (Ed.) L.G. Copping. Royal Soc. Chem. Cambridge, UK p. 27–49.

Plant Protection, Diversity and Conservation (2011) *Pages* **448–456**
Editors: **D. Prasad & Satya Kumar**
Published by: **BIOTECH BOOKS, NEW DELHI**

Chapter 33

Fungicidal Application for Control of Alternaria Blight in Chickpea

Versha Misra
Botany Department, D.G. College, Kanpur – 208 001, U.P.

The legumes (popularly knows as pulses) all belonging to family Fabaceae, are next only to cereals in their importance as human food especially in India where the people derive most of their protein requirements from these crops. Pulses are highly valuable since they not only supply proteins but also carbohydrates, fats vitamins and minerals. In India pulses account for about 20 per cent of the total area under food grain crops.

The green revolution is responsible for decline in area under pulse crop due to unstable and low yields. Among the major problems known to limit yields, biotic stresses play pivotal role. However under biotic stresses, diseases and insect pests are considered more important, Around 7-8 per cent crop is annually lost due to ravages of diseases alone costing nearly Rs. 1000 crores to the National exchequer. Majority of diseases attacking Bengal gram are caused by fungi and these include blight wilt and rust diseases.

A new leaf spot and blight disease caused by *Alternaria alternata* has been observed on Bengal gram crop for the last few years in agricultural farms in many places of Uttar Pradesh including Kanpur.

The foliar blight disease of chickpea caused by *Alternaria alternata* has been observed in severe form in many places of Uttar Pradesh during rabi season. Therefore, survey was conducted for two years at various places to know the severity of this disease. The data given in Table 33.1 showed that the disease was prevalent in all the areas surveyed showing its widespread occurrence in nature. In the year 2004 the intensity of disease varied 9-76 to 26.40 per cent maximum 26.40 per cent, recorded at C.S.A. University Research farm, Kanpur followed by agriculture field Unnao. Lowest 9.76 per cent intensity of disease was found of crop research farm Mauranipur, Jhansi.

Table 33.1: Prevalance and Severity of Leaf Spot and Blight Disease of Chickpea by *A. alternata* in Different Place

Sl.No.	Location	Average Intensity 2004	Disease (Per cent) 2005
1.	C.S.A. University Research Farm, Kanpur	26.40	19.80
2.	Indian Institute of Pulses Research, Kanpur	24.86	23.03
3.	Agriculture Fields, Unnao	25.36	18.00
4.	N.D. University Farm Masodha. Faizabad	12.80	8.73
5.	Agriculture Field Tirwa, Kannauj	17.43	25.60
6.	Crop Research Farm Mauranipur, Jhansi	9.76	6.33
7.	Agriculture Field, Shahjahanpur	13.46	8.73
8.	Agriculture Field, Hardoi	17.43	25.60
	General Mean+	18.43	16.97
	SE Mean +	1.75	1.49
	CD at 5 per cent	3.69	2.08

The leaves of chickpea showed rounded brownish spots and blightening from upward tip to basal part. The affected leaves appeared chlorotic which latter withered. The lesions of leaflets and pods were smaller rounded and brownish in colour but were dark brown over the stems. Severely infected pods showed many circular black dots. The disease symptoms started from upper leaves and gradually decreased downward showing chlorotic zone around infection spots.

The heavily infected leaves of chickpea with leaf spot and blight disease were collected in the crop season from Indian Institute of Pulses Research Kanpur, and brought to the laboratory for isolation of the pathogen.

For isolation the affected leaves were cut into small pieces and thoroughly washed unsterilized water and were surface sterilized by 0.01 per cent $HgCl_2$ (Mercuric Chloride) solution for 10 seconds. The pieces were washed immediately in order to remove traces of $HgCl_2$. The pieces were than blotted and placed over sterilized potato dextrose agar medium in 4" diameter petridishes, under aseptic condition. After 72 hour of incubation at 30°c the isolation was done by transferring superficial mycelium growing around infected pieces. Further purification of culture was done by repeated transfer hyphal tips and single spore culture technique.

The pathogen isolated was indentified as *Alternaria alternata* (fr) Keissiar by the author and further confirmed a C.S.A. University of Agriculture and Technology Kanpur of the basis of the character present in the fungus.

The screening of available germplasm of varieties of chickpea against Alternaria alternate under artificial epiphytotic condition of infection in the field of Indian Institute of Pulses Research Farm, Kanpur during rabi crop season during 2004-2005 and 2005-2006 65 germplasm accessions of chickpea were screened using susceptible checks cultivar under natural epiphytotic condition. They were usually sown in randomized block design with three replicates in a paired row in plots of 2.5 meter length and 3.0 meter width. One infector row after every fifth line of test germplasm was kept in order to build up good inoculum potential. The plot was irrigated from time to time. Disease intensity was assessed visually on the basis of percentage leaf area affected by the leaf spot disease at the time of harvest of crop by randomly selecting 50 leaves from each replication of 10 plants. The varieties and cultures were graded in different categories based on standard evaluation system of Khurana. *et al.* (1997)

Table 33.2

Grade	Percentage Leaf Area Affected	Reaction
0	Nil	Free
1	Upto 5 per cent	Resistant (R)
2	Upto 10 per cent	Moderately resistant (MR)
3	Upto 20 per cent	Moderately susceptible (MS)
4	Upto 30 per cent	Susceptible (S)
5	More than 30 per cent	Highly susceptible (HS)

The use of resistant varieties is considered to be the best method of disease control. Therefore The studies were carried out to find out the source of resistant against the leaf blight disease. The available 65 varieties/cultures of chickpea were screened for varietal reaction against the disease under natural condition during rabi season 2004 the results obtained have been presented in Table 33.3.

Table 33.3: Reaction of Chickpea Varieties to *A. alternnata* Under Natural Condition

Reaction	Chickpea Varieties
Highly resistant (No Infection)	ICCCC 13
Number-01	
Resistant (1-5 per cent Infection)	C235 F496, G 543G588, GL No. 89027, GNG 146, GNG 469 (Samrat)
Number-14	GPF2 (GF 89-36), JG 315, PBGI, Pusa 244 (BG244), Pusa 261 (BG 261) Pusa 413 (BGM 413)
Moderately resistant (5.1-10 per cent infection)	Awarodhi, BG 256 BG 362, BGM 408, FG 22, FG 133, Gaurav, GG 575
Number-23	GL 83055, GL 83103, GL 84297, GMG 737, GN 90062, GPF 90062, GPF 70022, H 76-2, H-82-2 JG62, JG 317, JG 326, KBK 13, Pragat, Pusa 372 (BG372)
Moderately Susceptible (10.01-20 per cent Infection)	GG 1111. GN Gaurav, GNG 1103, GMG 1107, GMG 1166, ICCC 166
Number-11	KBK-1 KGD 1168 Pant 114, Pusa 209, WCG 10
Susceptible (20.1-30 per cent Infection)	C 104, C 214 GL 76-9, H 208 Hare chhole 1, ICCC 10, JG 79-29, JG
Number-12	1265, K850, L 144 (Kabuli), L 550 (Kabuli), RSG 256
Highly Susceptible (>30 per cent Infection)	HMS 6, H 76-49, ICC 17M Radhey
Number-04	
Total number of varieties 65	

From the results it was clear that out of the 65 germplasm screened one was found to be immune, 14 resistant 23 moderately resistant, 11 moderately susceptible, 12 susceptible and 4 highly susceptible.

The control of foliar disease of chickpea was tried with the help of chemical treatment. These chemicals applied on the crop plants only protected them from subsequent infection but did not cure or stop the disease after it had started. During present study a number of fungitoxicants after their laboratory screening were used to control the blight disease of chickpea through seed treatment and foliar spraying methods.

Laboratory Screening of Fungitoxicants

Eighteen fungitoxicants belonging to different chemical groups were screened in the laboratory by "poisoned food technique" to select the effective ones for the control of foliar blight of chickpea caused by *Alternaria alternata*. The average diameter of the fungal colony of each fungitoxicants were Recorded and percent inhibition over control was calculated. The results have been given in Table 33.4.

Table 33.4: Inhibitory Effect of Fungicidal Compounds on the Growth of *A. alternata in vitro*

Sl.No.	*Fungitoxicants*	*Dose (Per cent)*	*Averrage Diameter of Fungal Colony (mm)*	*Percent Inhibition in Diameter Over Control*
1.	Indofil M-45	0.2	0.00	100.00
2.	Indofil Z-78	0.2	0.00	100.00
3.	Thiram	0.2	0.00	100.00
4.	Captan	0.2	0.00	100.00
5.	Calxin	0.05	0.00	100.00
6.	Agrosan	0.3	0.00	100.00
7.	Panolil	0.15	0.00	100.00
8.	Ceresan dry	0.3	0.00	100.00
9.	Vitavax	0.2	0.00	100.00
10.	Bavistin	0.2	7.00	90.63
11.	Difolatan	0.2	7.86	89.48
12.	Karathana	0.2	8.66	88.41
13.	Blitox-50	0.2	20.00	68.74
14.	Kavach	0.2	25.50	65.87
15.	Benlate	0.1	43.70	41.52
16.	Ridomil MZ	0.2	45.63	38.94
17.	Score	0.05	60.33	19.26
18.	Contaf	0.05	68.10	8.87
19.	Control		74.73	
	SE Mean ±	0.25		
	CD at 5 per cent		0.49	

The results showed that all the fungitoxicants were significantly superior to control in inhibiting growth of the fungus. Indofil M-45 indofil Z-78 Thiram Captan, Calixin, Agrosan Penolil, Ceresan dry and Vitavax proved highly effective because they inhibited the growth of the fungus completely (100 Per cent) others including Bavistin diflolatan and karathane were equally good showing inhibition of

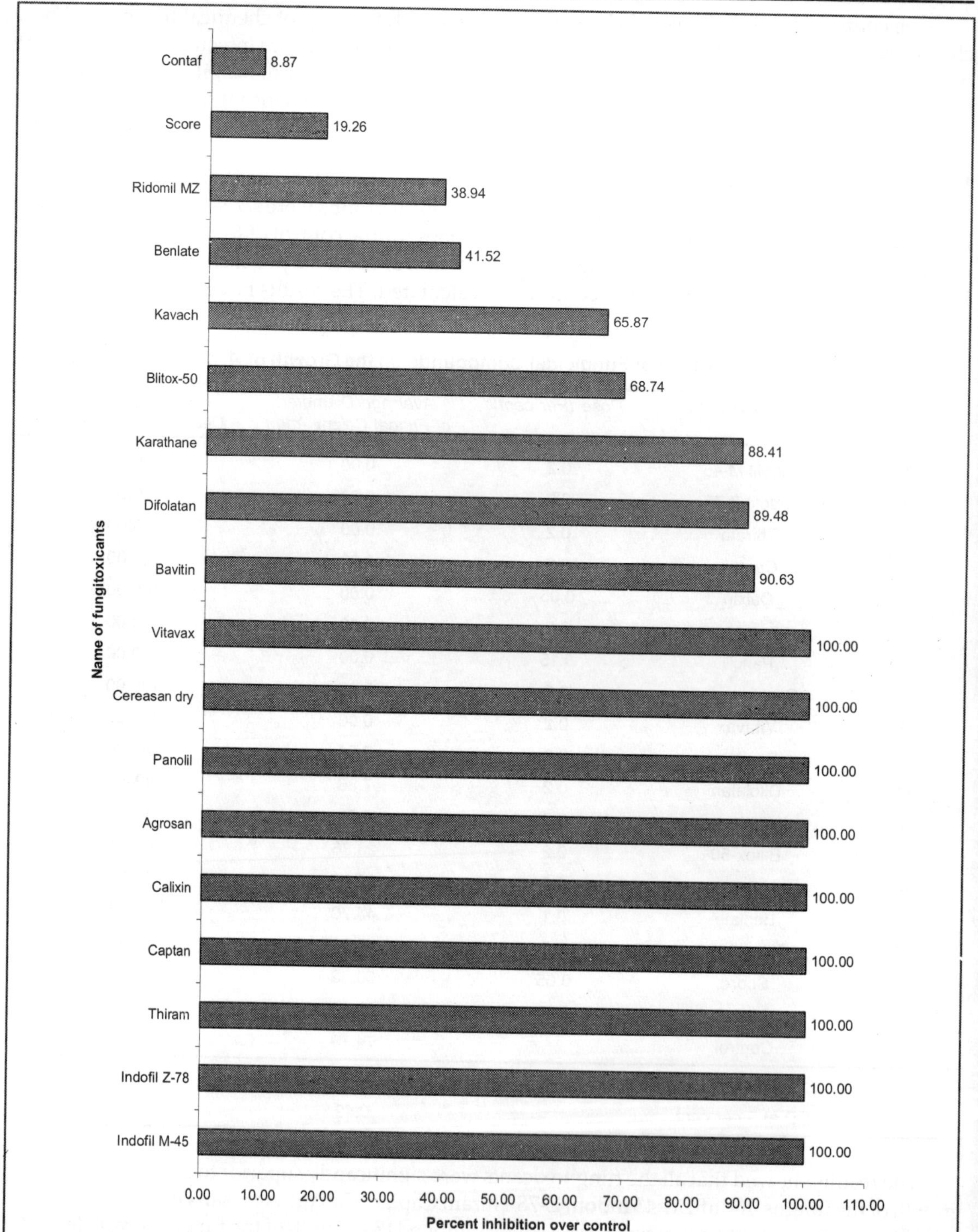

Figure 33.1: Inhibitary Effect of Different Fungitoxicants on the Growth of *A. alternata in vitro*

88.41-90.63 percent over control Besides this Blitox-50 Kavach, Benlate and Ridomil MZ were effective showing 38.94-68.74 per cent inhibition over control Remaining fungitoxicants Score and contaf were partially effective showing 8.87-19.26 per cent inhibition over control.

On statistical analysis it was revealed that the fungitoxicants namely Indofil M-45 Indofil Z-78 Thiram, Captan, Calixin, Agrosan, Penolil, Ceresan dry, Vitavax were at par with each other in completely inhibiting of radial growth of the fungus. Other fungitoxicants differed significantly from each other in inhibiting the growth of the fungus. However the treatment of Bavistin was significantly superior to rest of the treatments in percent inhibiton in diameter over control followed by difolatan and Karathane, Score showed least inhibition in diameter over control in comparison to others.

Effect of Seed Dressing Fungitoxicants on Seed Germination, Seed Borne Infection and Yield of Chickpea

The seed dressing fungitoxicants found effective in laboratory screening were assessed for their efficacy on seed germination and yield of chickpea in pot cultures. The data were statistically analysed and given in Table 33.5.

Table 33.5: Effect of Seed Dressing Fungitoxicants on the Seed Germination and Yield of Chickpea in Pot Culture

S.No.	*Fungi Toxicants*	*Dose (Per cent)*	*Seed Germination (Per cent)*	*Per cent Increase in Germination over Control*	*Disease Intensity (Per cent)*	*Per cent Decrease in Disease Intensity Over Control*	*Yield (gm/pot)*	*Per cent increase in yield Over Control*
1.	Indofil M-45	0.2	84.50	17.14	16.10	45.66	34.76	23.43
2.	Bavistin	0.2	84.13	16.63	16.90	42.96	34.06	20.95
3.	Captan	0.2	81.96	13.62	18.76	36.65	33.56	19.17
4.	Ceresan dry	0.3	79.43	10.12	19.20	35.20	33.63	19.42
5.	Thiram	0.2	81.90	13.54	22.86	22.54	32.13	14.09
6.	Agrosan	0.3	75.86	5.17	25.70	13.26	31.43	11.61
7.	Panolil	0.15	76.87	6.28	25.96	12.38	30.43	8.06
8.	Calixin	0.05	79.20	9.80	26.46	10.69	30.23	7.35
9.	Vitavax	0.2	78.13	8.31	26.86	9.34	29.43	4.50
10.	Karathane	0.2	75.00	3.97	27.50	7.18	28.90	2.62
11.	Control	–	72.07	–	29.03	–	28.16	–
	SE Mean		1.07	0.36			0.23	
	CD at 5 per cent		2.21				0.48	

The data revealed that seed dressing fungitoxicants studied during present investigation showed positive response of germination of chickpea seeds showing increase of 3.97-17.14 per cent over control. Maximum seed germination was recoreded in case of Indofil M-45 (84.50 percent) and Bavistin (84.13 percent) followed by captan (81.96 percent) and Thiram (81.90 per cent) Cerasandry, Agroson, Panolil, Calixin and Vitavax showed average seed germination (75.86-79-43) and were found statistically insignificant Karathane was relatively least effective (75.0 per cent) but were superior to control.

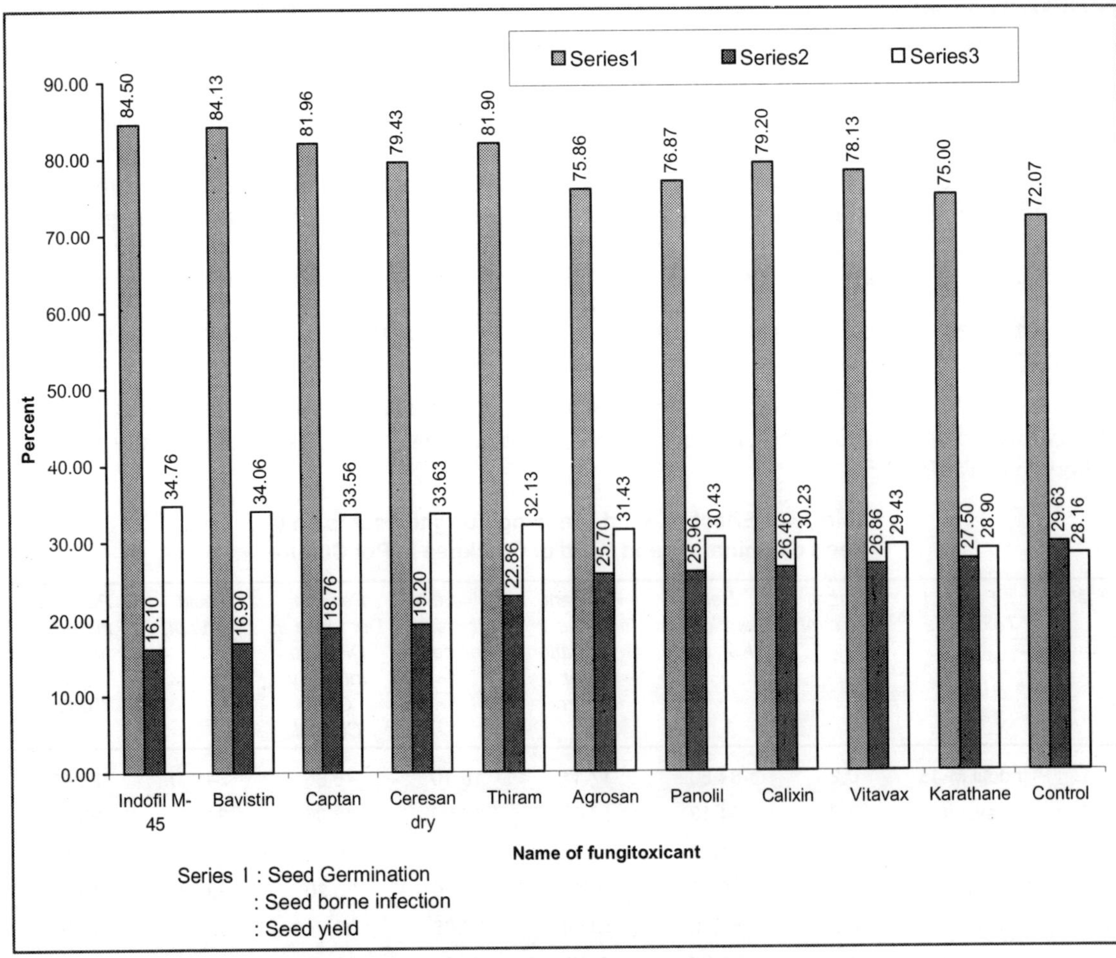

Figure 33.2: Effect of Seed Dressing Fungicides on the Germination of Seed, Seed Borne, Infection and Yield of Chickpea

Table 33.6 further showed that all the fungitoxicants taken in persent investigation improved seed yield of chickpea (2.62-23.43 percent) in comparison to control Maximum increase in seed yield was however recorded in case of Indofil M-45 (23.43 per cent) followed by Bavistin (20.95 per cent) captan (19.17 per cent) and ceresan dry (19.42 per cent). The variation however in latter three were statistically insignificant. Thiram was found superior than rest of fungitoxicants. Vitavax and Karathane gave relatively lower yield.

Effect of Spray of Fungitoxicants on the Intensity of Foliar Disease and Yield of Chickpea in the Field

The results showed that all the fungitoxicants tested in present study were significantly superior over control in disease intensity. Among the fungicides used Indofil M-45 proved to be best showing 55.14 per cent decrease in disease severity over control followed by Difolatan giving 48.83 per cent decrease in disease. Other two fungicides namely Indofile Z-78 and Bavistin showed 40.86 per cent

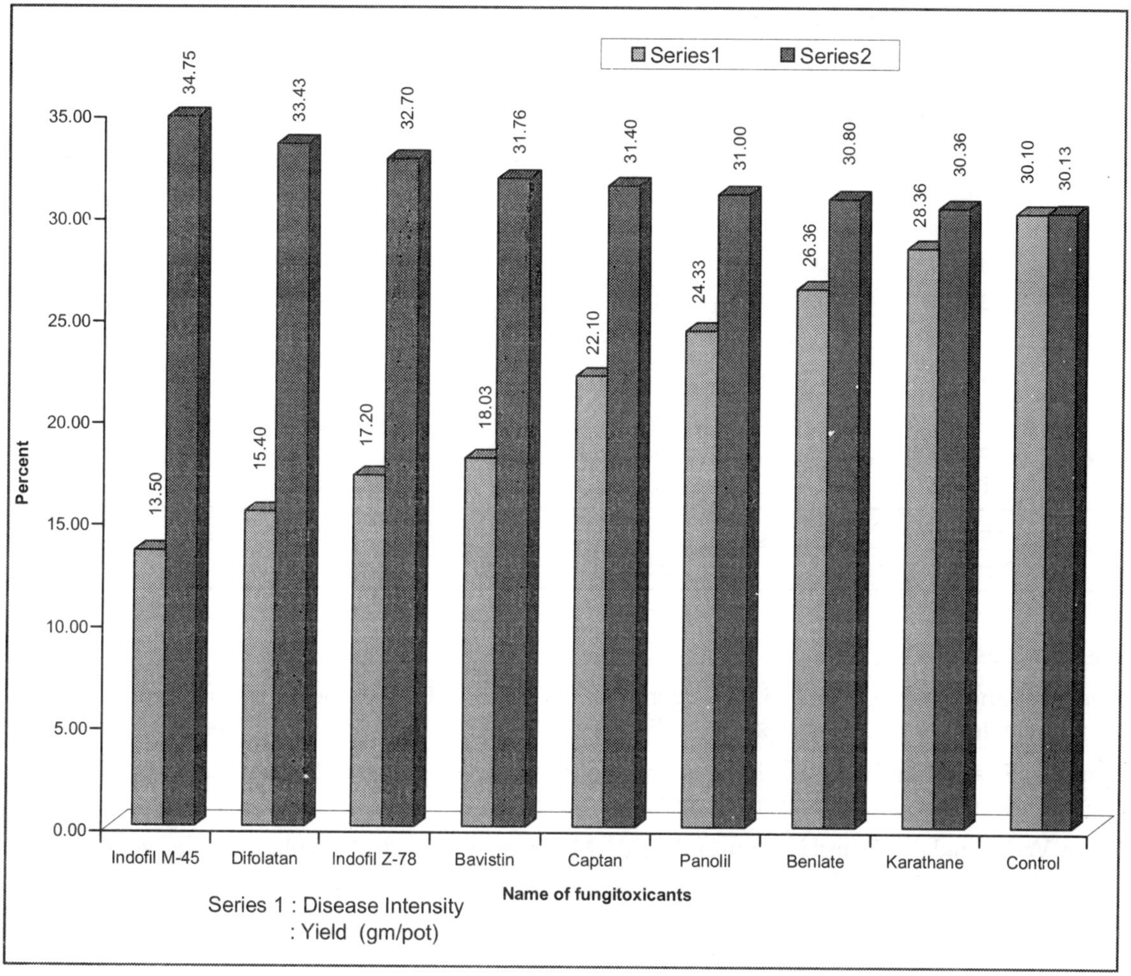

Figure 33.3: Effect of Fungitoxicants on the Intensity of Disease and Yield of Chickpea Under Field Conditions

and 40.09 per cent in disease over control respectively captan panolil and Benlate were poorly effective showing 12.42-26.57 per cent decrease in disease over control, Karthane was least effective showing only 4.88 per cent in disease over control. Statistically the disease intensity observed in different fungitoxicants were significant and the treatment of Indofil M-45 was significantly superior than others in controlling the disease. However, the variation in disease intensity observed between Indofil Z-78 and Bavistin were statistically insignificant.

Table 33.6 further revealed that all the treatment proved effective in increasing yield of chickpea. The pots treated with Indofil M-45 gave best yield showing 15.36 per cent increase in yield with over control followed by difolatan., Indofil Z-78 Bavistin and captan showing 4.21-10.95 percent increase in yield over control. Other two fungicides namely Panolil and Benlate showed 2.22-2.88 per cent increase yield over control, and they were found statically insignificant with each other. The lowest seed yield 0.76 per cent was observed in case of Karathane.

Table 33.6: Effect of Fungitoxicants on the Intensity of Disease and Yield of Chickpea Under Field Conditions

Sl.No.	Fungitoxicants	Dose (Per cent)	Disease Intensity (Per cent)	Per cent Decrease in Disease Intensity Over Control	Yield (gm/pot)	Per cent Increase in yield Over Control
1.	Indofil M -45	0.2	13.50	55.14	34.76	15.36
2.	Difolatan	0.2	15.40	48.83	33.43	10.95
3.	Indofil Z-78	0.2	17.20	40.86	32.70	8.52
4.	Bavistin	0.2	18.03	40.09	31.76	4.21
5.	Captan	0.2	22.10	26.57	31.40	5.40
6.	Panolil	0.15	24.33	19.16	31.00	2.88
7.	Benlate	0.1	26.36	12.42	30.80	2.22
8.	Karathane	0.2	28.36	4.88	30.36	0.76
9.	Control	-	30.10	-	30.13	=
	SE Mean +		0.48		0.14	
	CD at 5 per cent		0.99		0.29	

References

Kumar, P. and Singh, D.V., 2003. Management of *Alternaria* blight of radish by foliar application of fungicides. *Ann. Pl. Protec. Sci.*, 11: 175–176.

Sumitha, R. and Nimbkar, N., 2009. Efficacy of fungicides against *Alternaria* leaf spot disease in safflower. *Ann. Pl. Protec. Sci.*, 17: 507–508.

Ratan, Ved, Shukla, H.P. and Narain, Udit, 2003. Management of *Alternaria* blight of Marigold. *Ann. Pl. Protec. Sci.*, 11: 172–173.

Kushwaha, K.P.S. and Narain, Udit, 2001. Evaluation of fungicides against *Alternaria* leaf blight of Pigeon Pea. *Ann. Pl. Protec. Sci.*, 9: 330–332.

Raut, S.P. and Patil, R.N., 2005. Field evaluation of fungicides botanicals and *Trichoderma viride* for management of root rot and wilt of tomato. *Ann. Pl. Protec. Sci.*, 13: 231–232

Singh, R.B.and Singh, R.N., 2004. Management of *Alternaria* blight of linseed. *Ann. Pl. Protec. Sci.*, 12: 305–309.

Tetarwal, M.L. and Rai, P.K., 2007. Management of *Alternaria* blight of Senna (*Cassia angustifolia*) through fungicides and plant extracts. *Ann. Pl. Protec. Sci.*, 15: 165–169.

Plant Protection, Diversity and Conservation (2011) *Pages* **457–460**
Editors: **D. Prasad & Satya Kumar**
Published by: **BIOTECH BOOKS, NEW DELHI**

Chapter 34

Management of *Colletotrichum* Blight of Urdbean

Versha Misra[1], K.C. Srivastava[2] and Neena Srivastava[3]
[1]*D.G.P.G. College, Kanpur – 208 001*
[2]*D.A.V. College Kanpur – 208 013*
[3]*D.S.N. College, Unnao, U.P.*

Urdbean (*Vigna mungo* L.) is one of the most important pulse crop of India and grown in rainfed condition during rainy season and on residual moisture in winter season in eastern and southern parts of the country. Urdbean along with mungbean,lentil,lethyrus, rajmash and field pea share 42 per cent of the total pulse acreage and contribute 37 per cent of the national pulse basket (Ali *et al.*, 2002). The total area in the country under the crop has increased frome 18.4 lac hac in 1965-66 to 33.3 lac hac in 2004-05. The production has increased from 5.5 lac m tonn. in 1965-66 to 14.1 lac m tonn. in 2004-05. Maharastra and M.P. have been leading states producing 21.4 per cent and 20.3 per cent of total production in Urdbean in the country (Wealth of India, 1982).

Disease Problem

In spite of cultivation of Urdbean in large area, its genetic potential of productivity is not realized owing to various biotic constraints which include diseases. The extent of annual loss in pulse crops including Urdbean is to the tune of 15-30 per cent pf total loss, amounting to nearly 484 crores of rupees. In case of disease spectrum of the crop, blight disease caused by *Colletotrichum truncatum* is most important being endemic in high rain fall areas. High yielding varieties of Urdbean such as T-9, T-27, T-65 and K-10 have been found to suffer from this disease for the past several years.

The association of Urdbean seeds with *C. truncatum* was confirmed quite earlier (Agarwal *et al.*, 1972; Saxena and Sinha, 1977; Singh and Shukla, 1988). The fungus is a destructive leaf spotting pathogen and normally preceeds the attack of *Cercospora* sp. under high rainfall conditions (Kaushal,

1985). The pathogenicity of the fungus has also been confirmed on healthy Urdbean leaves within 5 days (Singh *et al.*, 1978).

The disease is of wide occurrence throughout Uttar Pradesh (Mishra, 2008), H.P. (Rana and Kaushal, 2004). Gurdaspur, Panjab (Bains *et al.*, 1989) and Assam (Ruthaiah and Sarma, 2004). Differential behaviour of the pethogen on Urdbean showed great pathogenic variations. The large host range of the pathogen may be a reason of its non-specific nature. The other reason of this variation may be a diversity of germplasm grown by the farmers in different areas. The germplasm get adopted to local ecogeographical conditions due to which they become genetically different. The pathogen seems to have acquired parasitic fitness of these ecogeographical conditions. Because of highly non-specialized nature of the pathogen, there seems a need to look for resistance in wide array of Urdbean germplasm.

Management Strategy

Like other pulse crops, the protective packages of Urdbean from diseases have not received much attention. Evaluation of germplasm and chemical control methods are the general and well adopted approaches. Besides these, crop rotation, proper spacing and megre use of fungicides are among the measures being practiced by traditional farmers, though some new cultivars have been recently developed with resistance to blight disease of Urdbean but their effective utilization has been vary limited. However, there is still scope for exploitation of the means of disease management such as cultural methods and use of bio-agents in combination with host plant resistance.

Resistant Varieties

Screening of disease resistant varieties of Urdbean is the best method for management of blight disease. But this method could not gain popularity due to vast host range, competitive saprophytic survival ability and presence of pathogen's variability in isolates of the pathogen. Only two varieties of Urdbean *viz.*112 and Br-10 proved resistant against blight disease (Mishra, 2008). Most of the varieties tested showed however moderately to highly susceptible reactions (Singh and Shukla, 1986).

Cultural Management

Time of sowing, weeding, crop rotation, wider spacing and sanitation are the important cultural methods which have a direct effect on incidence of blight disease of Urdbean (Srivastava and Mishra, 2008). Sowing the Urdbean crop between April-June (Allen, 1972), 30 June (Mittal, 1999), 2 July (Rana and Kaushal, 2005) and 30 July (Srivastava and Mishra, 2008) have been recommended for minimum disease incidence. Higher disease severity has been observed in inter-row spacing (Srivastava and Mishra, 2008). Disinfection of agricultural instruments and tools along with destruction of infected plants keep the disease at minimum level (Ali and Kumar, 2006). Hot water treatment of Urdbean seeds at 50°C for 5 minutes has been recommended (Srivastava and Mishra, 2008) and also at 58°C for 15 minutes (Singh, 1977).

Chemical Management

The use of fungicides is quite effective and can not be replaced by any other method of disease control but continuous use of fungicides pose the residual toxicity in the plants and soil and cause environmental pollution. Seed treatment and foliar sprays have greater scope for managing blight disease of Urdbean and maximizing yield. There fore effective strategies are required for the management of the pathogen by judicious use of fungicides so that there is maximization of advantages and minimization of disadvantages. Large number of fungicides have been screened *in vitro* against *C. truncatum* (Venugopal *et al.*, 2007; Wealth of India, 1982). Among these, Bavistin, Benlate, Cantaf, Calixin, Kitazine, Topsin-M and Vitavax were most effective.

Seed Treatment

Seed-borne infection play an important role in dissemination of pathogen as well as the initiation of the disease. The fungicidal seed treatment has been found very effective to eradicate seed-borne infection of *C. truncatum* in Urdbean and to increase the germination of seeds with reduction in blight disease development. Among seed-dressing fungicides, Bavistim (0.1 per cent) and Benlate (0.1 per cent) gave more than 80 per cent disease control (Joshi and Tripathi, 2002; Srivastava and Mishra, 2008).

Foliar Spray

Colletotrichum truncatum causes blight disease in Urdbean. Combination of seed treatment and subsequent spray of fungicides or foliar spray alone have been highly effective in control of blight disease of Urdbean. Agallol, Cerasan wet, Captan and Thiram have proved better to prevent blight disease in Urdbean (Suhag, 1975). Carbendazim and Benomyl are successfully used to prevent *C. truncatum* infection in Urdbean crop (Shailbala and Tripathi, 2004; Singh and Prasad, 1993). Four sprays of Indofil M-45 (0.2 per cent) at 30 days interval gives minimum severity of blight disease in Urdbean and maximam grain yield (Joshi and Tripathi, 2002). Ziram (0.2 per cent) proved effective in management of blight diseade in Urdbean (Ali and Kumar, 2006). Foliar spray of Bavistim (0.2 per cent) or Benlate reduced blight disease of Urdbean caused by *C. truncatum* to 76 per cent and increased 76 per cent of grain yield (Srivastava and Mishra, 2008).

References

Agarwal, V.K., Mathur, S.B. and Neergard, P., 1972. *Ind. Phytopath.*, 25:91–100.

Allen, R.N., 1972. *Agricultural Gazette of New South Wales*, 83: 244–246.

Ali, M., Tripathi, D.P. and Dixit, G.P., 2002. *AICRP ON MILLARP Front line Demonstration*, 1996–2001, IIPR Kanpur–24.

Ali, M. and Kumar, S., 2006. IIPR Kanpur.

Bains, S.S., Kaur, L., Dhaliwal, H.S., Gill, A.S. and Kaur, L., 1989. *Panjab Agri. Repr.*, 8(2):164–165.

Joshi, D. and Tripathi, H.S., 2002. *J. Mycol. Pl. Patho.*, 32(1): 52–55.

Kaushal, R.P., 1985. Geneties of resistance in Urdbean [*Vigna mungo* (L.) Hepper] to some major diseases occurring in H.P.. *Thesis*, H.P.K.U., Palampur, pp. 86.

Mishra, C.P., 2008. Pathological aspects and management of *Colletotrichum* blight of Urd (*Vigna mungo* Linn). *Ph.D. Thesis*, C.S.J.M., Univ., Kanpur.

Mittal, R.K., 1999. *J. Mycol. Pl. Pathol.*, 29(1): 120–130.

Rana, U. and Kaushal, R.P., 2004. *J. Mycol. Pl. Pathol.*, 34(2): 649–652.

Rana, U. and Kaushal, R.P., 2005. *Ind. Phytopath.*, 58(3): 374.

Ruthaiah, Y. and Sarma, S.K., 2004. *J. Mycol. Pl. Pathol.*, 34(2): 176–178.

Saxena, R.M. and Sinha, S., 1977. *Curr. Sci.*, 46: 716

Shailbala and Tripathi, H.S., 2004. *J. Mycol. Pl. Pathol.*, 34(3): 851–852.

Singh, H.N.P. and Prasad, M.M., 1993. *Lett Applied Microbiol.*, 17: 260–271

Singh, V. 1977. Studies on *Colletotrichum* corda associated with Urad (*Phaseolus mungo*) seeds. *M. Sc.(Ag.) Thesis*, J.N.K.V.V., Jabalpur.

Singh, R.R. and Shukla, P., 1986. *Ind. J. Mycol. Pl. Pathol.*, 16(2): 172–174.

Singh, B.M., Sood, A.K. and Saharan, G.S., 1978. *Ind. Phytopath.*, 31: 100–101.

Singh, R.R. and Shukla, P., 1988. *Farm Sci. J.*, 3(2): 184–185.

Suhag, L.S., 1975. *J. Mycol. Pl. Pathol.*, 5: 165–168.

Srivastava, K.C. and Mishra, C.P., 2008. *Ann. Pl. Protec. Sci.*, 16(1): 139–142.

Venugopal, C., Gopi, K., Srinivasalu, Y. and Vamana, K., 2007. *Ann. Pl. Protec. Sci.*, 15: 521–523.

Wealth of India, 1982. *A Dictionary of Indian Raw Material and Industrial Products, Vol. 10*, Publication and Information Directorate, CSIR, New Delhi.

Plant Protection, Diversity and Conservation (2011) *Pages* **461–480**
Editors: **D. Prasad & Satya Kumar**
Published by: **BIOTECH BOOKS, NEW DELHI**

Chapter 35

Current Status of Sesamum and their Disease Management

Sushma Nema[1] and S.K. Tripathi[2]
[1]J.N.K.V.V., College of Agriculture, Jabalpur, M.P.
[2]J.N.K.V.V., College of Agriculture, Rewa – 486 001, M.P.

Sesamum (*Sesamum indicum* L.), also known as sesame, til and gingerly, is an important and ancient oil-yielding crop. Africa is considered to be the primary centre of origin of this crop because of the presence of its diverse wild species in that continent. India is the secondary centre of origin and another secondary centre is Japan.Sesame is described as the most ancient oilseed crop in the world. It is widely cultivated in tropical and sub tropical parts of the world. The plant is cultivated all over India in the plains and upto an altitude of 1200 meters

Economic Importance

The sesamum seed is a rich source of edible oil. Its oil content generally varies from 46 to 52 per cent. Its grains may be eaten fried, mixed with sugar or in the form of sweetmeats. The tila seeds (Sesame seeds) contain many vitamins like thiamine, niacin, riboflavin, nicotinic acid, pantothenic, folic and ascorbic acids, choline, inositol, pyridoxine, vitamin A, and tocopherol. Sugars present are glucose, sucrose, galactose, planteose and raffinose. The fatty acids in the seeds are myristic, palmitic, stearic, arachidic, oleic, linoelic, hexadecenoic and linoceric acids. Sesamin isolated from seed oil and three stereoisomeric forms of sesamin, *viz.* sesamin, asarinin and epiasarinin. The leaves afforded a new flavone glycoside pedalin. A new antioxidant sesamol and tocopherol were obtained from seeds. The seeds and seed oil is used for medicinal purpose. As the seed oil (sesame oil) has various properties, it is the most widely used oil, by Ayurvedic practitioners and pharmacy. Externally, the massage with tila oil reduces the dryness of the skin and alleviates the vata dosha. The local application of the pulp of tila seeds promotes the wounds healing and relieves the pains in haemorrhoids also. The roots and leaves are emollient and a decoction of them forms a good hair wash, which promotes the hair growth, blacken them and bestows texture. To strengthen the teeth and gums and to help receding gums the til oil is massaged on gums or the seeds are chewed. The paste of ground seeds of tila can also be used as a face pack in dry skin. Tila oil is widely used in Panchakarma therapy for snehana (oleation), basti (enema) and nasya (nasal therapy). In medicated oil sesame oil forms a fat soluble medium. In Ayurvedic pharmacy, tila oil shares the major portion of oil preparations as the base. The massage with the combination of sesame oil and nutmeg oil is effective in swollen joints. The whole body massage with tila oil is salutary in obesity as well as in emaciation. It is extremely beneficial in vata diseases like paralysis, facial palsy etc. Internally, tila oil is used in vast range of diseases. In bleeding piles, the seeds are given along with butter. The sees powder, butter and rock candy is recommended in nonspecific ulcerative colitis. The seeds mashed with milk are given to nourishing mothers to augment the quantity of breast milk. The same preparation given along with rock candy is benevolent in Dysurea. In urinary calculi, the burnt ash of tila plant is given along with honey. Tila oil is a valuable remedy for dysmenorrheal and amenorrhea. It is also salutary in general, mental and sexual debility. It works well in hiccup and asthma as a snehana (oleation).

Properties

Tila is pungent, bitter, astringent and sweet in taste, pungent in the post digestive effect and has hot potency. It alleviates all the three doshas, predominantly, vata. It possesses heavy and oily attributes. It has special potency as a hair tonic. The chief properties of tila oil are a hair tonic, galactogtogue, an appetizer general tonic digestant and it is used in skin diseases, wounds, anorexia and dental diseases (Bhavaprakash Nighantu) Sesamum oil is used as a cooking-oil in southern India. It is also used for anointing the body, for manufacturing perfumed oils and for medicinal purposes. Sesamum-cake is a rich source of protein, carbohydrates and mineral nutrients, such as calcium and phosphorus. The cake is edible and is eaten avidly by working classes. It is also a valuable and nutritious feed for milch cattle.

It had earned a poetic label "Queen of oilseeds" because seeds have high quality polyunsaturated stable fatty acids and antioxidant agents, sesamin, sesmolin. Moreover, seeds are rich source of edible oil (48-55 per cent) and protein (20-28 per cent) consisting both methionine and tryptophane, vitamin (niacine) and minerals (Ca and P) (Bed *et al.*, 1985).

Botanical Description

The genus Sesamum belongs to the family Pedaliaceae which has a wide distribution, covering tropical Africa, Madagascar, Arabia, India, Sri Lanka, tropical Australia and a few of the eastern islands of the Malayan Archipelago. Ayurvedic Samhita mention that the white seeds yield more oil, where as, black seeds have better medicinal value.

Sesamum is an erect and annual herb. The plant is simple or branched. The leaves are opposite below and alternate above. The leaves numerous, alternate, very variable in form. Ovate or oblong-lanceolate, usually tapering at the base. The flowers are zygomorphic, solitary, occasionally two or three together, axillary, short- pedicelled, borne on the upper part of the stem or branches geniculate. The tube, bent downwords and is dialated above the oblique base. The calyx is small, and five sectioned, 0.5-0.6 cm long. The corolla is tubular, campanulate, approximately 3 cm long, with an applicate margin and slightly bilabiate, the upper lip is entire and the lower one is tripartite. There are four functional stamens and often one is sterile, didynamous. The antliers are sagittate. The ovary is superior, bilocular, but may be subdivided by false internal walls. The fruit is a capsule, erect and oblong. The flowers are white, pink or purplish, shortly stalked, solitary in the axils of the leaves and hairy. The fruits, quadrangular compressed capsules, narrowly oblong, with rounded angles. The seeds numerous, flattened, rounded triangular or oval in outline The capsule contains numerous small ovate seeds. The testa may be smooth or reticulate and may be white, The seed is dicotyledonous, albuminous and oleaginousigian

Distribution and Production

India, China Sudan, Mexico, Turkey, Burma and Pakistan are the important sesamum producing countries. India ranks first, both in the area and production of sesamum in the world. The annual area put under it in India is about 2-5 million hectares (45 per cent of the world hectarage) and the total production is nearly 52 thousand tonnes. Sesamum is grown on 21 lakh hectares in only eight states, *viz.* Uttar Pradesh (673,000), Rajasthan (562,000), Madhya Pradesh (345,000), Andhra Pradesh (237,000) Maharashtra (139,000), Gujarat (118,000), Tamil Nadu (117,000) and Orissa (103,000). Among other states only Karnataka has a sizable area (68,000) under sesamum.

Climate and Soil

Sesamum grows in the plains and at elevations up to 1,200 m. It cannot stand frost, continuous heavy rain or prolonged drought. It is grown in sandy-loam to heavy black soils. Most of the crop is confined to lighter soils and its cultivation in heavy soils is limited to certain regions. The pH of sesamum growing soils ranges from 5.5 to 8.2. The sesamum growing soils are shallow to medium in Andhra Pradesh, Orissa, Andhra Pradesh and Tamil Nadu and are deep in Rajasthan, Gujarat and Maharashtra.

Mixed Cropping

Sesamum is grown both as pure and mixed with other crops, whereas the semi-rabi and summer crops are taken as pure. The common component crops are red-gram (arhar) in Andhra Pradesh, bajra and red-gram in Gujarat, red-gram and jowar in Madhya Pradesh. Cotton, groundnut and red-gram in Maharashtra, and black-gram, green-gram and horse-gram in Orissa.

Cultivation

In India, seasmum is grown in three seasons, *viz.* kharif, semi-rabi and summer. The kharif crop occupies over 70 per cent of the area, whereas the semi-rabi and summer crops occupy 20 per cent and 10 per cent area respectively. The kharif sesamum is sown in June-July with the onset of the monsoon and is harvested in December-January. The kharif and semi-rabi crops are entirely rainfed, whereas summer crop is grown under irrigation. The yield of the *kharif* crop is poor, whereas those of the semi-rabi and summer crops are high, as they are grown in rich soils and under better management. The preparatory cultivation for the kharif crop is usually not thorough. For the semi-*rabi* and summer crops, the land is ploughed and harrowed repeatedly to secure a clean and fine seedbed. Except in Maharashtra and Gujarat, where line-sowing is practised, in all other states it is sown broadcast. The recommended spacings for the line-sown crop in different regions, however, range from 25 to 35 cm between the rows and 10 to 20 cm between the plants in the row. The seed being small, it is often mixed with sand to ensure its even distribution and the drill is operated rather shallow to avoid deep sowing. After sowing, the seed is covered lightly with a brush harrow. The seed-rate varies from 3 to 5 kg per hectare. The seed should be treated with Captan or Thiram at the rate of 3 g per kg of seed to control seed-borne diseases.

The rain fed crop is weeded and hoed once or twice and the irrigated crop is weeded often. Wherever the weed control with mechanical means is a problem, chemical weed control with pre-emergence application of Lasso at the rate of 3 litres/ha is recommended. The crop is generally recommended for different regions as 25-30 kg of W, 20-30 kg of P_2O_5 and 0-20 kg of K_2O per hectare. The crop is harvested when the leaves, stems and capsules begin to turn yellow and the lower leaves start shedding. To prevent the shedding of grains, the crop should not be allowed to become a drip in the field. The ripe plants are cut, carried to the threshing-yard, stacked for a week in the sun with the cut-ends downwards and well shaken or beaten to take out the grains from the dry capsules. Winnowing and cleaning completes the process.

Varieties

The varieties differ not only in respect of the season of growth and the time of maturity but also in the number of capsules formed in the leaf axils, the habit of growth (Single-stemmed or branched) and the size, colour and oil content of the seeds. The seeds may be white, dull white, light brown dark brown, red or black and their oil content varies from 48-52 per cent.

Important Factors for Obtaining High Yields

1. A suitable variety must he chosen for the tract and the season.
2. The crop should be sown in lines.
3. The application of fertilizers, inter-cultivation and plant-protection measures must be taken up at the appropriate time.
4. Harvesting should be done after the crop matures but must not be delayed up to the stage when the capsules shed seeds.

Sesame suffers due to diseases caused by fungi bacteria, viruses, nematodes and phytoplasma making the host weak resulting in losses both in quantity and quality. The pathogens are both seed borne and soil borne. The important diseases of sesame are damping off (*Pythium aphanidermatum, Thielavia terricola var. minor*); foot rot (*Sclerotium rolfsii*); *wilt* (*Fusarium oxysporum f. sp. sesami*); root rot, stem rot, charcoal rot, blight, seedling rot [(*Rhizoctonia bataticola*) (*Macrophomina phaseolina*)]; Phytophthora blight (*Phytophthora nicotianae var. parasitica*); Alternaria blight (*Alternaria sesami*); Cercospora leaf spot (*Cercospora sesami and C.sesamicola*); Corynespora blight (*Corynespora cassicola*); *Botryosphaeria blight* (*Botryosphaeria ribis*); *Cylinderosporium* leaf spot (*Cylinderosporium sesami*); anthracnose (*Colletotrichum sp.*); powdery mildew (*Leveillula taurica, Oidium sp., Sphaerotheca fuliginea*); Phoma leaf blight, blight, stem black (*Phoma sesamina, P. variosporeae*); bacterial leaf spot (*Pseudomonas syringae pv. sesami*); bacterial blight (*Xanthomonas campestris pv. sesami*) Leaf curl (*Nicotiana virus* 10); Phyllody (Phytoplasma) etc. Several other microorganisms have also been found associated with sesame seeds.

The incidence of important diseases vary from state to state based on agro-climatic situations. The diseases of sesame and their management are summarized below:

Fungal Diseases

More than 72 fungi are responsible for causing diseases in sesame. Several such fungi have been observed associated with seed and soil which play a vital role in disease cycle.

Damping off (*Pythium aphanidermatum* (Eds.) *Fitz.; Thielavia terricola* var. *minor* Rayss and Borut.)

Damping off of seedlings is important under high soil moisture both at pre and post-emergence. The radicle and plumule get infected before emergence and no seedlings are formed. At post-emergence stage, the stem close to the soil becomes water soaked, shrinks, rots and the seedling topples down. The pathogens are soil borne *T. terricola* also causes root rot.

Foot rot (*Sclerotium rolfsii* Sacc.)

Mishra and Khare (1970) first reported the disease from Jabalpur. The disease has been observed at different sesame growing areas. The basal part of the stem develops brown to black lesion due to necrosis of tissues. The affected stem shrivels and the plant dies. White mycelium is seen around the diseased part which develops mustard like sclerotia. The disease is severe under high humidity and high temperature. Leaf infection has also been observed.

Wilt (*Fusarium oxysporum f. sp. sesami (Zapro.) Castellani*)

The disease was first reported by Butler (1918). The disease is common in all the sesame growing areas. Loss upto 20 per cent has been observed due to wilt. As the diseased plants die before maturity heavy losses are possible. The leaves of diseased plants turn yellowish, droop down and dry. The vascular bundles turn brown and this brown discolouration starts from roots and gets extended upto the tip.

The fungus is soil borne as well as seed borne. The seed borne inoculum leads to infection of seedling at very early stage, resulting in mortality, depending upon the inoculum potential. The soil borne inoculum attacks and enters through finer roots.

Root Rot, Stem Rot, Charcoal Rot, Blight, Seedling Rot (*Macrophomina phaseolina* (Maubl.) Ashby (Pycnidial stage); *Rhizoctonia bataticola* Taub. Butler) (Mycelial and sclerotial stage)

These diseases are very important as infection occurs from seed germination and emergence to adult plant stage. The losses incurred are even upto 100 per cent. Murugesan *et al.* (1978) observed 1.8 kg yield loss per hectare at every one per cent increase in disease intensity. *M. phaseolina* causes seedling rot, stalk rot, root rot, charcoal rot and blight in sesame (Ali Ahmad and Saidawi, 1998). It is of major importance in West Bengal (Anonymous, 2001). The infected seeds are discoloured, on germination, seed rot occurs or the seedlings get damaged and dry. Brown dark spots develop on the stem which turn into lesions. The stem breaks. A number of pycnidia are formed in the lesion on the stem. Sclerotia develop in the tissues of diseased root and stem. On splitting it appears as if charcoal powder has bean sprinkled on the tissues, hence the disease is called charcoal rot. Black spots of various sizes are formed on the leaves which coalesce. The infected leaves are dropped off. Similar spots are also formed on capsules and sclerotia as well as pycnidia develop on them. Blackish hyphae are seen on the seed surface, associated with seed coat.

The pathogen is both seed and soil borne. The sclerotia survive for many years in soil. The seed borne infection is both externally and internally associated. Soil and seed borne inoculum is responsible for recurrence of the disease and seed borne inoculum helps in long distance transmission. The spread of the disease is through movement of workers, animals, water and equipments used in cultural management in the field. Seed to plant infection is systemic as well as local through seed borne inoculum and infected inert matter. Plant to seed transmission is through capsule. The pathogen has a very wide host range. High temperature and humidity favour the disease development. Phosphorus in soil checks the disease severity. Appressoria formation has bean observed on the epidermis of infected seedlings which helped the mechanical and chemical penetration. The hyphae extend inter and intracellularly from roots to stem, fruits and seeds. Gupta (1992) negatively correlated the production of volatile aldehyde compounds with the extent of seed infection. The infection of sesame plants was positively correlated with percentage of infected seeds. Singh and Singh (1990) examined sesame seed samples from 18 districts of Rajasthan and found 108 samples to be infected with 1-67 per cent incidence. Microsclerotia of the pathogen were present in the seed. Gupta and Cheema (1990) positively correlated the number of microselerotia on sesame seeds with plant infection and negatively with seed germination, dry matter production and root and shoot length of seedlings. Kushi and Khare (1979) examined several seed samples of 21 sesame varieties and found 17 fungi associated of which *M. phaseolina* was the most important along with *Corynespora cossicola* and *A. sesami*. Karunanithi *et al.* (1999) established three groups of *M. phaseolina* isolates based on colony characters and virulence.

Phytophthora Blight (*Phytophthora nicotianae var. parasitica* Dastur)

The disease was first observed by Dastur in India. It is the most serious disease in northen India described by Butler (1918). High humidity and heavy rains increase the disease severity resulting in even 80 per cent seedling mortality (Singh *et al.*, 1976). The plants get diseased at all stages of crop growth. The disease starts from the base of the plants and gradually moves up covering the stem which turns black; Infloresence, flowers and fruits are also attacked by the fungus. The first symptom of the disease appears as brown, water soaked spots on leaves and seedlings in very early stage In old plant, leaves develop water soaked spots which coalesce and give blight appearance. Many times the disease is so severe that nothing is left in the field if attack is at seedling stage and continuous rains occur for about a week. The partially affected plants produce shrivelled smaller seeds with brown discolourations, severely infected plant dry up

Figure 35.1: Phytophthora Blight

The pathogen survives in soil during the unfavorable period in the form of dormant mycelium and or in the form of chlamydospores/oospores. It is seed borne, the mycelium remains in embryo. The recurrence of the disease is through these two sources. Once the plants are attacked, sporangia are produced by the mycelium which spread the disease. Zoospores are formed in the sporangium which are oozed out and infect the host plants. High humidity and 28-30°C temperature favour the disease spread. Heavy soil and severe rains are responsible for high disease intensity (Kumar *et.al.*,1963). Low lying fields with poor or no drainage get the disease with very high severity.

The disease is both seed and soil borne, hence effective sanitation, rotation with other suitable crops and seed treatment are essential. Resistant sources are also available. In SriLanka Pathirana (1992) obtained a high yielding mutant tolerant to Phytophthora blight by Gamma irradiation of 450 and 600 Gy.

Alternaria Blight (*Alternaria sesami* Zimm.)

Figure 35.2: Alternaria Blight

The disease was first reported by Mohanti and Behera (1958) in Orissa, now the disease is prevalent in all sesame growing states in the country. Both typical leaf spot and blight symptoms are produced by the fungus. Losses upto 50 per cent have been reported due to this disease. The disease occurs at all the stages of crop growth however, Ojiambo *et al.* (1999) observed the plants to be least susceptible at 4 weeks and most susceptible at 8-10 weeks age. Dark brown spots develop on cotyledons. Water soaked, circular or irregular brown spots develop on leaf. Brown stripes are formed on stem. Capsules are also infected resulting in seed infection.

The pathogen is seed borne and the fungus survives in seed coat as well as in embryo. The recurrence of the disease is due to this inoculum. The cotyledons are infected on which conidia are produced which attack the foliage of other plants through air and the disease gets spread. The pathogen also survives in crop residues in soil and helps in disease recurrence. Temperature 20-30°C and pH 5-7 favour the disease development *A. sesami* produced PGTE,PTE, PG enzymes invitro (Rajpurohit and Prasad 1982).Cultural studies *viz.* effect of different media, temp.,level of pH, Effect of Carbon and Nitrogen on growth and sporulation of *A. sesami* were carried out in detail (Rajpurohit and Prasad 1982, Rajpurohit *et al.*, 1983). The possibility of races has been expected by some workers. A description of morphology and taxonomy of Sesamum *Alternaria* given by Rajpurohit 1981.

For the detection of *A. sesami* associated with seed, oat meal agar medium is the best. The seed infection has been observed from 9 to 25 per cent. Seed to plant transmission of the pathogen is important as 8 per cent infected seeds resulting in high incidence of the disease in field. According to Ojiambo *et al.* (2000) the tolerance level should be less than 2 per cent. In Karnataka it causes yield loss of 0.1 to 5.7 g seed per 100 capsules (Siddaramaiah, 1981).Seed infection reduces viability of seeds.

Rao and Vijayalakshmi (2000) tested nutritional requirements of *A. sesami* and its ability to produce phytotoxic and antibacterial metabolites. Chloroform extracts of the fungus exhibited antibacterial activity. The culture filtrate contained tenuazonic acid.

The pathogen survives in seeds as well as in soil on residues for more than one year as mycelium or chlamydospores Excessive rainfall favours occurrence and spread of the disease (Maiti *et al.*, 1988) Cumulative rains, rainy days and relative humidity for many days positively correlated whereas maximum and minimum temperature had negative co-relation with Alternaria leaf spot (Naik *et al.*, 2003).The time of plant to seed infection has been determined by Ojiambo *et al.* (1999) in Kenya, which was maximum in plants inoculated between 8 - 10 weeks age and least when inoculated at 4, 6 and 12 weeks age.

Cercospora Leaf Spot (*Cercospora sesami* Zimm. and C. *sesamicola*)

The disease is prevalent in all sesame-growing areas but more severe and important in Madhya Pradesh, Uttar Pradesh, Orissa, Tamil Nadu and Karnataka. Losses upto 60 per cent have been reported due to the disease. *C. sesami* produces conidia which attack both the sides of leaves by forming water soaked spots which later become circular or irregular. Initially the leaf spots are whitish to light brown, later turning dark brown. The leaf spots coalesce and occupy a large area of lamina giving blight appearance. The infected leaves dry and drop off.The disease spread to petioles, stem and pods on which linear, dark coloured, deep seated lesions are formed. *C. sesamicola* produces brown angular spots (Mohanti, 1958).

The pathogen is both seed and soil borne survives in the plant debris in soil as well as in and on seed for more than one year and acts as primary inoculum in the recurrence of the disease. The secondary invasion or spread of the pathogen is through conidia from plant to plant and field to field by wind. Plant to seed infection is through capsule.

Corynespora Blight [*Corynespora cassiicola* (Berk. and Curt.) Wei]

The disease was first observed at Bhubaneshwar. It has been reported from different sesame growing areas in this country. All the plant parts except root are attacked but stem and leaves are badly affected. On leaves brown specks are formed which later turn into large spots. At advanced stage of infection the leaves curl and drop off. On stem, purple brown specks develop which turn into elongated, scattered, irregular lesions, which enlarge at both the ends. Lesions turn into canker with straw coloured central area surrounded by dark brown margin. The infected stem bends and cracks. When both stem and leaves get infected plant dies prematurely. On capsules circular to elongate, minute, sunken, black spots develop which cover the whole surface. The infected seeds turn brown and become light. The disease reduces the yield. Low temperature and high humidity favour the disease development

The pathogen is seed and soil borne. It is internally and externally seed borne. The fungus survives in soil in infected crop residues. Both the sources are responsible for recurrence of the disease. The spread occurs through conidia formed after the primary inoculum infects plants, gets established and produces conidia. The pathogen has several other hosts like pulses, tomato etc. also on which it causes diseases, survives and attacks sesame. *C. cassicola* is highly seed borne. In seed stored at 26± 2°C and 50 per cent RH, the infection decreased from 89.8 to 8 per cent in seven months and in ten months the fungus was fully inactive (Navas and Subero, 1995).

Botryosphaeria Blight [*Botryosphaeria ribis* (Gross) Duggar]

The disease was first observed by Shukla and Vyas (1977) at Jabalpur. The disease appears towards crop maturity and causes heavy losses in yield. All above ground plant parts are infected and brownish specks are formed which later become irregular lesions. They coalesce and enlarge 2-25 mm in length. On leaves the infection starts from the margin, the spots are brown but later the central part turns black. The spots even cover the whole leaf. On capsules small, black sunken spots appear which increase under high humidity and temperature. The seeds inside the capsule get infected and turn brown.

The disease is both soil borne and seed borne. The pathogen survives on diseased plant refuge in soil and is associated with seed externally and internally. These sources act as primary inoculum for recurrence of the disease. Ascospores and conidia formed on diseased parts spread the disease through air.

Figure 35.3: Cercospora Leaf Spot

Figure 35.4: Cercospora Leaf Spot

Figure 35.5: Corynespora Blight

Cylindrosporium Leaf Spot (*Cylindrosporium sesami* Hansford)

The disease has been reported in 1965 from Gujarat. Water soaked spots are formed on leaves which later become darker and angular limited by veins. Necrosis occurs. Black acervuli of the fungus are formed on the upper surface in these spots and conidia are formed in large number. For the growth of the fungus 27°C temperature is the best. Sclerotia are formed which get mixed with the seed as inert matter and are responsible for recurrence of the disease alongwith those present in the infected crop residues in soil. The spread of the disease is through conidia formed on infected plant parts of the standing crop.

Anthracnose (*Colletotrichum* sp.)

Anthracnose has been reported from several sesame growing countries. In India, it was first reported from Uttar Pradesh in 1953. Infection occurs at all stages of crop growth and all the above ground parts attacked. Water soaked spots appear which turn brown and enlarge irregularly. The margins are dark brown. Black acervuli develop on the diseased portions. Stem also develops symptoms. The cortex is separated and woody part gets exposed. The infected plants dry. Capsules are either not formed on diseased plants or they are very few. The pathogen is seed borne and also survives in infected plant parts in soil and they help in recurrence of the disease. The spread is through conidia.

Powdery Mildew (*Leveillula taurica, Oidium* sp., *Sphaerotheca fuliginia*)

Powdery mildew is an important in all the sesame growing areas. The disease is severe under dry conditions. Small circular areas with whitish powder appear on leaves, later stem, leaves and capsules get fully covered with whitish powdery mass. The leaves start drying and fall off. The seeds produced

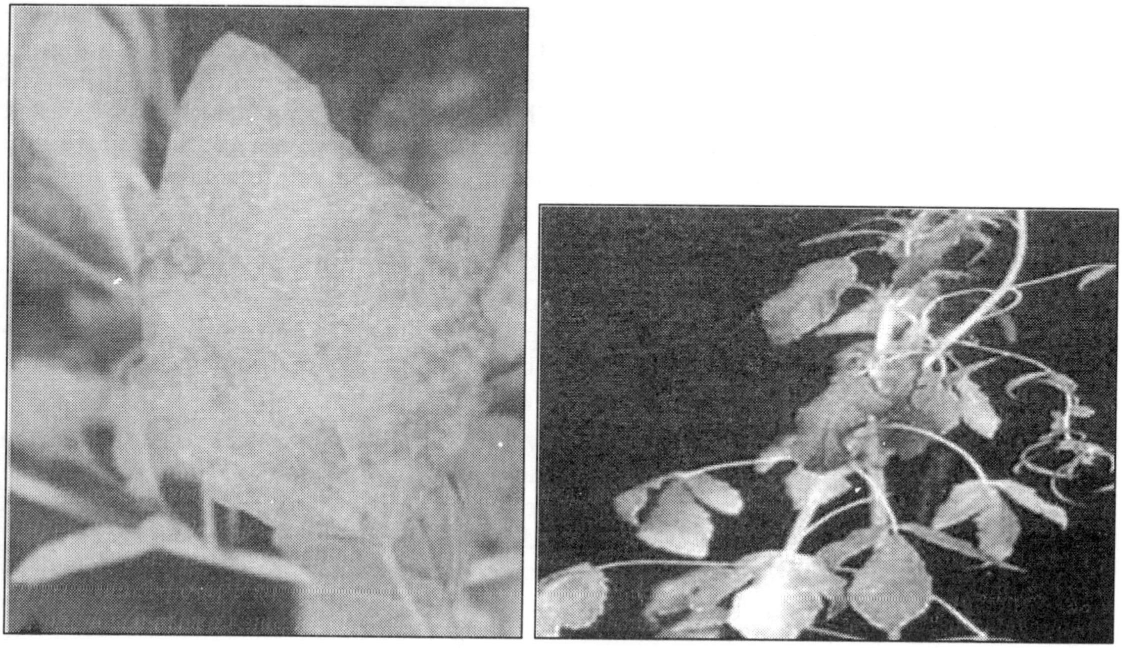

Figure 35.6: Powdery Mildew

from diseased plants are small and wrinkled. The disease is not seed borne. The pathogen has a wide host range and is always present in nature.

Rajpurohit (1993) has reported powdery mildew caused by *Erysiphe orontii* from Rajasthan. Ganesh *et al.* (1992) have found *Erysiphe cichoracearum* D.C. to be the casual fungus of powdery mildew in Tamil Nadu.

Phoma Leaf Blight and Black Stem (*Phoma sesamina* Sacc., *Phoma variosporeae Shreemali*)

Shreemali (1978) has reported *P.variosporeae* and *P. sesamina* from various places in India. The ashy spots surrounded by brown margin are formed on the leaf margins, which later cover entire leaf. The stem also develop similar symptoms. Black pycnidia are formed on the infected portions.

Bacterial

Bacterial Leaf Spot (*Pseudomaons syringae* pv. *sesami* Malkoff.)

The disease occurs in all the sesame growing areas of the world. In India the disease was first reported by Durgapal and Rao (1967) and is prevalent in Rajasthan, Uttar Pradesh, Madhya Pradesh, Tamil Nadu, Andhra Pradesh etc. The disease is also called black rot. The pathogen is seed borne. The pathogen has two physiological races. The most suited temperature for the growth is 30°C. Seed to plant transmission of the pathogen is systemic as well as local. Stomata and wounds are the entry points for the bacterium in disease spread. High temperature and frequent rain showers favour the disease development.

The crop is attacked at all stages of crop growth but it is more devastating at seedling stage. The disease starts from petiole, moves to leaf blade and small angular brown to black spots with dark margins are formed. They are confined by the veins. Bacterial exudates can be observed on the lower surface of the leaf. Under high humidity and temperature, the spots increase and coalesce. The diseased leaves drop off. On stem, black lesions are formed which girdle the whole stem and the plant dies. The capsules become black and seeds are not formed in them, but seeds, are formed in partially infected capsules which are discarded. The pathogen is internally and externally seed borne. Pandya *et al.* (2000) observed a decrease in N,P,K, Ca and increase in Mg content in infected leaves as disease intensity increased.

Bacterial Blight (*Xanthomonas campestris* pv. *sesami* Sabet and Dowson)

Sabet and Dowson (1960) reported the disease from Sudan first. At Jabalpur 10-25 per cent plants were found diseased in local and exotic varieties. The pathogen is highly seed borne and the recurrence of the disease is through infected seeds. The seed to plant transmission is both systemic and local. The spread of the disease is through air borne inoculum and rain splashes.

When infected seeds are sown, the colyledonary leaves exhibit small water soaked spots on the lower surface. These spots turn black. Under severe infection, the cotyledons dry and drop off. Typical dark brown to black angular spots appear on leaves which coalesce and give blight appearance. Bacterial ooze comes out from the affected parts under favourable conditions. Stem is also infected forming chlorotic spots which later turn dark brown and exude bacterial ooze. Capsules also develop redish brown raised spots (Wasnikar *et al.*, 1991). The bacterium also attacks certain weeds on which it survives till next season. Younger leaves are more susceptible. High humidity and 20-26°C temperature favour the disease. Higher dose of N increases the disease severity. Poor seed germination and damping off have been observed in Assam due to the bacterium by Phookan and Hazarika (1994).

Viral

Leaf Curl (*Nicotiana Virus* 10)

Leaf curl is a very harmful disease of sesame occurring in many sesame growing countries and sesame growing regions in India. As the diseased plant dies before flowering, 60-100 per cent losses have been observed. The virus has several hosts like tomato, tobacco, papaya etc. and is transmitted by white fly (*Bemisia tabaci Gen.*) hence the recurrence and spread of the disease take place through white fly obtaining the virus from such sources.

MLO

Phyllody (*Mycoplasma Like Organism- Phytoplasma*)

Phyllody is one of the most serious and wide spread disease of sesame and occurs in sesame growing states, Rajasthan, Madhya Pradesh, Tamil Nadu, Orissa, Maharashtra, Uttar Pradesh, Gujarat, Andhra Pradesh, Haryana, Punjab every year. The disease incidence ranges 1 to 100 per cent. Considerable losses are caused due to the disease. The disease was earlier considered to be due to physiological and genetic disorder, then due to virus and finally MLO. Murugesan *et al.* (1978) reported 8.36 kg/ha loss in yield on each one per cent increase in disease incidence. No capsules are formed on severely infected plants.

The disease symptoms of phyllody get manifested at flowering which get changed in leafy structures with abundant vegetative growth. Petals enlarge and turn green. The filaments of stamens are flattened and become green. Anthers also turn green and have no functional pollen. Ovules in the ovary get small petiole like growth. The diseased plants develop more branches with smaller leaves with clear veins.

The disease is transmitted by leaf hopper *Orosius albicinctus*, it has 15-23 day incubation period. The MLO survives in the leaf hopper throughout its life. It attacks various crops as well as weeds (82 plant species) throughout the year and is transmitted to sesame plant as the leaf hopper is also present. The MLO is transmitted fast by the leaf hopper at 25-40°C. Increased Indole Acetic Acid (IAA) due to infection is responsible for proliferation of vegetative and reproductive parts.

Application of N (15-45 kg/ha) imparted resistance to phyllody when coupled with medium dosage of P (15-30 kg/ha) (Borkar and Krishna, 2000).

Figure 35.7: Phyllody

Other Diseases

Besides above diseases, there are many other diseases which are not of great importance at present but close eye is needed so that they do not become limiting factor in sesame production.

Choanophora Wet Rot (*Choanephora cucurbitarum* Thax.)

The disease becomes severe in heavy rains. Small water soaked spots are formed on leaves which coalesce. Under high humidity condition whitish hyphae cover the diseased parts.

Drechslera Blight (*Drechslera sesami* (Miyake) Richa and Frasier)

The disease is prevalent in all the sesame growing areas. Small brownish elongated spots (0.2 x 2.0 cm) are formed on leaves and (1x4 cm) on stem. Capsules are also infected in which the seeds get spoiled. The spots are of V shape and later give blight appearance. *D. sesami* is seed borne.

Myrothecium Leaf Spot (*Myrothecium roridum* Tode ex Fries)

Small light brown circular spots are formed with darker margin and ashy central portion. Black spordochia are formed in the centre of the spot. The diseased portion is blown off by wind and shot hole symptoms develop. The pathogen is seed borne.

Synchytrium Gall Disease (*Synchytrium sesamicola*)

The disease has been reported from Bihar and Uttar Pradesh. Galls are formed due to hypertrophied host tissues.

Cyst nematode disease (*Heterodera* sp.)

The diseased plants remain stunted and roots do not develop fully. Capsules are either not formed or if formed, they remain small with small shrivelled seeds. The disease was first reported from Rajasthan.

Seed-borne Microorganisms

Several scientists (Kushi and Khare, 1978; Singh and Singh, 1978; Mahatma *et al.*, 2001) have examined the sesame seeds of different varieties and locations and observed the following fungi and bacteria associated with seeds. They are internally seed borne, externally seed borne, extraembryal, intraembryal, concomitant contaminant, mixed with the seed as inert matter or associated with inert matter *viz. Alternaria alternata, A. longissima, A. sesami, A. sesamicola, Aspergillus flavus, A. niger, A. ochraceus, A. tamarii, A.terreus, Beltrama rhombica, Botrytis cinerea, Cercospora sesami, C. sesamicola, Chaetoniium globosum, Cladosporium herbarum, Colletotrichum dematium, Corynespora cassicola, Curvularia lunata, Cylindrosporium sesami, Drechslera hawaiensis, D.sesami, Epicocum nigram, Fusarium moniliforme, F. oxysporum, F.semitectum, F.solani, Macrophomina phaseolina, Mucar haematis, Myrothecium roridum, Penicillium chrysogenum, P. citrinum, P. purpurogenum, Phoma sp., Phytophthora parasitica, Pleospora infectoria, Pseudomonas syringae, Rhizopus nigricans, R. oryzae, R. stolonifer, Setophaeria sp., Stemphylium sp., Syncephalasturm racemosum, Trichoderma hamatum, T. viride, Ulocladium sp., Xanthomonas campestris* etc.

Management

Cultural Practices

It is necessary to follow suitable phytosanitary measures, crop rotations in endemic areas, proper date of planting, integrated fertilizer management, application of required micro-nutrients, desired plant population, removal of weeds etc.Sesame grown as mixed or inter cropped with moth or green gram 1:1 gave less incidence of Macrophomina stem and root rot of sesame (Rajpurohit 2003).

Chemical Control

Seed treatment with suitable chemicals like Thiram (3g/kg), Bavistin (2g/kg), Captan (3g/kg), Thiram + Bavistin (2g + 1g) should be done before planting(Rajpurohit 2005). Foliar fungal diseases can be checked by the spray of Mancozeb (0.2 per cent), Zineb (0.2 per cent), Copper oxychloride (0.2 per cent). Soil treatment with Benlate or Rizolax-T (0.04-0.1 per cent) has been recommended (Gabr *et al.*, 1998). Powdery mildew can be checked by spraying sulfex (0.3 per cent) (Rajpurohit 1993). Efficacy of plant product and new chemicals against *A. sesami* reported by Rajpurohit 2004.

To control bacterial diseases seed treatment with Streptocycline (250-1000 ppm) is recommended. For hot water treatment seeds are dipped in water at 51°C for 10 minutes then dipped for 6-9 hours in water having 0.025 per cent Agrimycin 100. The foliar infection can be checked by spraying streptocycline.

For the control of leaf curl white fly needs to be checked by the application of 0.1 per cent metasystox as spray or any other suitable insecticide. Foliar spray of Mancozeb +Methyl demeton reduced Alternaria blight, Phyllody and leaf curl diseases of sesame(Rajpurohit 2004) Spray of 0.05 per cent. Dimethoate, Fenthion and Monocrotophos has been recommended to control *Orosius albicinctus* the transmitting insect of phyllody MLO. Soil application of Phorate granules 11kg/ha combined with 0.03 per cent Dimethoate spray checks the disease. Management of Macrophomina stem and root rot through fungicidal seed treatment and varietal resistance reported by Rajpurohit 1997.

Integrated Disease Management

Integrated disease management module comprised seed treatment with Thiram 0.2 per cent + Carbendazim 50 WP 0.1 per cent + foliar spray of Mancozeb 0.2 per cent + Endosulfan 0.07 per cent at 30-45 DAS and 45-55 DAS reduced *Macrophomina* stem and root rot from 24.54 to 11.93 per cent, *Alternaria* leaf spot 41.73 to 10.76 per cent, Phyllody from 3.20 to 0.64 per cent, Leaf curl 3.3 to 0.47 per cent. This module recorded seed yield of 693 kg/ha with 2.92 BC ratio (Rajpurohit and Bishnoi 2004).

Biological Control

Three isolates of *Bacillus subtilis* suppressed wilt and root rot diseases when applied in artificially infested soil (Gabr *et al.*, 1998). Seed treatment with *Trichoderma viride* (0.4 per cent) was effective in checking *M. phaseolina* (Rajpurohit 1995). From the experiments conducted under All India Coordinated Project on Sesame, it was inferred that seed treatment with *T. viride or T. harzianum* (0.4 per cent) and Carbendazim 25 SD (0.25 per cent) + Thiram (0.15 per cent) or seed treatment with *T. viride* (0.4 per cent) +soil application of *T. viride* @2.5kg/ha minimized the incidence of Macrophomina stem and root rot; Seed treatment with *Pseudmonas fluorescens* (0.4 per cent) and Contaf (0.05 per cent) effectively managed Cercospora leaf spot incidence (Anonymous, 2001).

Table 35.1: Varieties/Lines of Sesame Resistant to Different Diseases

Disease	*Resistant Varieties/Lines*	*Reference*
Wilt	JT-7, N-32, MB-12	Vyas *et al.* (1981)
Macrophomina diseases	AT-9, HT-16, RT-46, ST-58, Gwalior-5, ES-25, ES-105, ES-124, IS-156, IS-190,Line-144, 885, BAG-59, 21,65,9,5,11	Gupta (1995) Saharan and Chand (1988) Li *et al.* (1991)
	RT-46, JTS-8, SI-249, SI-254-1,EC-370932, EC-370929,EC-351832,ES379-3-84,IS-97, IS-100 B,IS-646-1	Arriel *et al.* (1998) Anonymous (2001) Rajpurohit *et al.* (2003)
Phytophthora blight	75A/1-1/2-1,M-3-1, M-3-2, No.2-39, No.66-193, Spring borok	Gemawat and Prasad (1964) Saharan and Chand (1988) Sasikumar and Sardana (1988)
Alternaria blight	No.-2, No.-4, J-77, E-8,EC-370929, EC-370932	Saharan and Chand (1988) Rajpurohit and Solanki {2002)
Cercospora leaf spot	IS-4, 14, 15, 21, 29, No.-41, No.-41A, No.-41B, No-128, No.128°,L-36, L-45, RSB-1, HT-1, SI-2373, SI-1501, SI-1032, X-79-6, TMV-6, S-1683, BIC-7-2, Sidhi-54, Rewa-114, Seoni Malwa,65B58, 60/2/3-1-8B, 69B-392, 73A-96B	Kushwaha and Kaushal (1970) Vyas *et al.* (1981) Saharan and Chand (1988) Tripathi *et al.* (1994) Poswal and Misari (1994)
Powdery mildew	CO-1, BS-129, BS-518-6G, X-74-24, SI-3170, 59-1-1, IS-401,OMT 30 and DORS 101	Ganesh *et al.* (1992) Shadakshari *et al.* (1981) Rajpurohit (1993)
Synchytrium gall	JT-66-276, SP-70-23	Variar *et al.* (1981)
Bacterial leaf spot	Maporal, Ajimo Atar S-5, Morada, Almora white, C-24, C-87, C-96, M-31,NP-8, NP-23, Punjab-1, SSM-34, SSM-44,SSM-49, EC-4094, EC-13536, EC-14536, EC-20783, EC-20785, EC-20787, N-63-177-74-145, TC-325, T-58	Urdaneta and Mazzani (1976) Duragapal *et al.* (1969)
Bacterial blight	T-58,Tozi-3, S-76F2-22, K-112	Saharan and Chand (1988) Abdal Rahim and Adam (1990)
Phyllody	RT-46, RT-54, RT-103, RT-125, VRI-1	Anonymous (2001)

Resistance

Resistance source is the most important to control the diseases. Selim *et al.* (1976a) have reported susceptibility in mature plants to *M. phaseolina* to be dominant over tolerance, controlled by 1,2 or 3 pairs of genes. Selim *et al.* (1976 b) have reported that in mature plants the tolerance to wilt (*F. oxysporium f.sp. sesami*) is governed by 1 or 2 dominant pairs of genes and in one cross it was governed by 3 pairs of genes. Tolerance was recessive. The resistance to powdery mildew (*Erysiphe cichoracearum*) is controlled by two major complimentary genes (Krishnaswami *et al.*, 1983). Mishra *et al.* (1994) while studying the inheritance of leaf curl disease observed both additive x additive and/or additive x dominance types of gene interaction. The resistance to phyllody disease in *Sesame alatum* is due to insect resistance (Parani *et al.*, 1996). The available genotypes resistant to single disease is given in Table 35.1 and to multiple diseases in Table 35.2.

Table 35.2: Sesame Varieties/Lines Resistant to Multiple Diseases

Diseases	*Varieties/Lines*	*Reference*
Phyllody + wilt	Mulayanum	Mehetre *et al.* (1994)
Powdery mildew + Bacterial leaf spot	ES-277	Shadakshari *et al.* (1989)
Phytophthora blight + Alternaria blight	Suwonkkae	Lee *et al.* (1992)
Phytophthora blight + Alternaria blight + Cercospora blight	*Sesame radiatum*	Malaguti (1973)
Fusarium wilt + Macrophomina rot + *Rhizoctonia solani* root rot	87-AN, Golmarmara, Muganli, Ozberk-82, 9/10-1-2, 1/10-2-1	Alac *et al.* (1994)
Powdery mildew + Bacterial wilt + Cercospora blight + Fusarium stem rot	RAU-8817-4, CO-1, AT-6	Ram *et al.* (1995)
Macrophomina rot + Bacterial spot + Phyllody + Powdery mildew	Rajeshwari	Gangakishan *et al.* (1989)

Abd El, Moneem *et al.* (1997) evaluated 112 somaclones obtained from epicotyl segments of variety Giza 25 grown in tissue culture. S_2 clones had higher resistance to root rot, charcoal rot and wilt diseases. The resistance is attributed to increased contents of either peroxidase or phenolic compounds in roots.

References

Anonymous., 2001. Annual Progress Report Sesame and Niger 2000–2001. All India Coordinated Research Project (Sesame and Niger), J.N. Agricultural University Campus, Jabalpur, M.P. pp.152.

Arriel, N.H.C., Moreira, J. de, A.N., Guedes, A.R., Costa, I.T., Da, Tarso, F.P. de, da Costa, I.T. and da, Tarso, F.P., 1998. Agronomic and botanical evaluation of the live collection of sesame germplasm at Embrapa Algodao. Pesquisa em Andamento, Embraba Algodao. No. 91 pp.9.

Atac, A., Cetin, V. And Ipkin, B., 1994. Studies on root and crown rot disease of sesame. Bitki Koruma Bulletin 34: 43–54.

Bedigian, D., Seiogler, D.S.andHarlan, J. R., 1985.Sesamia, sesamolia and the origin of sesame.*Biochem. Syst. Ecology,13: 133–149.*

Borkar, S.G. and Krishna, A., 2000. Effect of fertilizers on development of phyllody in Sesame. *Indian J. Mycol. Pl. Pathol.* 30: 422–423.

Butler, E.J.1918. Fungi and Disease in Plants. Thaker Spink and Co. Calcutta. pp.547.

Durgapal, J.C. and Rao, Y.P., 1967. Bacterial leaf spot of *Sesame orientale in India. Indian Phytopath.* 20: 178–179.

Durgapal, J.C., Patel, P.N. and Rao, Y.P., 1969. Resistance in crops to bacterial disease in India. Evaluation of Sesame for resistance to bacterial leaf spot incited by *Pseudomonas sesami Indian Phytopath.* 22: 292–294.

Gabr, M.R., Hussein, N.A., Saleh, O.I. and Khalil, M.A., 1998. Susceptibility of certain varieties and genotypes and control of wilt and root rot diseases of sesame attributed to *Fusarium oxysporum* f. sp. *sesami* and *Macrophomina phaseolina. Egyptian Journal of Microbiology* 33: 403–428.

Ganesh, S.K., Kandasamy, G., Kousalya, G., Sivarman, M.R. and Gangadharan, K., 1992. Field evaluation of (*Erysiphe cichoracearum* D.C.). *Madras Agricultural Journal* 79: 115–117.

Gangakishan, A., Jagadishwar, K., Raju, C.S. and Sudarshanum, A., 1989. Rajeshwari– a high yielding white seeded variety of sesame for Andhra Pradesh. *Indian Farming* 39(3): 7–8.

Gemavat, P.D. and Prasad, N., 1964. Further studies on *Phytophthora* blight of Sesame. *Indian Phytopath.* 17: 273–283.

Gupta, I.J. and Cheema, H.S., 1990. Effect of microsclerotia of *Macrophomina phaseolina* and seed dressers on germination and vigour of sesame seed. *Seed Research* 18: 169–172.

Gupta, L.J., 1992. Change in the production of volatile aldehyde compounds in *Macrophomina* infected sesame (*Sesamum indicum*) seed. *Seed Research* 20: 27–30.

Gupta, R.B.L.1995. Evaluation of sesame genotypes for resistance to root rot and oozing. *Indian Phytopathology* 48; 194–195.

Karunanithi, K. Muthusamy, M. and Seetharaman, K., 1999. The cultural and pathogenic variability among the isolates of *Macrophomina phaseolina* causing root rot of sesame. *Plant Disease Research* 14: 113–117.

Krishinaswami, S., Rao, A.V., Appadurai, R. and Venkatarao, A., 1983. A preliminary study on the inheritance of powdery mildew disease in Sesame. Proc. National Seminar on breeding crop plants for resistance to pests and diseases. TNAU, Coimbatore.

Kushi, K.K. and Khare, M.N., 1979. Seed borne fungi of sesame and their significance. *Seed Research* 7: 48–53,

Kushwaha, U.S. and Kaushal, P.K., 1970. Reaction of Sesame varieties to *Cercospora* leaf spot in M.P. Mysore J. Agric. Sci. 4: 228–230.

Lee, J.I., Kang, C.W., Lee, B.H. and Park, R.K.1992 A nonbranching good quality and high yielding sesame variety 'Suwonkkae'. Research Report R.D.A.U.& I.C. 34: 91–96.

Li, L.L., Wang. S.Y., Fang, X.P., Hyang, Z.H., Wang, S.T., Li, M.L. and Cui, M.Q., 1991. Identification of *Macrophomina phaseolina* resistant germplasm of sesame in China. *Oil crops of China*. No. 1: 3–6.

Mahatma, L., Singh, S.D. and Lodha, P.C., 2001. Pathogenic seed mycoflora of sesame (*Sesamum indicum L.*). *J. Mycol. Pl. Pathol.* 31: 377–379.

Malaguti, G., 1973. Leaf diseases of sesame (*Sesamum indicum L.*) in Venezuela. Rev. de la Fac. de. *Agronomia* 7: 109–125.

Mahetre. S.S., Ghatge, R.D. and Lad, S.K., 1994. Wild *Sesame mulayanum*: a source of multiple disease resistance. *Annals of Agricultural Research* 15: 243–244.

Mishra, A.K., Tiwari, R.C., Pathak, K.N., Ali, S.A. and Ghuraya, R.S., 1994. Inheritance of leaf curl disease in *Sesamum indicum* L. *Crop Research* Hisar 8: 131–136.

Mishra, R.P. and Khare., M.N., 1970. A sclerotial leaf and stem rot of mesta. *Indian Phytopath.* 23: 706–708.

Mohanti, N.N., 1958. *Cercospora* leaf spot of sesame. *Indian Phytopath.* 11: 186–187.

Mohanti, N.N. and Behera, B.C.1958. Blight of sesame (*Sesame orientale* L.) caused by *Alternaria sesami* (Kawamara) Comb. *Current Science* 27: 492–493.

Murugesan, M., Shanmugam, N., Menon, P.P.V., Arokiaraj, A., Dhamo, K.P. and Kochubabu, M., 1978. Statistical arrangement of yield loss of Sesame due to insect pests and diseases. *Madras Agric.* J. 65: 290–295.

Navas, M. and Subero, L.J., 1995. Survival of *Corynespora cassicola* in sesame seeds under two storage temperatures. *Anales de Botanica Agricola* 2: 16–19.

Ojiambo, P.S., Ayiecho, P.O. and Nyabundi, J.O., 1999. Effect of plant age on sesame infection by *Alternaria* leaf spot. *African Crop Science Journal* 7: 91–96.

Ojiambo, P.S., Ayiecho, P.O., Narla, R.D. and Mibey, R.K., 2000. Tolerance level of *Alternaria sesami* and the effect of seed infection on yield of sesame in Kenya. Experimental Agriculture 36: 335–342.

Pandya, R.K., Triphathi, M.L., Reetu Singh and Singh, R., 2000. Change in mineral content in sesame leaves infected by *Pseudomonas syringae* pv. *sesami* (Malkoff). *Crop Research*– Hisar 20: 140–144.

Parani, M., Singh, K.N., Rangasamy, S.R.S. and Ramalingam, R.S., 1996. A study on mechanism of phyllody disease resistance in *Sesame alatum* Thonn. *Current Science* 70: 86–89.

Pathirana, R., 1992. Gamma ray induced field tolerance to *Phytophthora* blight in sesame. *Plant Breeding* 108: 314–319.

Phookan, A.K. and Hazarika, R., 1994. Damping off of Sesame caused by *Xanthomonas campestris* pv. *sesami* in Assam. *Indian Phytopath.* 47: 249–250.

Poswal, M.A.T. and Misari, S.M., 1994. Resistance of sesame cultivars to *Cercospora* leaf spot induced by *Cercospora sesami* Zimin. *Discovery and Innovation* 6: 66–70.

Rajpurohit,T.S 1981.Morphology and taxonomy of sesamum *Alternaria*. Madras Agricultural Journal.68(10): 696–697.

Rajpurohit,T.S. and N.Prasad 1982 Effect of different media, temperature and levels of pH on the growth and sporulation of *Alternaria sesami.Geobios Vol.9(3)142–144.*

Rajpurohit,T.S.and N.Prasad 1982.Production of pectinolytic enzymes by *Alternaria sesami in vitro.* Indian.J.of Mycol.&Pl.Path.Vol 12(2): 220–221.

Rajpurohit,T.S.N.Prasad and P.D.Gemawat1984.Evaluation of fungicides against *Alternaria sesami in vitro.* Pesticides special issue on fungicides Jan: : 57.

Rajpurohit, T.S., 1993. Occurrence, varietal reaction and chemical control of new powdery mildew (*Erysiphe orontii* Cast) of sesame. *Indian J. Mycol. Pl. Pathol.* 23: 207–209.

Rajpurohit,T.S., 1997.Management of Macrophomina stem and root rot of sesame through fungicidal seed treatment and varietal resistance. Indian.J.of Mycol.&Pl.Path.Vol. 27 No.1Page 98.

Rajpurohit,T.S., 1998.Integrated management of *Phytophthora* blight, *Cercospora* leaf spot and *Alternaria* leaf spot of sesame.J.of Mycol.&Pl.Path.VoI 28(1): 68.

Rajpurohit,T.S., 1999.Trichoderma viride Biocontrol agent effective against *Macrophomina* stem and root rot of sesame.4th Agricultural Science Congress,Feb.21–24,Jaipur.Abstract P.297.

Rajpurohit,T.S and Z.S.Solanki2002.Identification of resistance sources of *Alternaria* leaf spot of sesame.Asian congress of Mycology and Plant Pathology oct.1–4,University of Mysore.Page 207.

Rajpurohit,T.S and Z.S.Solanki and Ahuja, D.B., 2003. Screening of promising sesame genotypes against stem and root rot (*Macrophomina phaseolina*)under sick plot.NationalSeminar on Strees management in oilseeds for attaining self reliance in vegetable oils Indian Society of Oilseeds, DOR, Hyderabad, 28–30 Jan 2003.

Rajpurohit, T.S. 2002.Influenced of intercropping and mixed cropping sesame with pearl millet,greengram and mothbean on the incidence of stem and root (*Macrophomina phaseolina*)of sesame Sesamum and Safflower news letter.No.–17Page 40–41.

Rajpurohit,T.S. 2003. Studies on fungicides plant products and bioagents as spray in sesame disease management. Journal of Eco–physiology 6(3–4): 87–88.

Rajpurohit,T.S.2004.Studies on plant products and new chemicals as spray against *Alternaria* blight of sesame.Gebios,31: 64–65.

Rajpurohit,T.S.2004. Efficacy of foliar spray on incidence of diseases of sesame. Sesamum and Safflower news letter.No.–19Page 67–69.

Rajpurohit,T.S.2004 Evaluation of plant products and fungicides and insecticides against *Alternaria* blight, Phyllody and leaf curl diseases of sesame.Indian Phytopathology Vol. 57(3): 340–341.

Rajpurohit,T.S.and H.R.Bishnoi 2004 Integrated disease management in sesame (*Sesamum indicum* L.)Paper presented in 26 Annual conf.& Symp.ISMPP.Goa university Goa, 7–9 Abstract.pp 144–145.

Rajpurohit,T.S., 2005 Seed treatment and Bioagent for the management of stem and root rot of sesame. Sesamum and Safflower news letter.(Under publication)

Ram, T., Ansari, M.M. and Gangwar, B., 1993. Evaluation of sesame varieties in rainfed conditions for better yield and resistance to diseases in a rice based cropping system. In Proceedings of Symposium on farming systems for sustained productivity in humid tropics. Ed. Rao, G.C., Central Agricultural Research Institute, Port Blair. pp. 119–122.

Rao, N.R. and Vijayalakshmi, M., 2000. Studies on *Alternaria sesami* pathogenic to sesame. *Microbiological Research* 155: 129–131.

Rao, R. and Pavgi, M.S., 1975. Two sclerotial diseases of crops in Varanasi. *Indian Phytopath.* 28: 401–404

Sabet, K.A. and Dowson, W.J., 1960. Bacterial leaf spot of sesame (*Sesame orientale* D.). *Phytopath. Z.* 37: 252–258.

Saharan, G.S. and Chand, J.N., 1988. Tilhani Faslon ke Rog. Commission for Scientific and Technical Terminology, Government of India, New Delhi. pp 47–83.

Sasikumar, B. and Sardana, S., 1988. Siping borok (Jhum til). Source of some valuable genes. *Journal of Oilseeds Research* 5: 183–184.

Selim, A.K., Serry, M.S., Omran, A.O. and Satour, M.M., 1976 (a). Breeding for disease resistance in Sesame, *Sesamum indicum* L. II. Inheritance of resistance to root and stem rot disease caused by *Sclerotium bataticola* Taub. *Egyptian Journal of Phytopathology* 8: 15–18.

Selim, A.K., Serry, M.S., Satour, M.M. and Al Ahmar, B.A., 1976 (b). Breedidng for disease resistance in sesame, *Sesamum indicum* L. III Inheritance of resistance to *Fusarium oxysporum* Schlecht. *Egyptian Journal of Phytopathology* 8: 19–24.

Shadakshari, Y.G., Virupakshappa, K. and Siddaramaiah, A.C., 1989. Reactions of Sesame genotypes against powdery mildew and bacterial leaf spot. *Current Research* 18: 4–5.

Shreemali, J.L., 1978. Two new species of Phoma from India. *Indian J. Mycol. Pl. Pathol.* 8: 220–221.

Shukla, B.N. and Vyas, S.C., 1997. Two new diseases of Sesame (*Sesamum indicum* L.). J.N.K.V.V. Res. J. 11: 142–143.

Singh, B.P., Shukla, B.N. and Kaushal, P.K., 1976. Evaluation of Sesame varieties for their susceptibility to *Phytophthora parasitica* Dastur at Jabalpur, M.P. J.N.K.V.V. Res. J. 10: 76–77.

Singh, T. and Singh, D., 1990. Incidence of *Macrophomina phaseolina* (*Tassi*) Goid in sesame seeds of Rajasthan. *Indian J. Mycol, Pl. Pathol.* 20: 111–114.

Tripathi, S.K. Tiwari, K.L. and Patel, R.P., 1976. Assessment of stability in resistance to *Cercospora* leaf spot in Sesame. *Agricultural Science Digest Karnal* 16: 233–236.

Urdaneta, U.R. and Mazzani, B., 1976. Varietal differences in susceptibility to bacteriosis in sesame (*Sesamum indicum L.*) in Venezuela. Ag*ronomia Tropical* 26: 311–320.

Variar, M., Pavgi, M.S. and Variar, M., 1981. Varietal reaction of sesame to *Synchytrium* gall disease. *Indian Phytopath.* 34: 430–433.

Vyas, S.C., Prasad, K.V.V. and Khare, M.N., 1981. Diseases of Sesame and Niger and their Control. Department of Plant Pathology, J.N.K.V.V, Jabalpur, M.P. Bull. pp.16.

Wasnikar, A.R., Khare, M.N., Deshmukh, M.R. and Yasin, M., 1991. Bacterial blight of sesame. *Current Research* 20: 183.

Plant Protection, Diversity and Conservation (2011) *Pages* **481–490**
Editors: **D. Prasad & Satya Kumar**
Published by: **BIOTECH BOOKS, NEW DELHI**

Chapter 36
Trends in Sustainable Weed Management

Ozair Chaudhry[1]* and Rene Van Acker[2]**

[1]*Director, Ontario Institute of Agrologists and Instructor, Albert Campbell Collegiate Institute, Toronto, Canada*
[2]*Associate Dean External and /Professor, Ontario Agriculture College, University of Guelph, Canada*

The agricultural model of production promoted during the green revolution lead to tremendous gains in crop yield via the use of highly bred crop varieties and the increased use of crop production inputs including fertilizer and pesticides (Dobbs 2006). As the world's population continues to grow at a rapid pace there remains a need for substantive crop yield increases in agriculture achieved through the use of agricultural technology including pesticides (Godfray *et al.*, 2010; Tonukari and Omotor 2010; Simmons and Kappes 2010). The yield success of the inaugural green revolution has however came at some cost, in some cases to the environment and to farmer livelihoods (Dobbs 2006), and there have been calls for more sustainable approaches to achieving increased yields (Gomez-Casero, *et al.*, 2010; Van Acker 2008) including a reduction in pesticide use and diversified and integrated approaches to weed management (Nazarko et.al. 2005).

Weeds as Pests

Weed interference is responsible for one third of crop yield loss worldwide (Oerke 2006). In the United States, weed infestations are estimated to cause losses amounting to US $ 34.7 billion annually (WSSA.net 2008). In Canada, 10-30 percent of the total crop production input costs are spent on herbicides (Derksen *et al.*, 2002) and potential crop yield losses in Canada due to weed interference vary from 10% to as high as 53 % (Figure 36.1) (OMAFRA 2009). Unlike diseases or insects, weed pest infestations tend to be stable and relatively predictable from year to year. The infestations also tend to

E-mail: *ozair.chaudhry@tdsb.on.ca; **vanacker@uoguelph.ca

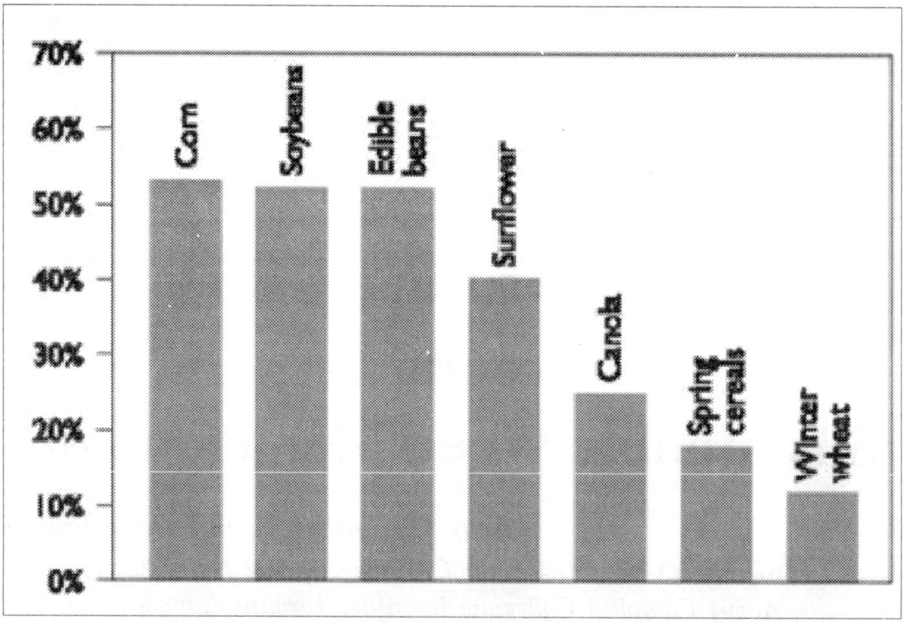

Figure 36.1: Typical Yield Losses of Different Field Crops Due to Weed-Crop Competition
***Source*: OMAFRA, 2009.**

be comprised of a wide variety of species existing as a field-based complex (Nazarko *et al.*, 2005). These factors pose difficulties for management because there are multiple species and the weed complex tends to produce persistent populations via dormant seed banks. To some extent this is why highly efficacious herbicide use remains the primary means of weed management in modern farming systems (Nazarko *et al.*, 2005).

However, over the past decade, the range of herbicides used has narrowed, in particular in regions where genetically modified herbicide tolerant (GM-HT) crops have been widely adopted (Sheridan 2010). But, the repeated and continuous use of herbicides has lead to the development of herbicide resistant weeds. Currently, 319 weed species worldwide have been documented to be herbicide resistant, and 8 different herbicide groups have resistant weed biotypes (Figure 36.2) (Heap 2010; Webster 2009.,Chaudhry 2010., Hager 2010).

Pressures for Changes to Weed Management

Recognized climate change will impact weed management. Increasing global temperature is predicted to cause more rapid species extinctions and species shifts (Stern 2004), and will likely change the types and range of weed species we will be managing (Chaudhry 2008). In addition, there is a desire among many societies to reduce reliance on synthetic chemical pesticides including herbicides (Kudsk 2008) and this has lead to pesticide bans in some jurisdictions. The rise in the popularity of organic foods mirrors some of these sentiments. Additionally, herbicide resistant weeds are causing many farmers to consider alternative weed control measures, and in some cases to develop farming systems that are inherently more resilient to weeds.

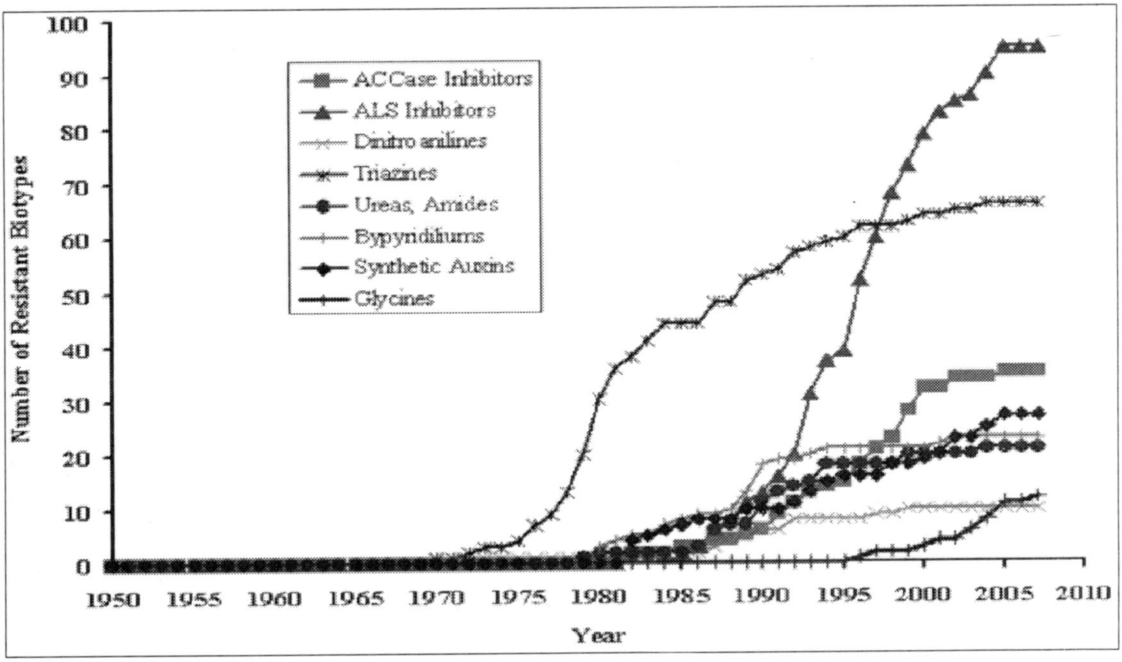

Figure 36.2: Trends in the Number of Herbicide Resistant Weeds for Several Herbicide Classes. *Source*: Heap, 2007.

Strategies and Tactics for Reducing Herbicide Use

There are a number of approaches for reducing herbicide use and reliance on herbicides in agriculture. These include; increasing herbicide efficiency, innovations in mechanical weed control, innovations in biological control and systems design.

Increasing Herbicide Efficiency

Spray Delivery Adjustments

Reducing herbicide dose while maintaining efficacy is challenging but can be achieved by a variety of means including innovations in nozzle type, delivery speed, boom height, nozzle spacing, directed vs. open spray, electrostatic charging of spray droplets (Nazarko *et al.*, 2005). Kudsk 2008 reported that the evolution of hydraulic nozzles greatly minimizes (and can sometimes eliminate) spray drift associated with flat fan nozzles. Reducing boom height can also facilitate further reductions in herbicide dose. He also observed that angling of nozzles increased the efficacy of foliar-active graminicides applied at early growth stages. However, because angling of nozzles can increase the risk of drift it has to be accompanied by a reduction in boom height from 50-25 cm.

Adjuvants/Surfactants

Selective spray solution additives may help enhance herbicide deposition, absorption and spreading ability on leaves when applied post emergence. Adjuvants are often used to reduce the required herbicide dose by increasing the ability of molecules to reach the target site of action within the plant more rapidly and reliably. For example, weeds have been shown to retain more active ingredients as oil adjuvant concentration increased from 0.25-4 % v/v (Chaudhry *et al.*, 1989, Nalewaja *et al.*, 1988). And even for low dose chemistries such as sulfonylurea herbicides, enhanced efficacy has been achieved with the use of surfactants especially when spray conditions and water types are variable (Top Crop Manager 2001)

Herbicide Formulation

There has been a steady evolution of herbicide formulations to improve efficacy and allow for reduced dose. Recently, BASF introduced "Polymer Herbicide" technology where minute quantities of active ingredient can be encapsulated in a water based polymer. This polymer coating allows for slow release of the herbicide providing longer term control, better handling and greater application flexibility for preemergence soil applied herbicides in all tillage and environmental conditions. This method also offers targeted release and superior surface release (Top Crop Manager, 2009).

Use of Domesticated Chemicals

In organic systems and in jurisdictions where pesticide bans are in place there is a great interest in the use of domesticated chemicals for weed management. For example, a blend of grain vinegar and clove oil applied @636L/ha provided 100% control of redroot pigweed seedlings and 73 % control of velvetleaf seedlings in field trials (Evans 2009). Application timing is critical to maximize efficiency with these chemicals. Their use in non-crop areas is promising but selectivity in cropped (or sodded) areas is still a tremendous challenge.

Nanoherbicides

Nanotechnology has made it possible to distribute nanoparticles (herbicidal molecules) into plant vascular system and guide them to reach specific sites of action. Nanoencapsulation of herbicides could also provide better penetration through leaf cuticle tissue (Perez-de-Laque and Rubiales, 2009).

Site Specific Assessment for Weed Management

Distribution of weeds and their composition in an area requires weed mapping by either field scouting or by remote sensing Given the typical distribution of weeds there is great potential in reducing overall herbicide use through site-specific application, in particular if there is technology that can allow for the discrimination of weeds from crops (Kudsk 2008). In the United States, a robotic system has been developed that consists of a machine vision system, a controlled illumination chamber and a precision chemical applicator. It was able to distinguish weeds from cotton and apply herbicides only to weeds with 80-90% reliability (Lamm, *et al.*, 2002). Similarly, in Europe, weed scientists have made tremendous advancements in developing means of distinguishing weedy monocots from monocot crops using differences in hyperspectral signatures (Gomez-Casero, *et al.*, 2010). Weed-detecting sprayer technology has been commercialized that does not require weed mapping prior to herbicide application. These sprayers are however, specific only to non-crop areas (fallow and pre-plant or post-harvest) but they can provide herbicide use reductions of 19-60% in these situations (Blackshaw *et al.*, 1998a and 1998b).

Innovations in Mechanical Weed Control

Innovative Mechanical Weeding Tools

Pre-plant tillage and post-crop emergence tillage has been a common practice for creating a clean seedbed and providing crops a competitive advantage over later emerging weeds. Post-emergence mechanical weed control has traditionally been between rows in row crops with some minor exceptions (including blind tillage) and there often remains a requirement for herbicide use in these systems, especially over the row where the weed interference can cause tremendous yield (Nazarko *et al.*, 2005). There have been some advancements in mechanical weed control designed to remove weeds from within the crop row and these include torsion finger and brush weeders as well as weeders that employ compressed air (the "Pneumat" weeder) to blow weeds away (Figures 36.3a, b, c, d) (Van Der-Weide, et.al. 2008).

Forcella (2009) has explored the possibility of using compressed air and an abrasive agent to physically control weeds. He found that granulated nut shells delivered by sand blaster @ 517 Kpa at a distance 30-60 cm from *Chenopodium album* seedlings reduced weed weight by 23-86 % and 100 % with one and two blasts respectively at the 1-4 leaf stage.

Innovations in Biological Control

Several organisms (mammals, insects, fungi, bacteria, and nematodes) can prey on weed plants and provide a range of control levels of weeds (Nazarko *et al.*, 2005). In countries like Canada, where changes in legislation facilitate the registration of bioherbicides there has been an increase in evaluations and registrations of biopesticides (Figure 36.4) (Bailey *et al.*, 2010).

The first bioherbicide used in Canada was *Collectotrichum gloeosporioides* in 1992 for round-leaved malva (*Malva pusilla*) control. During 1972-2008, the Pest Management Regulatory Agency (PMRA) in Canada approved and registered 24 microbial active pesticides with 83 formulations but only 4 of these were herbicides (Bailey *et al.*, 2009). High price, storage challenges and field application requirements are all limits to the commercialization of bioherbicides. In addition, these agents tend to be species specific and are of limited in typical multiple weed species infestations (Nazarko *et al.*, 2005). With a need for chemical pesticide alternatives in home lawns due either to residential pesticide bans or consumer demand there has been considerable advancement in the development of bioherbcides for lawn weeds such as dandelion (*Taraxacum officinale*) (Schnick *et al.*, 2002). *Sclerotinia minor* has been reported to provide up to 100 % control of plantain (*Plantago major*) if applied on 3-6 week old plants (Abu-Dieyeh, et *al.*, 2010). Pesticide bans may be a significant driver for further bioherbicide development, and may allow for products that are competitive with chemical herbicides.

Systems Design

Ultimately, system design is the foundation for effective and sustainable weed control, and farming systems that include a greater amount of diversity (e.g. crop rotation) are more robust with an inherently greater ability to resist the invasion, proliferation and interference of weeds (Derksen *et al.*, 2002; Nazarko *et al.*, 2005). Innovations in weed management techniques will aid in our ability to reduce herbicide use, in particular in situations where system diversity is not as practical an option (e.g. home lawns). Over the past 3 decades, increasing farm size, narrowing margins on-farm and increasing cash flow immediacy have all worked to limit farmer's opportunities to build greater diversity into their farming systems. Premiums paid for organic products have helped to allow some organic farmers to develop more robust farming systems (Dobbs 2006). The pressures related to reductions in pesticide use are being met by farmers at the same time as growing concerns over the impact that climate change will have on agricultural production (Stern 2004; Chaudhry 2008) and this may create additional challenges, but it may also create additional incentive to develop resilient agricultural systems and systems based capacity to reduce herbicide use while maintaining or increasing productivity and yield.

Figure 36.3a: Harrowing Torsion Weeder in Maize

Figure 36.3b: Harrowing Torsion Finger Weeder in Cabbage

Figure 36.3c: Pneumat Compressed Air Weed Control Device

Figure 36.3d: Sal Radis Intelligent Weeder

Source: Van DerWeide *et al.,* 2008., Coutier, *et al.,* 2007.

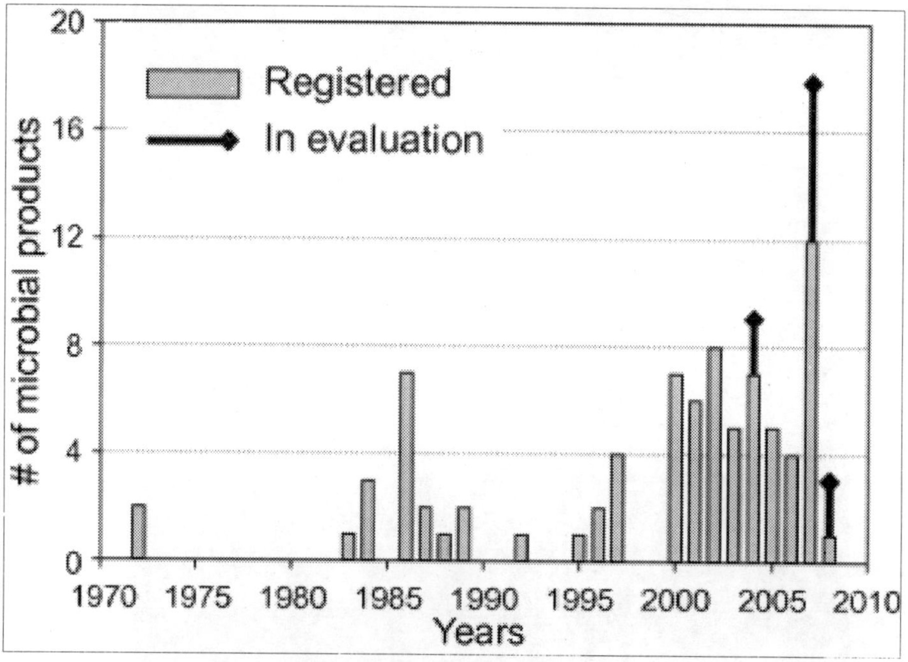

Figure 36.4: Microbial Products Registered Under Regulatory Evaluation in Canada from 1972-2008
***Source*: Bailey *et al.*, 2010.**

Conclusion

Sustainability is the ultimate driver for future efforts in weed management. Weed management development has evolved significantly over the past 2 decades and is substantively more diverse in approach versus efforts during the green revolution. Chemical herbicides will always be a core tool for weed management but sustainability requires a diversity of options and weed populations, especially in the context of climate change, can only be sustainably controlled if there is a diversity of strategies and tactics available. In this paper we have briefly highlighted some of the advancements in weed management and have demonstrated the trend towards the creation of a broader diversity of weed management options.

References

Abu-Dieyeh, M. H., Shaheen, I. and Watson, A.K. 2010. Effect of plant age and turf grass composition on the efficacy of the *Sclerotina minor* granular biopesticide on broad leaf plantation and prostrate knock weed. Biocontrol Science and Technology. 20(2): 213-226.

Blackshaw, R. E., Molnar, L.J. and Lindwall, C. W. 1998a. Merits of weed sensing sprayer to control weeds in conservation fallow and cropping system. Weed Sci. 46: 120-126.

Blackshaw, R. E., Molnar, L.J. and Lindwall, C. W. 1998b. Factors affecting the operation of the weed sensing detect spray system. Weed Sci. 46:127-131.

Bailey, K. L., Boyetchko, S. M. and Langle, T.2010. Social and economic drivers shaping the future of biological control: A Canadian perspective on the factors affecting the development and use of microbial biopesticides. Biological Control. 52(3): 221-229.

Courtier, D. C.,Van der Weide, R. Y., Peruzzi, A. and Leblanc, M.L. 2007. Mechanical Weed Management. Book Chapter In: Nonchemical Weed Management Principles, Concepts and Technology. CABI Int'l. Pub. pp: 111-134.

Chaudhry, O., Rana, M. A., and Khan, M. A. 1989. Activity of foliar applied Fluazifop butyl as influenced by surfactant, relative humidity and season of planting of sunflower. Pak. J. of Weed Res. 4 (1): 25-31.

Chaudhry, O., Khan, A. R., Aslam, M. and Rana, M. A. 1993. Performance of herbicides in spring and autumn planting of soybean (Glycine max L.). Pak. J. of Ag. Res. 14(1): 29-37.

Chaudhry, O. 2008. Review Article: Futuristic vision on climate change adversity and herbicide technology. Ann. Plant Protec. Sci. 16 (1): 1-5.

Chaudhry, O. 2010. Herbicide-Resistance and Weed-Resistance Management. Book Chapter In: Potential Plant Protection Strategies. I.K. International Publishing House Pvt. Ltd. New Delhi 110-016. ISBN 978-938-0578-293. Courtesy Posted on US/WSSA home page.

Derksen, D. A., Anderson, R.L., Blackshaw, R. E and Maxwell, B. 2002. Weed dynamics and management strategies for cropping system in the northern great plains. Agron. J. 94: 174-185.

Dobbs, T. L. 2006. Challenges facing a second green revolution: Expanding the reach of organic agriculture. Crop Management. DOI: 10.1094/CM-2006-0921-02 (accessed 14/06/10 online at http://www.plantmanagementnetwork.org/pub/cm/symposium/organics/Dobbs/)

Evans ,G.J., Bellinder, R. R. and Goffinet, M. C. 2009. Herbicidal effects of vinegar and clove oil products on redroot pigweed (*Amaranthus retroflexus*) and velvetleaf (*Abutilon theophrasti*). Weed Tech. 23 (2): 292-299.

Forcella, F. 2009. Potential use of abrasive air-propelled agricultural residues for weed control. Weed Res. 49: 334-345

Godfray, H. C. J., Bedlington, J. R., Crute, R., Hadad, L., Lawrence, D., Muir, J. F., Pretty, J., Robinson, S., Sandy, M. and Toulmin, C. 2010. Food Security: The challenge of feeding a billion people. Science 327: 812-818.

Gomez-Casero, M. T., Castillejo-Gonzalez., Garcia-Ferrer, A., Pena-Barragan, J.M., Jurado-Exposito, M.,Garcia-Torres, L. and Lopez-Granados, F. 2010.Agronomy for Sustainable Development. Agron. J.DOI 10.1051/agro/2009052.

Hager, H. 2010. Weed Resistance to Herbicide: A disturbing Trend. Top Crop Manager (East) 35 (6):38-39.

Heap, I. M. 2007. International survey of herbicide resistant weeds on line WSSA. http.www.weedscience.org.In.asp.dated6/7/2008.

Kudsk, P. 2008. Recent advances in weed management. Zemdirbyste-Agriculture 95 (3): 103-109.

Lamm, R. D., Slaughter, D. C. and Giles, D.K. 2002. Precision weed control system for cotton. Transaction of the ASABE. 45(1) : 231-238.

Liaghat, S and Balasundram, S. K. 2010. A review: The role of sensing in precision agriculture. Am. J. Ag. and Biol. Sci. 5 (1): 50-55.

Nalewaja, J. D. Gross, G. R. and Tann, R. S. 1988. Spray retention is affected by spray Parameters, species, adjuvants. In: Pesticide Formulations System. Eighteenth Vol. ASTM STP 1347 Eds. American Soc. For testing and materials.

Nazarko, O. M., Van Acker, R. C. and Entz, M. H. 2005. Strategies and tactics of herbicide use Reduction in field crops in Canada: A review. Can. J. of Plant Sci. (85): 457-479.

Oerke, E. C. 2006. Crop losses to pests. J. of Ag. Sci. 144: 31-43.

OMAFRA. 2009. Weed Control: crop yield losses due to weeds. http://www.omafra.gov.on.ca/english/crops/pub811/12crop.htm

Perez-de-Luque, A. and Rubiales, D. 2009. Mini review: Nanotechnology for parasitic plant control. Pest Management Sci. 65 (5): 540-545.

Schnick, P. J., Stewart-Wade, S. M. and Boland, B. J. 2002. 2,4-D and *Sclerotinia minor to* control common dandelion. Weed Sci. 50: 173-178.

Stern, N. 2004. The Economics of Climate Change. Executive summary. http://www.hm.treasury.gov./media/8Ac/F/7/Executive_summary.pdf., http://www.news.bbc.ca.uk1/hi/business6098362.htm.

Sheridan, C. 2010. Report blames GM crops for herbicide spike, downplays pesticide reduction. Nature Biotechnology 28: 112-113.

Simmons, K. and Kappes, S.M. 2010. Using Genetic Tools to Combat Hunger. Agricultural Research. 58(5): 2.

Top Crop Manager. 2001. Choosing the Right Surfactant. http://www.topcropmanager.com/content.view.2646/38/

Top Crop Manager. 2009. Innovation leads to grower benefit. http://www.topcropmanager.com/index.php?option=com_content&task=view&id=4247

Tonukari, N.J and Omotor, D.G. 2010. Standard Review: Biotechnology and food security in developing countries. Biotech and Molecular Biology Reviews. 5(1): 13-23.

Van Der Weide, R. Y., Bleeker, P. O., Achten, V. T. J. M., Lotz, L. A. P., Fogelberg, F., and Melander, B. 2008. Innovation in mechanical weed control in crop rows. Weed Res. 48 (3): 215-224.

Webster, T. M. and Sosnoskie, L. M. 2010. The loss of glyphosate efficiency: a changing weed spectrum in Georgia cotton. Weed Sci. 58:73-79.

Weed Science Society of America 2008. http://www.wssa.net/WSSA/PressRoom/wssawildfires.htm

Plant Protection, Diversity and Conservation (2011) *Pages* **491–498**
Editors: **D. Prasad & Satya Kumar**
Published by: **BIOTECH BOOKS, NEW DELHI**

Chapter 37

The Pumpkin Beetle and its Integrated Management: A Review

S.K. Mandal

Department of Entomology and Agricultural Zoology,
Rajendra Agricultural University, Pusa (Samastipur) – 848 125, Bihar

The pumpkin beetle is a major pest of cucurbitaceous crops (Lall, 1964 and Menon *et al.*, 1972). Red pumpkin beetle, *Aulacophora foveicollis* Lucas, blue pumpkin beetle, *A. intermedia* Jacoby and grey pumpkin beetle, *A. cincta* Fab. are three major species of pumpkin beetle which occupy prominent status in India as serious pest of cucurbits (Mishra, 1983). According to Rawat and Mishra (1967), Shaheen (1973), Sandhu and Bhalla (1974), Butani and Verma (1977), Shinde and Purohit (1978), Mavi and Bajwa (1984), Pawar *et al.* (1984), Luthra *et al.* (1995), Suliman (1997) Ciampolini *et al.* (1998) and Johri *et al.* (2004), the red pumpkin beetle *Raphidopalpa foveicollis* Lucas. is most destructive pest throughout the cropping period of cucurbits.

Distribution

A. foveicollis is widely distributed throughout South-east Asia, Australia, Southern Europe, Africa and Mediterranean region (Butani and Jotwani, 1983).

Many workers reported that the red pumpkin beetle is a serioLis pest in different parts of states in all over India *viz.*, Karnataka (Venkataryappa and Govindan, 1978), Madhya Pradesh (Shinde and Purohit, 1978), Assam (Debi *et al.*, 1980), Kerala (Gopalkrishnan *et al.*, 1983), Maharashtra (Pawar *et al.*, 1984) and Manipur (Shriram and Pathak, 1984).

Host Plants

Cucurbits are by far the most important host plant of the red pumpkin beetle (Nath, 1963, 1964; Nath and Thakur, 1965; Bogawat and Pandey, 1967; Singh and Krishna, 1970; Pal *et al.*, 1978; Shinde and Purohit, 1978; Gopal Krishnan *et al.*, 1983; Shri Ram and Pathak, 1984; Sandhu and Grewal, 1985 and Johri and Johri, 2003). However a few alternate host plants have been studied by Lall (1964),

Bogawat and Pandey (1967), Pruthi (1969) and Shinde and Purohit (1978). As compared to the red pumpkin beetle, *A. foveicollis* has a wide range of host plants (Table 37.1).

Table 37.1: Host Plant of *Aulacophora foveicollis*

Scientific Name	*Common Name*	*References*
Lycopersicon esculentum L.	Tomato	Rawat and Mishra, 1967
Raphanus sativus L.	Radish	Butani and Verma, 1977
Cucumis melo L.	Kharbuj	Shinde and Purohit, 1978
Momordica charantia L.	Karela	Gopalkrishnan *et al.*, 1983
Cucumis sativus L.	Kheera	Shri Ram and Pathak, 1984
Calotropis gigentia	–	Verma, 1984
Thavatia nerifolia	–	Verma, 1984
Citrullus vulgaris Schard	Watermelon	Singh, 2000

The qualitative and quantitative aspects of food determine the extent of relationship between insect and host, which ultimately governs the possible host range and their influence on the development of insect. Being cosmopolitan and polyphagous in nature of red pumpkin beetle, *A. foveicollis* Lucas a series of different cucurbit hosts (Hussain and Sah, 1926; Nath, 1964 and Pareek and Kavadia, 1988). Therefore, first preference was to *Cucurbita maxima* followed by *Cucumis sativus, Citrullus vulgaris* var. *fistulosus, Luffa cylindrical* and *Laginaria vulgaris* (Khan and Hajela, 1984). It is evident that the growth index (G.I.) values increased with the increase in the suitability of the host. Johri and Johri (2003) also observed the maximum growth index of *A. foveicollis* on pumpkin followed by cucumber, melon, musk melon, water melon, bottle gourd, round gourd. However, lowest growth index was obtained on sponge gourd and ridge gourd. Thus, the highest growth index value of *A. foveicollis* was done on different cucurbit hosts to high nutritive value and high food intake both by the adults and grubs.

Symptoms of Damage

The eggs are laid in moist soil around the host. The grubs feed on roots, underground portion of stems and bore into the fruits that in contact with the soil (Kadam and Patel, 1957). Its larval form feeds on the roots and stems while adult beetle feeds mainly on leaves (Saini, 1959). The damage is followed by secondary infection by saprophytic fungi resulting in rotting of the underground plant parts and ripe fruits whereas, the unripe damaged fruits gradually get wrinkled and dry up which becoming unfit for human consumption. Adult beetles feed voraciously on leaf lamina making irregular holes. They prefer young seedlings and tender leaves as well as damage at this stage may even kill the seedlings. Again Shinde and Purohit (1978) and Pareek and Kavadia (1988) were reported that the grubs feed on the roots while the adults make characteristics per formation on the leaves and flowers and cause considerable reduction in yield.

Seasonal Abundance

Beetle population was significantly higher in the crops sown in June and July. This might be due to peak activity of the pest during rainy season (Butani, 1975). Again he reported in 1975b the beetle was active from March to October and the peak period of activity is April to June. Minimum plant damage (19.5 per cent) was recorded in 20th September sown crop followed by August. This might be due to over wintering of the pest in winter days (Butani and Jotwani, 1984). The peak activity period for red pumpkin beetle is mid March to April and August to September (Sharma, 1997). The maximum yield was recorded during summer crop with 33.6 per cent plant damage followed by rainy season crop. Though, the winter sown crop recorded with minimum plant damage but yield was significantly low (Borah, 1999). Johri and Johri (2003) found the peak infestation (47.49 per cent) of red pumpkin beetle was in the month of July and lowest (3.92 per cent) in the month of February. Again they reported that tile highest damage potential by the beetle was in the germinating cucurbit plants in March to May.

Losses

R. foveicollis and *R. intermedia* damage the cucurbits plant by eating to the foliage and fruits 80 to 100 per cent and 40 to 88 per cent, respectively (Verma, 1984) while Purwar (2005) observed 12 to 72 per cent losses the yield due to red pumpkin beetle.

Life-Cycle

Immature stages of these species are mostly similar. Eggs are elongated in shape and yellowish-brown in colour. Grubs are 10 to 16 mm long and creamy-yellow in colour. Adult of *A. foveicollis* are 6 to 7 mm long with shiny reddish brown elytra, *A. cincta* is same in size having grayish-brown elytra, with dark-brown margin all round, while *A. intermedia* is comparatively smaller in size, 5 to 6 mm long and having shiny blue-black elytra. Eggs, grub, prepupal and pupal stages of *A. foveicollis* last for 5 to 8, 13 to 15, 2 to 5 and 7 to 15 days, respectively (Narayanan, 1953). Total life cycle takes 32 to 65 days and adults live for 15 to 30 days. There are 5 to 8 overlapping generations in a year. The adults hibernate from November to April under dry cucurbitaceous creepers, grasses, weeds and even in the soil. Mortality among the hibernating adult is quite high. Dutt and Dalapati (1977) also biological studies on *A. foveicollis* Locus in cucurbits.

Integrated Pest Management

The component of integrated pest management *viz.*, cultural, mechanical, biological and chemical measures play significance role to reduce the extent of damage to the crops.

Cultural and Mechanical Control

1. The crops heavily fertilized with nitrogen are more promote to its attack. Therefore, a balanced fertilization is suggestive against the pest.
2. Pest can be controlled mechanically by covering the fruits with paper bags (Gopalkrishnan *et al.*, 1983).
3. Wood-ash is dusted over the plants early in the morning to check the attack of the red pumpkin beetles (Luthra *et al.*, 1995). The author further suggested that intercultural operations should be performed before the veins begin spreading. Urea in three equal split doses should be applied to the plants when they are 25, 40 and 55 days old.
4. Sowing of the crops should be completed in the shortest possible period preferably in April to May. As far as possible, there should be synchronization of sowing time in a particular area. Late maturing varieties should never be sown after May. Cucumber cultivation from August to September proved uneconomic as the crop did not grow well in winter (Borah, 1997). Late planting in rainy and early sowing in summer seasons reduced the damage of the pest (Sharma, 1997).

5. Regularly collect and destroy the leaves and branches having egg masses and other visible stages of insects on them.
6. Adopt proper spacing, thinning and irrigation schedules for the crop. Hoeings and weedings should be done regularly in order to keep the soil loose and weed-free (Butani and Jotwani, 1977a).
7. Raking of soil: Frequent raking of soil expose to sun and destroy the eggs of red pumpkin beetle iaid in the soil (Sharma, 1997).
8. After harvesting the infested fields, these must be immediately deep ploughed to kill the grubs in the soil (Nayer *et al.*, 1976).
9. To protect the crop growing in full sun in a well-drained field, which has been deeply ploughed previously (Mishra, 1983).

Resistant Varieties

Varietal resistance is one of the important component in a pest management programme. Judicious integration of chemical control and varietal resistance is a practical technique to manage the insect pest of cucurbits. Because resistant varieties need fewer insecticide applications, the cost of chemical control is reduced and use of less insecticide favours the survival of natural enemies. Different resistant varieties of cucurbitaceous plants are detailed as under:

Musk melon	RM 43, Punjab Sunheri	Luthra *et al.*, 1995
	Durga Pura Madhu	
	MM-28, Punjab Sunehari	Gill, 2003
Cucumber	AAUC1, AAUC2, Diphu local	Borah, 1999
Bitter gourd	Kerela collection-1	Sharma, 1997
	Faizabad-1	
Bottle gourd	NB-29	Sharma, 1997
Water melon	Selection-1, Candy	Sharma, 1997

Biological Control

Braconids like *Opius fletcheri* Silv. and *Opius compensans* SHv. are pupal parasite of this pest (Srivastava, 1993).

Nematode Parasitoid

Stenernema sp. of nematode use in soil has been found effective against maggots of red pumpkin beetle (sharma, 1997).

Botanical Pesticides

Neem products have been found very effective as repellent for red pumpkin beetle (Sharma, 1997). Johri *et al.* (2004) reported that 100 ml of aqueous fruit extract of emetic nut, *Randia spinosa* Poir was most effective (16.0 to 49.2 beetles killed/plant) on cucurbits. Neem seed kernel extract, Tulsi leaves, leaf extract of Ramboans, each at 5 per cent and neem oil 10 per cent were highly effective as ovipositional deterrents (Parwar, 2005).

Judicious Use of Insecticides

The judicious application of insecticides only when necessary with the minimum of environmental disruption. For this method choice of insecticide and time of application are of almost importance. Use of insecticides with long persistence should never to be recommended. Vegetable growers will have to be educated on the minimum time between application of an action may have to be taken for the insecticides used for management of pests and efficient. Organisation may have to be set up to check the misuse of insecticides and enforce the law regarding tolerance limits. Since a few number of organa-synthetic Insecticides including HCH, endrin and parathion have been reported phytotoxic to various cucurbits (Banerjee and Chatterjee, 1955; Mookerjee and Wadhi, 1956 and Nagaraja Rao, 1959) and recently banned for further use. Great care need to be exercised in selecting proper insecticides.

A number of insecticides have been reported by different scientists as effective for the control of this pest (Bogawat *et al.*, 1969; Pruthi, 1969; Sharma, 1970; Kadyan *et al.*, 1971; Regupathy *et al.*, 1973; Sandhu and Bhala, 1973b; Kavadia *et al.*, 1974; Oebi *et al.*, 1980 and Bajwa and Havi, 1981). Sandhu and Bhala (1974; reported that fenitrothion, fenthion and monocrotophos @ 500 ml/ha proved effective for controlling the pest upto 10 days. Dusting with 5 per cent carbaryl or 4 per cent endosulfan or spraying with 0.2 per cent carbaryl or 0.05 per cent endosulfan as suggested by Butani and Verma (1977). Singh and Mishra (1977) observed that 0.1 per cent carbaryl effective while Krishnaiah *et al.* (1979) were found 0.2 per cent carbaryl 0.5 per cent phenthoate, 0.05 per cent phenthoate and 0.05 per cent leptophos very effective for controlling the pests. Spraying with carbaryl W.P. 0.075 per cent is recommended for the control of the pest (Anonymous, 1983). Some of the recently developed synthetic pyrethroids can also be recommended (Butani and Jotwani, 1984). Mavi and Bajwa (1984) reported that Acephate 50 EC and Quinalphos 25 EC @ 125 9 8.i./ha while phosalone 35 EC and Phoxin 50 EC @ 187.5 g a.i./ha were most effective and protected the crop from its attack for 10 days. Ciampolini (1998) observed that the good results have been obtained against he adults using insecticides especially pyrethroids. Borah (1999) observed that deltamethrin, cypermethrin and fenvalerate gave acceptable level of control of *Raphidopalpa foveicollis* on infesting cucumber.

Conclusion

The depleting demand of Indian vegetables in the international market due to objectionable level of insecticidal residues and high price is a fact that is harsh to admit. Indian vegetable growers are more acquainted with the traditional way of cultivation. So it is time for synchronization between the existing and present development in vegetable research for boosting up the production associated with low cost of cultivation and good quality product.

References

Anonymous, 1983. *Punjab Agricultural Handbook*, pp. 123.

Bajwa, D.S. and Mavi, G.S., 1981. Evaluation of granular insecticides with respect to the control of r dpumpkinbee

le, *Aulacophora foveicollis* on musk melon. *J. Res. P.A.U.*, 18(3): 292–299. • Banerjee, S.N. and Chatterjee, B.

., 1955. Studies on the phytotoxic action of BHC and DOT on the common cucurbitaceous plants in West Bengal *Proc*

edings of Indian Academy of Sciences, 41–8(5): 227–239.

Bogawat, J.K. and Pandey, S.N., 1967. Food preference in *Aulacophora* spp. *Indian J. Ent.*, 29(3): 349–352.

Bogawat, J.K., Kaul, C.L. and Srivastava, B.K., 1969. Food preference in *Aulacophora* spp. *Indian J. Ent.*, 29(4): 349–353.

Borah, R.K., 1999. Seasonally and varietal preference of red pumpkin beetle (*Raphidopalpa foveicollis*) on cucumber (*Cucumis sativus*) in Assam. *Indian J. Agric. Sci.*, 69(3): 180–181.

Butani, D.K., 1975. Insect pests of fruit crops and their control. *Pesticide*, 9(3): 40–42.

Butani, D.K., 1975b. Insect pests of fruit crops and their control–15: *Date Palm*, 9(3): 40–42.

Butani, D.K. and Verma, S., 1977a. Insect pests of vegetables and their control–Cucurbits. *Pesticides*, 11(3): 37–41.

Butani, D.K. and Verma, S., 1977. Pests of vegetables and their control–Cucurbits. *Pesticides*. 11(3): 37–41.

Butani, D.K. and Jotwani, M.G., 1983. Insect as limiting factor in vegetable production. *Pesticides*, 17(9): 6–13.

Butani, D.K. and Jotwani, M.G., 1984. Insect in vegetables–Cucurbits. *Pesticides*, 9(3): 40–42.

Clampolini, M., Regalin, R., Lunghini, D. and Chimienti, G., 1998. Observations on *Aulacophora foveicollis*–A species highly damaging to melon. *Rev. of Agril. Ent.*, 86(12): 1509.

Davi, R.K., Qureshi, G. and Sarma, S.I., 1980. Field evaluation of insecticides spray against *Aulacophora foveicollis*. *Pesticide*, 14(3): 7–8.

Dutt, N. and Dalapati, A., 1977. Biological studies on *Aulacophora foveicollis* Lucae. in cucurbits. *Indian J. Ent.* 30(1): 70–80.

Gill, C.K., 2003. Insect pests and cucumber mosaic virus on musk melon. *Insect Environment*, 9(3): 102–103.

Gopalakrishnan, P.K., Joseph, Salikutty, Gopalakrishnan, T.R. and Peter, K.V., 1983. A better bitter–gourd for Kerala. *Indian Hort.*, 28(3): 11–12.

Hussain, M.A. and Shah, S.A., 1926. The red pumpkin beetle, *Aulacophora abdominalis* Fb. (Coleoptera: Chrysomelidae) and its control with a short note on *A. atripennis* Fb. Mem. Dep. Agri. *Indian Ent. Ser.*, 9: 31–57.

Johri, R. and Johri, P.K., 2003. Seasonal biology abundance and intensity of attack of red pumpkin beetle, *Aulacophora foveicolis* Lucas on cucurbit crops at Kanpur in Uttar Pradesh. *J. Appl. Zool. Res.*, 14(2): 144–147.

Johri, R., Tiwari, D., Mourya, R. and Johri, P.K., 2004. Efficacy of fruit extract of emetic nut, *Randia spinosa* Pair against red pumpkin beetle infesting bottle. *J. Appl. Zool. Res.* 15(1): 64–65.

Kadam, M.V. and Patel, G.A., 1957. Pests of crucifers, chillies, onions, cucurbits and bhindi. In: Crop Pests and How to Fight them. *Pesticide*, 17(9): 6–13.

Kadyan, A.S., Kaushik, S.N. and Gupta, D.S., 1971. Phytotoxicity of some insecticides to muskmelons. *Indian J. Ent.*, 33(4): 463–466.

Kavadia, V.S., Allah, N., Doval, S.C. and Gupta, H.C.L., 1974. Control of red pumpkin beetle, *R. foveicollis* through certain insecticides and their effects on the yield of melon fruits. *Indian J. Pl. Prot.*, 62(1 and 2): 92–102.

Khan, M.Z. and Hajela, K.B., 1987. Studies on *Aulacophora foveicollis* Lucas (Coleoptera: Chrysomelidae)–food preference and extent of damage. *Indian J. Ent.*, 49(4): 457–459.

Krishnaiah, K., Tondon, P.L., Mathur, A.C. and Jagan Mohan, N., 1979. Evaluation of insecticides for the control of major insect pest of *cucurbits. Indian J. Ent.*, 4(4): 311–315.

Lall, B.S., 1964. Vegetable pests, entomology in India. *Ent. Soc., India*, pp. 187–211.

Luthra, J.P., Singh, H. and Singh, B., 1995. New muskmelon 'RM 43'. *Indian Hort.*, 40(2): 6–7.

Mavi, G.S. and Bajwa, D.S., 1984. Comparative efficacy of some emulsifiable concentrates applied with "fogair" for the control of red pumpkin beetle, *Aulacophora foveicollis* Lucas on musk melon. *J. Ent. Res.*, 8(2): 167–170.

Menon, M.G.R., Katiyar, K.N. and Kapoor, V.C., 1972. Taxonomic identities of some Indian species of the genus *Aulacophora chevralot* associated with cucurbitaceous plants (Coleoptera: Chrysomelidae Galerucinae. *Journal Natural History*, 6(1): 17–20.

Mishra, K.C., 1983. Large scale cultivation of small bitter gourd problems and possibilities. *Indian Hort.*, 28(3): 5–8.

Mookherjee, P.B. and Wadhi, S.R., 1956. Further studies on the phytotoxicity of modern organic insecticides. *Indian J. Ent.*, 18(4): 332–335.

Narayanan, E.S., 1953. The red pumpkin beetle and its control. *Indian Farming*, 3(2): 8–9.

Nath, P., 1963. Evaluation of cucurbits for resistance to red pumpkin beetle. *Q. Report. Hort. IARI*, pp. 11.

Nath, P., 1964. Resistance of cucurbits to the red pumpkin beetle. *Indian J. Hort.*, 21(1): 77–78.

Nath, P. and Thakur, M.R., 1965. Evaluation for red pumpkin beetle resistance in gourd. *Indian J. Hort.*, 22(3 and 4): 330–339.

Nagaraja Rao, K.R., 1959. Studies on phytotoxicity of certain systemic pesticides. *Indian J. Ent.*, 21(2): 89–92.

Nayar, K.K., Ananthakrishnan, T.N. and David, B.V., 1976. *General and Applied Entomology*. Tata McGraw Hill Publishing Company Ltd., New Delhi, pp. 589.

Pal, A.B., Srinivasan, K., Bharatan, G. and Chandravadana, M.V., 1978. Studies on the feeding behaviour of *Raphidopalpa foveicollis* on cucurbitacin. *J. Ent. Res.*, 2(2): 148–152.

Pareek, B.L. and Kavadia, V.S., 1988. Economic insecticidal control of two major pests of muskmelon, *Cucumis melo* in the pumpkin beetle, *Raphidopalpa* spp. and the fruitfly, *Dacus cucurbitae* in Rajasthan, India. *Tropical Pest Management*, 34(1): 15–18.

Pawar, D.B., Joi, M.B. and Sonone, H.H., 1984. Chemical control of red pumpkin beetle and fruitfly of musk melon by modern insecticides. *Ind. Hort.*, 32(5): 317–318.

Pruthi, H.S., 1969. *Textbook of Agricultural Entomology*. ICAR, New Delhi, pp. 649.

Purwar, J.P., 2005. Integrated inset pest management in vegetable crops. *Agrobios Newsletter*, 3(9): 25–27.

Rawat, R.R. and Mehra, C.K., 1967. Host preference of pumpkin beetle, *Raphidopalpa foveicollis* (Lucas) and *Aulacophora intermedia* Jacoby (*Atripennis* Fabr.. *JNKVV Res. J.*, 1: 127.

Regupathy, A., Subramaniam, T.R. and Santharam, G., 1973. Chemical control of *Aulacophora foveicollis* Lucas. *Madras Agric. J.*, 60(7): 612–613.

Saini, R.S., 1959. Studies on the food preference of Aulacophora. *Madhya Bharti*, 8: 55–57.

Sandhu, G.S. and Bhala, J.S., 1973. Evaluation of some new insecticides for the control of red pumpkin beetle, *Raphidopalpa foveicollis* Lucas on muskmelon. *Haryana J. Hort. Sci.*, 2(3 and 4): 147–155.

Sandhu, G.S. and Bhalla, J.S., 1974. Red pumpkin beetle: An important pest of cucurbits. *Punjab Hort. J.*, 14(1–2): 70–74.

Sandhu, G.S. and Grewal, S.S., 1985. Testing cucurbits through 'multiple choice' and 'no choice' testing under cages by infesting red pumpkin beetle adult. *Indian J. Ent.*, 47(2): 169–172.

Sharma, R P., 1970. Chemical control of *Aulacophora foveicollis* Lucas. *Madras Agric. J.*, 57(6): 316–349.

Sharma, S.S., 1997. Insect pest management of kharif vegetables. In: *Paper Presented in the Advanced Training Course on Insect Pest Management* at the Department of Entomology, CCS Haryana Agricultural University, Hisar, 10–29 March.

Shaheen, A.H., 1973. Biological studies on *Aulacophora foveicollis* Lucas in Egypt. *Agril. Res. Rev.*, 51(1): 91.

Shinde, C.B. and Purohit, M.L., 1978. Population study of red pumpkin beetle (*Aulacophora foveicollis* Lucas) a serious insect pest of cucurbits. *Science and Culture*, 44(7): 335.

Singh, G. and Misra, P.N., 1977. Note on the efficacy of some insecticidal dusts and sprays against red pumpkin beetle, *Raphidopalpa foveicollis* (Lucas). *J. Agric. Sci.*, 47(5): 257–258.

Shri Ram and Pathak, A.K., 1984. Occurrence and distribution of insect pest complex of vegetable in Manipur. *Bull. Ent.*, 25(2): 193–198.

Singh, A.K. and Krishna, S.S., 1970. Studies on the feeding behaviour of *Aulacophora foveicollis* on cucurbitacin. *J. Eco. Ent.*, 63(1): 333–335.

Singh, S.V., Misra, A., Bisen, R.S. and Malik, Y.P., 2000. Host preference of red pumpkin beetle, *Aulacophora foveicollis* and melon fruit fly, *Dacus cucurbitae*. *Indian J. Ent.*, 62(3): 242–246.

Srivasatava, K.P., 1983. *A Textbook of Applied Entomology*. Vol. II, Kalyani Publishers, New Delhi, pp. 99–114.

Suliman, A.M., Mohamed, Y.F., Daffalla, G.A. and Giha, O.H., 1997. Wilt diseases of melon (*Cucumis melo* L.) and associated organism in sudan. *R.A.E.*,'86(6): 711.

Verma, R.R., 1984. Two new alternative host plant of *Raphidopalpa* (=*Aulacophora*) *foveicollis* Lucas and R. (=*A*) *intermedia* Jacoby in Garhwal Hills. *Punjab Hort. J.*, 24(1–4): 197.

Venkataryappa, T. and Govindon, R., 1978. Record of *Aulacophora foveicollis* on brinjal. *Current Research*, 7(1): 17.

Plant Protection, Diversity and Conservation (2011) *Pages* **499–515**
Editors: **D. Prasad & Satya Kumar**
Published by: **BIOTECH BOOKS, NEW DELHI**

Chapter 38

Survey of Mango Fruit Borer, *Deanolis albizonalis* (Hampson) in Andhra Pradesh, Orissa, West Bengal, Karnataka States and Explorative Research on Management through IPM

A. Sujatha[1]* *and M.S.V. Chalam*[2]

[1]*Andhra Pradesh Horticultural University.*
Horticultural Research Station, Ambajipeta – 533 214, A.P.
[2]*Department of Entomology, Agricultural College, Bapatla*

The mango, *Mangifera indica* L., is the most important fruit crop of India. The crop is grown in an area of 12.3 lakh hectares, with an annual production of 10.99 million tonnes which accounts for 57.18 per cent of total world production and earning export of nearly Rs. 68.0 crores. Andhra Pradesh ranks second in India in respect of both area (2.4 lakh ha.) and production (2.9 million tonnes). There is a good potential for export of fresh fruit to other countries, but constraints like pest and disease problems restricted the expansion of export trade from India. The crop is attacked by more than 100 species of pests (Tandon and Verghese, 1985) among which mango leafhoppers, mango nut weevil, stem borer, red banded mango fruit borer/mango fruit borer were found to be major pests limiting the fruit production and export.

The incidence of mango fruit borer, *Deanolis albizonalis* (Hampson) was first reported in Andhra Pradesh and was found to be serious pest in certain locations (Zaheruddeen and Sujatha, 1993). This pest is appearing in serious proportions resulting in large scale devastation of fruits of all sizes, all

* E-mail: avvaru_001@rediffmail.com

varieties through out the season and causing annoyance to farmers as well as traders. Field estimation of damage revealed that, the fruit yields were reduced up to 50 per cent in the infested orchards causing severe economic loss to the mango growers. Primary information about the pest *i.e.*, larval behaviour, nature and symptoms of damage and biology of the pest were reported by Sujatha and Zaheruddeen (2002) and distribution in Andhra Pradesh was also reported by Sujatha and Zaheruddeen in the year 2006 (Plates 38.1–38.3]. In the present investigations, survey was carried out in four mango growing states *i.e.*, Andhra Pradesh, Orissa, West Bengal and Karnataka to know the occurrence, distribution pattern of the pest and extent of the damage caused by mango fruit borer [MFB] and more importantly to identify hot spot areas.

As this pest is an internal feeder of edible fruit, its management with eco-friendly approach is important. Even though, controlling of the borer is a difficult task, certain weak links in the biology and ecology of the pest would provide chances for attempting appropriate control measures. The exit of larvae from the fruit for migration/pupation; pupation in the dead wood available on the tree/ cracks and crevices/under bark; prepupal diapause in off-season under the bark or in dead wood are among the weak points in the biology which can be exploited for planning the control measures (Zaheruddeen and Sujatha, 1995). With an objective of working out a management strategy for this serious pest, a series of experiments on various aspects like sanitation and cultural methods and screening of effective pesticides were conducted both in the field and laboratory.

Survey and Identification of Hot Spot Areas

Roving survey was conducted in four major mango growing states *i.e.*, Andhra Pradesh, Orissa, West Bengal and Karnataka with a view to assess the damage, distribution pattern and preference to principle mango varieties. Data on fruit damage/occurrence of mango fruit borer was collected from 17 districts in Andhra Pradesh, 5 districts in Orissa, 4 districts in West Bengal and 6 districts in Karnataka covering 385 villages in Andhra Pradesh, 64 in Orissa, 42 in West Bengal and 58 villages in Karnataka. Based on the intensity of fruit damage *i.e.*, low, medium and high, districts were categorized and hotspot areas were located. Observations were also made on varietal preference of the pest based on the fruit damage in the field.

Periodical observations were taken up in the mango gardens of Gudapalli village of East Godavari district, to know the varietal preference of mango fruit borer. Four different varieties *viz.*, Bangalora, Baneshan, Suvarnarekha and Neelam along with a local variety were observed at fortnightly interval and recorded the fruit damage. Ten plants were randomly chosen from each variety to count the total number of fruits as well as damaged fruits.

While conducting field surveys certain parasitized pre-pupae/pupae of the pest were collected and maintained in the laboratory till the emergence of parasitoids. The parasitoids so emerged were got identified.

Development of IPM Against Mango Fruit Borer

Studies on the Impact of Sanitation

Collection and Destruction of Dead Wood and Infested Fruits

To know the impact of sanitation measures on fruit infestation of MFB, a study was conducted during the years 2004-05 and 2005-06. Six MFB infested gardens with different crop protection and agronomic practices were selected in East and West Godavari districts of Andhra Pradesh and monitored to know the incidence of mango fruit borer. Management practices included regular collection and destruction of infested fruits, collection and destruction of diapausing larvae/pre pupae under

Plate 38.1: Symptoms of Damage

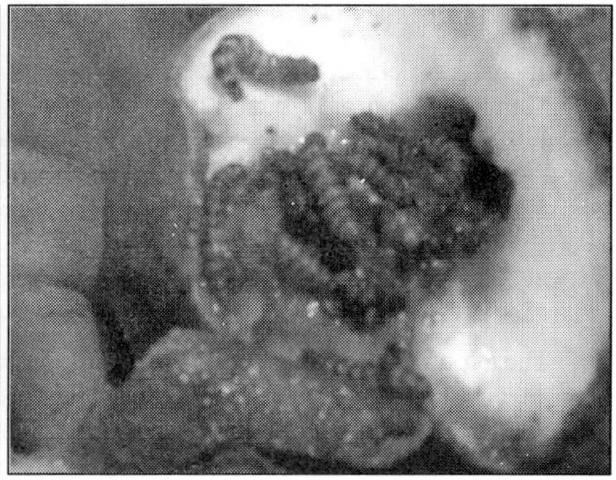

Plate 38.2: MFB Larvae Feeding in the Kernel

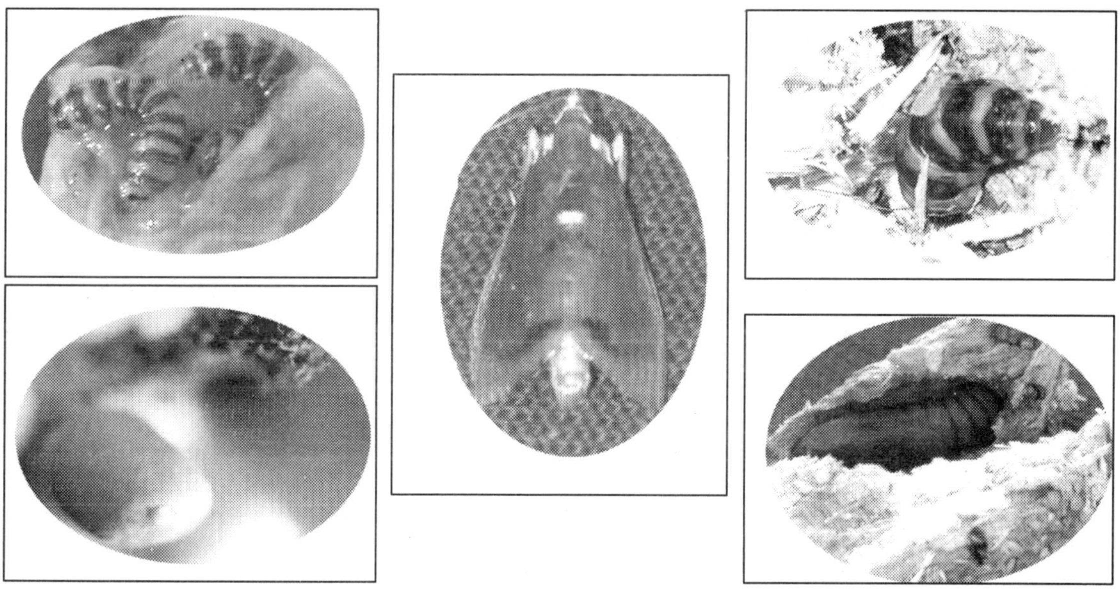

Plate 38.3: Life Stages of Mango Fruit Borer

the bark or cracks and crevices of the trunk and pruning of dead and unwanted wood in the off-season.

Collection and Destruction of Diapausing Pre Pupa

Generally the mango fruit borer larvae spend their entire off-season in diapausing stage under the bark of the tree which is a weak link in the biology of the pest. To exploit this opportunity for managing the pest, periodical visits were made to different gardens in East Godavari and West Godavari districts and different stages of MFB population from the bark were collected and destroyed. Data was collected on fruit damage in the ensuing season.

Cultural Methods

Studies on Artificial Niches to Attract Mango Fruit Borer Larvae for Pupation

Mango fruit borer larvae after completion of larval period comes out of the fruit in search of a suitable site for pupation and enter into dead wood available on tree or the cracks and crevices of the bark. As per the studies of Sujatha and Zaheruddeen (2002), mango fruit borer larvae prefer dead wood which is corky and soft, for pupation. Hence, artificial niches with the materials like mango wood, thermocool sheet and coconut husk were arranged to the branches of mango trees having infested fruits to attract the fruit borer larvae to pupate in these niches and these attract d larvae/pupae/diapausing larvae in the off season/season were destroyed in large numbers.

The experiment was conducted in the mango gardens in two different villages *viz.,* Gudapalli and Turpupalem of East Godavari district and in each garden ten trees were selected at random. In the first year (2004-05) artificial niches with materials like mango wood, thermocool sheet and coconut husk @ 5 nos. of each material were arranged to single tree in different branches having infested fruits early in the season. These artificial niches were observed at periodical intervals to count the number of larvae/pupae/diapausing larvae in each artificial niche till completion of season and also in off season.

Whereas in the 2nd year 2005-06, the experiment was conducted with mango wood pieces (which soft) in four gardens in four different villages of East Godavari district *viz.,* Gudapalli, Kathimanda, Turpupalem and Kannapuram villages in West Godavari district. Ten trees were randomly selected and ten mango wood pieces measuring about 50 cm were tied on the branches for each tree. These artificial niches were observed at regular intervals and number of larvae/pupae/diapausing larvae were counted till the next season.

Swabbing of Insecticides to the Trunk

As pre-pupal stage of mango fruit borer is present in cracks and crevices of mango tree trunk, a field experiment was conducted by swabbing the chemicals around the tree truck during the off season to target the diapausing/pupating larvae under the bark. The experiment was conducted in mango gardens of the village Kannapuram of West Godavari district during 2004-05 and 2005-06. Three treatments *i.e.,* Neem oil (50 ml), Monocrotophos 36 per cent SL (5 ml) and Chlorpyriphos 20 per cent EC (5 ml) per litre of water were applied to the trunk up to a height of 2 metres using a paint brush on Baneshan variety and each treatment was replicated thrice. The treatments were imposed in the month of November *i.e.,* during off season (2004-05); the data on the percentage fruit damage was recorded in the subsequent fruiting season (2005-06). The efficacy of the treatments was assessed by calculating the per cent fruit damage.

Screening of Effective Botanical/Chemical Pesticides Against Mango Fruit Borer

Evaluation of pesticides in the laboratory against mango fruit borer: A laboratory experiment was conducted to assess the efficacy of botanicals (neem formulations); in combination with insecticides against mango fruit borer. For this study, neem formulations *viz.,* Azadirachtin 1500, 3000, 10000 ppm were used alone as well as in combination with chlorpyriphos 20 per cent EC at different dosages.

Ten fruit borer infested mango fruits of the variety Baneshan were chosen and the spray solution was sprayed on fruits with one litre sprayer. Then the fruits were allowed for drying in shade and kept in plastic containers individually. Observations on mortality of MFB larvae were recorded after 24 hours by adopting destructive sampling method. Number of live larvae as well as dead larvae present in each treatment was counted and percentage mortality was arrived. An untreated control was also maintained to have a comparison with different treatments. The data obtained was subjected to Completely Randomized Design (CRD) analysis after $\sqrt{x+1}$ transformation.

	Treatment	Dosage
T_1	Azadirachtin 10000 ppm	3 ml/lit.
T_2	Azadirachtin 3000 ppm	4 ml/lit.
T_3	Azadirachtin 1500 ppm	7 ml/lit.
T_4	Neem oil	3 ml/lit.
T_5	Neem oil + Chlorpyriphos 20 per cent EC	3 ml + 1 ml/lit.
T_6	Azadirachtin 1500 PPM + Chlorpyriphos 20 per cent EC	7 ml + 1 ml/lit.
T_7	Azadirachtin 3000 PPM + Chlorpyriphos 20 per cent EC	4 ml + 1 ml/lit.
T_8	Azadirachtin 10000 PPM + Chlorpyriphos 20 per cent EC	3 ml + 1 ml/lit.
T_9	Carbaryl 50 per cent w.p.	3 g/lit.
T_{10}	Control	–

Field Evaluation of Pesticides Against Mango Fruit Borer

A field experiment was conducted during 2005-06 in the mango gardens of the village Kathimanda in East Godavari district to study the efficacy of insecticidal combination with neem formulations against mango fruit borer. Three different combinations *viz.,* Azadirachtin 1500 ppm (7 ml) + Chlorpyriphos 20 per cent EC (1 ml); Azadirachtin 1500 ppm (7 ml) + Dichlorvos 76 per cent EC (1 ml) and Chlorpyriphos 20 per cent EC (1 ml) + Monocrotophos 36 per cent SL (1 ml) per litre of water were sprayed in the pest infested garden. The experiment was laid out in Randomized Block Design (RBD) and eac treatment was replicated five times and untreated control was also maintained on Baneshan variety. During the season, the treatments were imposed and percentage mortality of mango fruit borer larvae was arrived.

Bio-assay Studies on 3rd Instar Larvae of Mango Fruit Borer

A laboratory study was conducted to test the susceptibility of 3rd instar larvae of mango fruit borer to different pesticides. Seven commonly used insecticides *i.e.,* Monocrotophos 36 per cent SL (1.5 ml/litre of water), Dichlorvos 76 per cent EC (1 ml/litre of water), Chlorpyriphos 20 per cent EC (2 ml/litre of water), Acephate 75 per cent SP (0.75 g/litre of water), Carbaryl 50 per cent WP (3 g/litre of water), Endosulfan 35 per cent EC (2ml/litre of water) and Azadirachtin 1500 ppm (7 ml/litre of water) were used at recommended concentrations by adopting residue film method. The insecticide solution was applied as thin film in a glass tube of size 6"x3" and allowed to air dry for about 30 minutes. The third instar larvae of mango fruit borer (10 larvae/treatment) were released into glass tube which was coated with thin film of pesticide. The mortality count of the larvae was recorded at 24 and 48 hours after exposure to insecticide. The live larvae were observed for further period to know whether they pass normal biological stages like pupation/moth emergence or not.

Identification of Hot Spots of the Pest

Among the 17 districts surveyed in Andhra Pradesh, the incidence of mango fruit borer were recorded in seven districts *i.e.,* Srikakulam, Vizianagaram, Visakhapatnam, East Godavari, West Godavari, Krishna and Khammam. Highest incidence of mango fruit borer (6 – 33 per cent fruit damage) was recorded in four districts *i.e.,* Visakhapatnam, East Godavari, West Godavari and Khammam districts of Andhra Pradesh. A total of 385 villages were surveyed from all the districts of

Andhra Pradesh, out of which; 67, 11, 23 villages recorded high, medium, low incidence of mango fruit borer, respectively (Table 38.1). The infestation level ranged from 1.0 to 35.0 per cent in Andhra Pradesh.

Table 38.1: Incidence of Mango Fruit Borer in Various Districts of Mango Growing States

	Name of the State		*Name of the District*	*No. of Villages Surveyed*	*Intensity of Incidence*
1.	Andhra Pradesh	1.	Srikakulam	16	L
		2.	Vizianagaram	24	M
		3.	Visakhapatnam	15	H
		4.	East Godavari	60	H
		5.	West Godavari	10	H
		6.	Krishna	23	M
		7.	Nellore	44	–
		8.	Chittoor	53	–
		9.	Kadapa	7	–
		10.	Kurnool	6	–
		11.	Khammam	8	H
		12.	Medak	21	–
		13.	Nalgonda	6	–
		14.	Adilabad	5	–
		15.	Mahbubnagar	2	–
		16.	Karimnagar	56	–
		17.	Nizamabad	29	–
			Total	385	
2.	Orissa	1.	Khurda	6	L
		2.	Ganjam	9	L
		3.	Puri	19	L
		4.	Rayagada	15	L
		5.	Koraput	15	L
			Total	64	
3.	West Bengal	1.	Malda	27	H
		2.	Murshidabad	8	H
		3.	Nadia	3	L
		4.	North 24 Paraganas	4	–
			Total	42	
4.	Karnataka	1.	Kolar	28	–
		2.	Bangalore Rural	8	–
		3.	Mandya	3	–
		4.	Mysore	6	–
		5.	Chitradurga	6	–
		6.	Dharwad	7	–
			Total	58	

Low: 1–5 per cent fruit damage; Medium: 6–15 per cent fruit damage; High: >15 per cent fruit damage; –: No incidence.

In Andhra Pradesh, four districts *viz.*, East Godavari, West Godavari, Visakhapatnam and Khammam has recorded highest percent of mango fruit borer infestation. A total of 93 villages were surveyed in the four districts, out of which 75 villages recorded the incidence of mango fruit borer ranging from medium to high level (6-33 per cent fruit damage) (Table 38.2). An informal meeting with the local farmers of mango orchards also revealed that mango fruit borer is a persistent pest, since a decade in these four districts of Andhra Pradesh. Based on the above mentioned information and observations it is notified that the four districts of Andhra Pradesh *viz.*, East Godavari, West Godavari, Visakhapatnam and Khammam are recognized as hot spot areas for mango fruit borer infestation (Figure 38.1). The presence of pest in the adjoining districts indicates that likely hood of pest spread to the adjoining places.

Survey in Orissa

The borer incidence was noticed in all the five districts surveyed but the incidence level is low (1–5 per cent fruit damage). Pest incidence was noticed at lower level in 44 villages out of 64 villages surveyed (Tables 38.1 and 38.2).

Table 38.2: Hot Spot Areas of Mango Fruit Borer in Andhra Pradesh and West Bengal

Sl.No.	*Name of the State*		*Name of the District (Hot spot of MFB)*	*No. of Villages Surveyed*	*No. of Villages having MFB Incidence*	*Percentage of MFB Incidence Observed*
1.	Andhra Pradesh	1.	East Godavari	60	51	6 – 33
		2.	West Godavari	10	10	18 – 21
		3.	Visakhapatnam	15	9	6 – 20
		4.	Khammam	8	5	16 – 20
2.	West Bengal	1.	Malda	27	27	10 – 22
		2.	Murshidabad	8	8	10 - 22

Low: 1–5 per cent fruit damage; Medium: 6–15 per cent fruit damage; High: >15 per cent fruit damage.

Survey in West Bengal

A roving survey was conducted in the four major mango growing districts of West Bengal. Mango fruit borer incidence was recorded from all the four districts surveyed *i.e.*, Malda, Murshidabad, Nadia and North 24 Paraganas. The districts, Malda and Murshidabad recorded a high per cent infestation (10–22 per cent fruit damage), where as Nadia has recorded a low infestation level (1-4 per cent fruit damage). Totally 42 number of villages were surveyed in the four districts in West Bengal, out of which 20,15, 3 villages recorded high, medium and low incidence of mango fruit borer, respectively (Tables 38.1 and 38.2).

In West Bengal, two districts *viz.*, Malda and Murshidabad recorded high per cent (22 per cent) of mango fruit borer incidence. A total of 35 villages were surveyed in these two districts which have recorded the incidence mango fruit borer. Sahoo and Jha (2006) also reported the wide spread occurrence of mango fruit borer in Malda and Murshidabad districts of West Bengal. Hence, the districts *viz.*, Malda and Murshidabad of West Bengal can be considered as hot spot areas for mango fruit borer incidence (Figure 38.2).

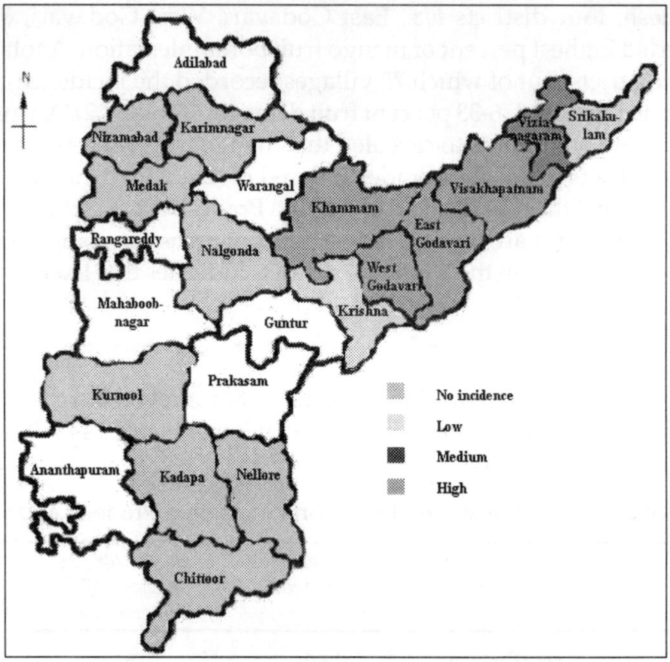

Figure 38.1: Hot Spot Areas of MFB in Andhra Pradesh

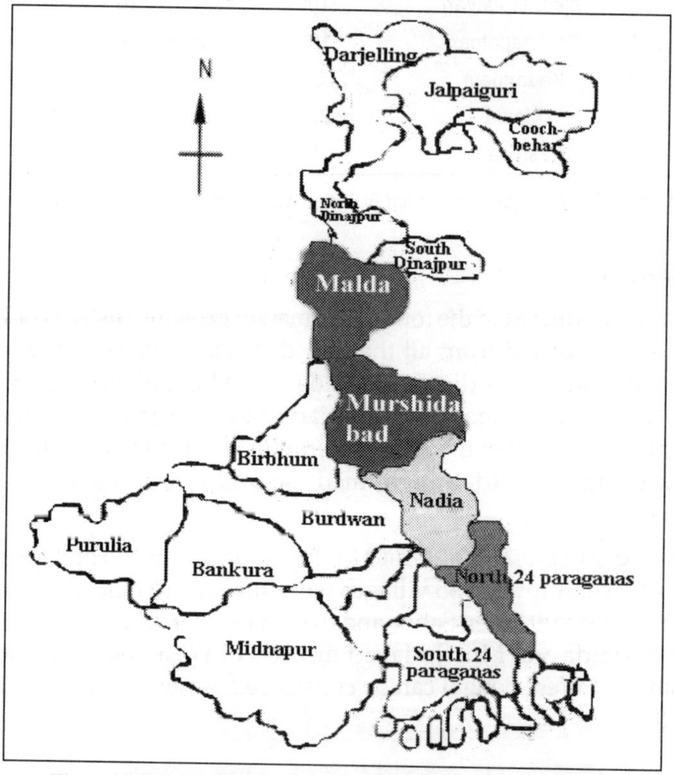

Figure 38.2: Hot Spot Areas of MFB in West Bengal

In case of Orissa, the pest damage was classified as low and no hot spot area is recognized from Orissa. From the observations, it was noticed that no systematic orchard management or plant protection measures were being followed in Orissa state.

Survey in Karnataka

A roving survey was conducted in six mango growing districts of Karnataka *i.e.*, Kolar, Bangalore rural, Mandya, Mysore, Chitradurga and Dharwad. A total of 58 numbers of villages were surveyed from six districts of Karnataka and from the survey, it was confirmed that, no fruit borer incidence was present in Karnataka state (Table 38.1).

From the survey results, it was found that highest fruit borer incidence was recorded in Andhra Pradesh (4 districts) and West Bengal (2 districts) and low incidence (5 districts) in Orissa state.

Persistant severe incidence of mango fruit borer in Andhra Pradesh and West Bengal states might be due to the indiscriminate usage of pesticides especially synthetic pyrethroids, lack of proper knowledge about the pest ecology and biology. Hence, there is every need to create awareness among the farming community about mango fruit borer pest identification, symptoms of damage, biology, ecology, pest cycles, critical stages of the pest, peak occurrences, management of the pest through litreature distribution, conducting seminars, kisan melas, interaction programmes *etc.*

Collection of Natural Enemies of the Pest and their Identification

In course of survey and periodical monitoring of the pest in mango gardens of different locations, a few natural enemies of pupae and prepupae of mango fruit borer were collected from the trunk region of mango tree and were maintained in the laboratory for observation. Few hymenopteran parasitoids were emerged from field collected prepupae and pupae and they were identified as:

Sl.No.	*Name of the Parasitoid*	*Family/Order*	*Subfamily/Tribe*
1.	*Brachymeria lassus* (Walker)	Chalcididae/Hymenoptera	Chalcidinae/Brachymerini
2.	*Brachymeria apicicornis* (Cameron)	-do-	-do-
3.	*Oxycoryphe* sp. Kriechbaumer	-do-	Haltichellinae/Haltichellini

Most of the parasitoids identified belong to super family chalcidoidea of family chalcididae (order: Hymenoptera). *Brachymeria lassus* is a black chalcidid wasp with short ovoid abdomen and a smooth first tergite. *Brachymeria lassus* generally parasitizes older larvae of different lepidopterons. The female lays a single egg in the host's body and development of parasitoid until pupation takes place inside the host. The longevity of adult parasitoid is about 3 – 5 days. The adult parasitoid lives for about one month under laboratory conditions in East Coast (Andhra Pradesh). *Brachymeria apicicornis* is hither reported only from Kerala and in the present studies it was also reported from Andhra Pradesh. Just like *Brachymeria lassus* it also parasitizes older larvae and pupae of lepidopterons.

During field surveys, data was collected on pest population (Pre pupae/pupae/empty pupae) observed under bark. Based on the empty pupal case characters, moth emergence/parasite emergence was recorded. The parasitoid collected from the gardens of Gudapalli, Kannapuram and Odalarevu villages were identified as *Oxycoryphe* sp. which can parasitize both larvae and pupae of the host and highest natural parasitisation under field conditions *i.e.*, 30.9 per cent was recorded in the village Odalarevu followed by 22.0 per cent in Gudapalli village (Table 38.3).

Table 38.3: Parasitisation Particulars of *Oxycoryphe* sp. Under Field Conditions of A.P.

Sl.No.	Name of the Village	Pest Population Collected from the Bark				Per cent parasitisation
		Larvae	Empty Pupal Shells		Total population	
			Moth Emerged	Parasite Emerged		
1.	Gudapalli	30	69	28	127	22.0
2.	Kannapuram	6	40	6	52	11.5
3.	Odalarevu	6	23	13	42	30.9

Other Natural Enemies Observed in the Field

One coleopteran beetle was also collected from the field which was predacious on mango fruit borer; red ants were quiet commonly observed while carrying mango fruit borer larvae to their nests. Another important feature observed was where ever the red ants are there on the branches or inside the wood, no mango fruit borer prepupae were observed. Mantids (*Mantis religiosa* Linnaeus; *Hierodula coarctata* Westw.) and ear wigs *Acrania picta* (Guerin-Menibille); *Euborallia stali* (Dohrn); *E. femoralis* (Dohrn) are other predatory fauna were observed and collected from the bark where the pest spends its prepupal, pupal and diapause stages.

Sujatha *et al.* (2002) reported insects *viz.*, red ant, earwigs, mantids, coleopteran beetle and two fungal pathogens *i.e.*, *Metarhizium anisopliae, Beauveria bassiana* as natural enemies under field conditions. Present investigation confirms the earlier report by Sujatha *et al.*, [2002]. In addition to this list, the hymenopteran parasitoids were newly reported in the present study on this pest. Further studies on the exploitation of natural enemies need to be conducted for evaluation of potential bio-agents, their mass multiplication and field evaluation.

Varietal Preference of the Pest

From the data it is apparent that there is no possible varietal preference mechanism in mango fruit borer. Almost all varieties have recorded fruit damage (Table 38.4). The pest first appeared on Suvarnarekha which is an early variety and later on spread to other mid as well as late varieties. However detail studies with more number of varieties during the season are to be conducted to know the varietal preference of the pest. Insect growth ratios like approximate digestibility and food conversion efficiency are to be worked so as to ascertain the varietal preference.

Development of IPM Technology for the Management of *D. albizonalis*

Studies on the Impact of Sanitation

Collection and Destruction of Dead Wood and Infested Fruits

Initial appearance of the pest was occurred in the first week of February in the gardens and gradually reached to peak level in the month of March. The fruit borer damage was lowest (2.6 per cent - in April month) in the garden number

Table 38.4: Per cent Fruit Borer Infestation in Different Mango Varieties

Sl.No.	Name of the Variety	Percentage of Fruit Damage
1.	Suvarnarekha	6.2
2.	Baneshan	9.0
3.	Bangalora	5.1
4.	Neelam	4.7
5.	Local	3.7

IV (Kannapuram village in West Godavari district) in which timely management practices were implemented (Table 38.5). Management included regular collection and destruction of infested fruits, collection and destruction of diapausing larvae/pre pupae under the bark or cracks and crevices of the trunk and pruning of dead and unwanted wood in the off-season, which resulted in reduction of fruit damage or further spread of the pest. The orchard (Garden-V) in the same village has recorded damage level up to 20.50 per cent which was under neglected conditions. In garden No.1 and No.2 in East Godavari district also highest fruit borer damage *i.e.*, 31.6 per cent and 20.3 per cent respectively were recorded where proper IPM was not followed. This clearly emphasizes the impact of sanitation and crop husbandry on the fruit borer infestation. Hence, collection and destruction of infested plant material would be an effective ecofriendly approach which greatly reduces the pest load.

Collection and Destruction of Diapausing Pre Pupa

Highest population of diapausing larvae/pupae could be collected from the garden where severe infestation of MFB was recorded in previous year (Gudapalli village in EG district - 105 nos/tree) (Table 38.6). This clearly indicates that elimination of pest population through collection and destruction during off-season/season having profound impact on population buildup which results ultimately in fruit damage. Mechanical collection and destruction of the pest is one of the recommended IPM measures in managing any pest.

Table 38.5: Impact of Sanitation on MFB Damage (Average of two years – 2004-05 and 2005-06)

Sl.No.	*Particulars of Garden*	*Name of the Village*	*Per cent Fruit Damage by Mango Fruit Borer*			
			January	*February*	*March*	*April*
1.	Garden-I E.G. Dist.	Kathimanda	2.0	11.0	20.0	31.6
2.	Garden-II E.G. Dist.	Gudapalli	No incidence	2.5	14.5	20.3
3.	Garden-III # E.G. Dist.	Erusumanda	No incidence	2.0	7.0	7.0
4.	Garden-IV * W.G. Dist.	Kannapuram	No incidence	1.0	2.0	2.6
5.	Garden-V W.G. Dist.	Kannapuram	No incidence	5.0	14.4	20.5
6.	Garden-VI E.G. Dist.	Kathimanda	No incidence	13.0	22.0	33.4

#: Partial bearing; *: IPM implemented garden.

Cultural Methods

Studies on Artificial Niches to Attract Mango Fruit Borer Larva for Pupation

From the observations it was found that mango wood was preferred for pupation of mango fruit borer larvae rather than thermocool sheet and coconut husk as no of larvae/pupae were observed in thermocool sheet and coconut husk (Table 38.7).

As first year studies revealed that, mango wood is most preferred to larvae of MFB for using as artificial niche; further studies were planned in 2005-06 by using mango wood as artificial niche material.

Table 38.6: Mango Fruit Borer Larvae Collected from the Bark of Mango Trees

Sl.No.	*Name of the Village*	*No. of Diapausing Larvae/Pupae Collected*
1.	Turpupalem	24
2.	Kannapuram	36
3.	Errakoneru	7
4.	Dondavaka	5
5.	Rapaka	7
6.	A.Kothapalli	5
7.	Koshuru	5
8.	Gudapalli	105
9.	Kathimanda	57
10.	Maruvada	1
11.	Odalarevu	36

Table 38.7: Studies on Artificial Niches to Attract MFB Larvae for Pupation (2004-05)

Sl.No.	*Name of the village*	*No. of Artificial Niches*			*Pest Population Observed/Tree*		
		Mango Wood	*Thermocool Sheet*	*Coconut Husk*	*Mango Wood*	*Thermocool Sheet*	*Coconut Husk*
1.	Gudapalli	5	5	5	5	–	–
2.	Turpupalem	5	5	5	5	–	–

From the data, it was found that considerable number of mango fruit borer larvae were attracted to these artificial niches in all the four villages. The highest population of 63 number was recorded from all the ten trees in Kathimanda village followed by Kannapuram (55), Turpupalem (53) and Gudapalli (49) villages (Table 38.8).

From these studies, it is evident that mango fruit borer larvae/pupae/diapausing larvae can be trapped by using artificial niches on the trees during the season and destroyed. Hence, this technique can be used as one of the IPM steps for the control of mango fruit borer in the infested gardens.

Table 38.8: Studies on Artificial Niches to Attract MFB Larvae for Pupation (2005-06)

Name of the Village	*Pest Population Observed in Mango Wood/Tree*			
	Larvae	*Pupae*	*Empty pupae*	*Total*
Kannapuram	–	–	55	55
Gudapalli	3	–	46	49
Turpupalem	4	–	49	53
Kathimanda	11	–	52	63

Swabbing of Insecticides to the Trunk

Among the three chemicals tested, Chlorpyriphos 20 per cent EC proved superior by recording the lowest fruit damage (11.48) followed by Monocrotophos 36 per cent SL (12.71) and Neem oil (15.57). From these studies, it can be concluded that swabbing of the tree trunk with Chlorpyriphos 20 per cent EC (or) Monocrotophos 36 per cent SL @ 5 ml/lit. of water during the off season can reduce the per cent fruit damage during the ensuing fruiting season (Table 38.9). Neem Oil @ 50 ml/lt. of water can also be used for swabbing the trunk as it is botanical origin.

Table 38.9: Effect of Trunk Swabbing on Mango Fruit Borer Infestation

Sl.No.	*Treatment*	*Dose*	*Percentage of Fruit Damage*	*Per cent Decrease Over Control*
1.	Neem oil	50 ml/litre of water	15.57	36
2.	Monocrotophos 36 per cent SL	5 ml/litre of water	12.71	48
3.	Chlorpyriphos 20 per cent EC	5 ml/litre of water	11.48	53
4.	Control	–	24.45	–

Table 38.10: Laboratory Evaluation of Pesticides Against Mango Fruit Borer Larvae

	Treatment particulars	*Dose*	*Percentage Mortality*			*Treatment Mean*
			R1	*R2*	*R3*	
T_1	Azadirachtin 10000 ppm	3 ml/lit.	0 (1)	0 (1)	0 (1)	1
T_2	Azadirachtin 3000 ppm	4 ml/lit.	0 (1)	0 (1)	0 (1)	1
T_3	Azadirachtin 1500 ppm	7 ml/lit.	0 (1)	0 (1)	0 (1)	1
T_4	Neem oil	3 ml/lit.	0 (1)	23.18 (4.91)	0 (1)	2.3
T_5	Neem oil + Chlorpyriphos 20 per cent EC	3 ml + 1 ml/lit.	100 (10.04)	100 (10.04)	100 (10.04)	9.59*
T_6	Azadirachtin 1500 PPM + Chlorpyriphos 20 per cent EC	7 ml + 1 ml/lit.	100 (10.04)	100 (10.04)	100 (10.04)	10.04*
T_7	Azadirachtin 3000 PPM + Chlorpyriphos 20 per cent EC	4 ml + 1 ml/lit.	100 (10.04)	100 (10.04)	100 (10.04)	10.04*
T_8	Azadirachtin 10000 PPM + Chlorpyriphos 20 per cent EC	3 ml + 1 ml/lit.	100 (10.04)	100 (10.04)	100 (10.04)	10.04*
T_9	Carbaryl 50 per cent w.p.	3 g/lit.	0 (1)	50 (7.14)	17.69 (4.32)	4.15
T_{10}	Control	—	0 (1)	0 (1)	0 (1)	1
		CD [0.05 per cent]				7.15
		SED				3.43

* Sig Figures in parenthesis are $\sqrt{X+1}$ values

Screening of Effective Botanical/Chemical Pesticides Against Mango Fruit Borer

Evaluation of pesticides in the laboratory against mango fruit borer: From the data, it is clearly evident that combination of neem oil and neem formulations with chlorpyriphos 20 per cent EC was very effective which recorded 100 per cent mortality (Table 38.10) (T_5, T_6, T_7 and T_8) and are significantly superior over other treatments. However, all the treatments of neem formulations in combination with chlorpyriphos 20 per cent EC are on par with each other. The neem formulations alone could not produce significant mortality and are on par with control. Hence, it can be inferred that the combination of Azadirachtin formulations + chlorpyriphos 20 per cent EC can be used successfully as one of the components in the IPM programme of this pest rather than using insecticides or botanicals alone (Table 38.10).

Field evaluation of pesticides against mango fruit borer: From the results, it is evident that, Azadirachtin 1500 ppm + Chlorpyriphos 20 per cent EC [T_1] recorded highest mortality (34.47 per cent) of mango fruit borer larvae which is significantly superior over other treatments followed by Chlorpyriphos 20 per cent EC + Monocrotophos 36 per cent SL [T_3] which has recorded a mortality of (34.30 per cent) (Table 38.11). However, both these treatments were on par with each other. The next best treatment was Azadirachtin 1500 ppm + Dichlorvos 76 per cent EC [T_2] which has recorded a mortality of 28.98 per cent and was found on par with T_3. From this experiment, it is evident that a of Azadirachtin and Chlorpyriphos 20 per cent EC combination treatment was found to be more efficacious and can be incorporated into the IPM of the mango fruit borer. Sujatha *et al.*, [2001] reported neem formulations were effective in reducing the fruit borer damage, but continuous applications throughout the season were suggested to check the pest.

Table 38.11: Evaluation of Different Pesticides Against Mango Fruit Borer in the Field

	Treatment Particulars	*Dose*	*Per cent mortality*					*Treatment Mean*
			R1	*R2*	*R3*	*R4*	*R5*	
T_1	Azadirachtin 1500 ppm + Chlorpyriphos 20 per cent EC	7 ml + 1 ml/lt. of water	40.11 (41.47)	33.77 (30.87)	30.00 (25.03)	30.92 (26.39)	37.58 (37.18)	34.47 ab
T_2	Azadirachtin 1500 ppm + Dichlorvos 76 per cent EC	7 ml + 1 ml/lt. of water	35.49 (33.66)	23.97 (16.54)	26.71 (20.23)	26.35 (19.26)	32.39 (28.70)	28.98 bc
T_3	Chlorpyriphos 20 per cent EC + Monocrotophos 36 per cent SL	1 ml + 1 ml/lt. of water	30.20 (25.26)	33.96 (31.24)	43.97 (48.21)	32.20 (28.4)	31.18 (26.85)	34.30 ab
T_4	Control		5.26 (0.74)	7.49 (1.72)	8.13 (1.98)	8.13 (2.03)	5.68 (0.98)	6.93 d
		CD						5.40
		SEM						3.84

Figures in parentheses are arcsine transformed values.

From the above experiments it was observed that mortality was high in case of early instars where as nil in case of 3rd instar larvae. Hence, another laboratory experiment was conducted to know whether 3rd instar larva of mango fruit borer has developed any resistance.

Bio-assay Studies on 3rd Instar Larvae of Mango Fruit Borer

The data obtained 24 hours after exposure revealed that Monocrotophos 36 per cent SL produced highest kill (85.7 per cent) followed by Dichlorvos 76 per cent EC (71.4 per cent) and Endosulfan 35 per cent EC (43.0 per cent). Acephate 75 per cent SP and Carbaryl 50 per cent WP recorded 28.6 and 14.3 per cent mortality, respectively where as mortality was not observed in treatments of Chlorpyriphos 20 per cent EC and Azadirachtin 1500 ppm (Table 38.12).

Table 38.12: Bio-assay Studies on 3rd Instar Larvae of Mango Fruit Borer

Sl.No.	*Treatment Particulars*	*Dose*	*Per cent Mortality After*		*Per cent of Larvae Successfully Entered into Diapause/Pupation*
			24 Hours	*48 Hours*	
1.	Monocrotophos 36 per cent SL	1.5 ml/lt. of water	85.7	100.0	0.00
2.	Dichlorvos 76 per cent EC	1 ml/lt. of water	71.4	71.4	0.00
3.	Chlorpyriphos 20 per cent EC	2 ml/lt. of water	0.00	100.0	0.00
4.	Acephate 75 per cent S.P.	0.75 g/lt. of water	28.6	85.7	14.28
5.	Carbaryl 50 per cent W.P.	3 g/lt. of water	14.3	43.0	28.57
6.	Azadirachtin 1500 ppm	7 ml/lt. of water	0.0	0.0	14.28
7.	Endosulfan 35 per cent EC	2 ml/lt. of water	43.0	100.0	0.00
8.	Control	—	0.0	0.0	71.42

After 48 hours of exposure Monocrotophos 36 per cent SL, Chlorpyriphos 20 per cent EC and Endosulfan 35 per cent EC recorded 100.00 per cent mortality. These treatments are closely followed by Acephate 75 per cent S.P. (85.7 per cent) and Dichlorvos 76 per cent EC (71.4 per cent). Carbaryl 50 per cent W.P. resulted only about 43.0 per cent kill where as mortality was zero in Azadirachtin 1500 ppm and control treatments.

From the observations of live population from different treatments, it was recorded that live larvae in Acephate 75 per cent SP (14.28 per cent larvae) and Carbaryl 50 per cent WP (28.57 per cent larvae) treatments entered into diapause successfully. Though Azadirachtin 1500 ppm treatment recorded zero per cent mortality after 48 hours of exposure, only 14.28 per cent of larvae entered into diapause while the remaining larvae (85.72 per cent) were either malformed/deformed into larval/pupal intermediaries causing mortality. These deformities in biological stages, emphasizes the slow action and growth retardant effects of neem formulations on larvae of mango fruit borer.

This study clearly indicates that 3rd instar of mango fruit borer was susceptible to recommended doses of commonly used insecticides as 70-100 per cent mortality was recorded in all the insecticides tested but it was not vice-versa in the field conditions. The site of feeding of mango fruit borer larvae may offer protection thus creating a sort of pseudo/ecological resistance which could have contributed to failure of insecticide under field conditions. Hence, it can be suggested that timely application of insecticides targeting 1st and 2nd instar stage of the pest when fruits are in pea or marble size (before the advancement of stage of larva as well as fruit) would be beneficial.

Through the present investigations, survey conducted in four mango growing states *i.e.* Andhra Pradesh, Orissa, West Bengal and Karnataka revealed the presence of pest in Andhra Pradesh, Orissa and West Bengal. East Godavari, West Godavari, Visakhapatnam and Khammam districts of Andhra Pradesh and Malda and Murshidabad districts of West Bengal were identified as hot spot areas of the pest. From the survey pest samples, hymenopteran parasitoids *viz., Brachymeria lassus, Brachymeria apicicornis* and *Oxycoryphe* sp. were identified as parasitoids of the mango fruit borer larvae and pupae. One coleopteran beetle, two species of ear wings and two species of mantids, red ants *etc.*, were also recorded as natural enemy fauna of the pest. The presence of sex pheromone in mango fruit borer moths was also established through wind tunnel method keeping virgin female as pheromone source and found pheromone release activity is highest in one day old moths. Sujatha *et al.* (2002) confirmed the presence of sex pheromone in *D.albizonalis* through wind tunnel and electroantennogram (EAG) studies in the laboratory. Further detailed studies are needed regarding pheromone release and its exploitation in the management of mango fruit borer.

In the light of the investigations carried out, the following integrated pest management package can be suggested to keep the MFB under check.

- ☆ Collection and destruction of dried branches/dead wood in the mango orchards during off-season.
- ☆ Collection and destruction of diapausing pre-pupae from the cracks, crevices and knots on the trunk and branches of mango tree during off-season.
- ☆ Collection and destruction of infested fruits regularly starting from the incidence of pest prevents the spread of pest infestation.
- ☆ The mango fruit borer larvae/pupae/diapausing larvae can be trapped by using artificial niches on the trees during the season and destroyed.
- ☆ Swabbing of the chemical chlorpyriphos 20 per cent EC @ 5 ml/lit. of water would kill the pre-pupae/pupal population under the bark and helps in reduction of fruit damage.
- ☆ Spraying relatively safer insecticides such as chlorpyriphos 20 per cent EC (2.5 ml/l), dichlorvos 76 per cent EC (1.5 ml/l) or carbaryl 50 per cent W.P. (3g/l) in the second fortnight of January coinciding with the moth emergence/hatching of eggs of first brood in the gardens where the pest incidence was severe in previous year.
- ☆ Application of dichlorvos 76 per cent EC (1.5 ml/l) late in the evening (as the pest is nocturnal) when the fruits are of marble size to coincide with the migration of the larvae.
- ☆ Spraying of Azadirachtin 1500 ppm + chlorpyriphos 20 per cent EC @ 7 ml + 1 ml/litre of water is also efficacious.
- ☆ Sprays of insecticides to be made in the evening to coincide with the peak activity time of the pest.
- ☆ Application of insecticide should be done with a tractor mounted power sprayer to ensure uniform coverage of the trees. In addition, local quarantine is to be exercised by regulating the movements of the fruits in and out of the infested tracts in Andhra Pradesh and also to other states. The movement of damaged fruits should be strictly avoided, as the infestation can be seen externally, to prevent the spread of inoculum to neighboring places.

Acknowledgements

Thanks are due to the Agricultural and Processed Food Export Development Authority (APEDA), Government of India for sponsoring the project work on an important quarantine pest. Honourable Vice Chancellor and Director of Research of Acharya N.G. Ranga Agricultural University, Andhra Pradesh are greatly acknowledged for providing infrastructural facilities to carry out the project work and sanctioning administrative provisions relating to the project work. Dr. Manickavasagam, Professor in Entomology, Annamalai University, Chidambaram is also acknowledged for his help rendered in identification of natural enemies. Thanks are also due to Dr. B.S.Nandihalli, Associate Professor (Entomology), University of Agricultural Sciences, Dharwar, Karnataka; Horticulture Departmental staff, Karnataka; staff from APEDA, Regional Office, Bangalore; staff from APEDA, Regional Office, Kolkata; Department of Horticulture, West Bengal; staff from Bidhan Chandra Krishi Viswavidyalaya, Kalyani; Department of Horticulture, Orissa; and Dr. H.S. Singh, Senior Scientist, Central Horticultural Expt. Station (IIHR), Bhubaneswar for their help rendered while survey in the respective states.

References

Sahoo, S.K. and Jha, S., 2006. Fruit borer, a threat to mango cultivation. *Association for Advancement in Plant Protection–Newsletter*, 1(1): 5.

Sujatha, A. and Zaheruddeen, S.M., 2002. Biology of pyralid fruit borer *Deanolis albizonalis* (Hampson): A new pest of mango. *J. of Appl. Zool. Research*, 13(1): 1–5.

Sujatha, A. and Zaheruddeen, S.M., 2006. Survey on the occurrence of mango fruit borer, *Deanolis albizonalis* in different mango tracts of Andhra Pradesh and its preference to mango varieties. *Indian J. Appl. Ent.*, 20(1): 33–37.

Sujatha, A., Zaheruddeen, S.M., Arjuna Rao, P. and Nagalingam, B., 2002. Electroantennogram and wind tunnel studies on olfactory responses of mango fruit borer, *Deanolis albizonalis* (Hampson) to female sex pheromone. *Pest Management in Horticultural Ecosystems*, 8 (2): 83–89.

Sujatha, A., Zaheruddeen, S.M., Nagalingam, B. and Arjuna Rao, P., 2001. Efficacy of different neem formulations against *Deanolis albizonalis* (Hampson) – A new mango fruit borer. In: *Proc. of the 2nd National Symposium on IPM in Horticultural Crops, New Molecules, Biopesticides and Environment*, pp 3-5.

Tandon, P.L. and Verghese, A., 1985. *World List of Insect Mite and Other Pests of Mango*. Indian Institute of Horticultural Research, Bangalore, Technical Document No. 5.

Zaheruddeen, S.M. and Sujatha, A., 1995. Larval behaviour of pyralid mango fruit borer *Deanolis albizonalis* (Hampson). *J. of Appl. Zool. Research*, 6(2): 157.

Zaheruddeen, S.M. and Sujatha, A., 1993. Record of *Deanolis albizonalis* (Hampson) (Pyralidae: Lepidoptera) as mango fruit borer in Andhra Pradesh. *Journal of Bombay Natural History Society*, 90: 528.

Plant Protection, Diversity and Conservation (2011) *Pages* **516–524**
Editors: **D. Prasad & Satya Kumar**
Published by: **BIOTECH BOOKS, NEW DELHI**

Chapter 39

Pest Complex and their Succession on Maize with Special Reference to Lepidopteran Insect Pests

Ishtiyaq Ahmad*, R.M. Bhagat, Hafeez Ahmad, Devinder Sharma and Saurav Gupta

Division of Entomology, SK University of Agricultural Sciences and Technology–J, Chatha, Jammu – 180 006, J&K

Maize is an important staple food of people living in hilly and sub-mountainous regions of Himalayas in Jammu division, occupying an area of 221.93 thousand hectares with an average productivity of 17.33 q ha^{-1} (Anonymous, 2007). However, the average yield is far below as compared to the world's average yield 38q ha^{-1}. Arthropod pests are among the key factors contributing to low yield of maize. Globally losses due to insect pests of maize are reported to be approximately 9 per cent, which is equivalent to 52 million MT and worth approximately $5.7 billion annually (James, 2003). A variety of insect pests attack maize crop right from seedling to maturity. These insect pests includes: lepidopterous pests (borers, cutworms, armyworms, earworms, and grain moths) are the most damaging world over, followed by the coleoptera (root worms, wireworms, grubs, grain borers, and weevils). Next in importance is the group of insects that serve as carriers (vectors) for disease agents or pathogens, among which leafhoppers and aphids are of prime importance (Alejandro, 1987).

In India the crop is being attacked by about 139 species of insect pests with varying degree of damage to the maize crop. However, only about a dozen of these are quite serious (Siddiqui and Marwaha, 1993). Seventeen species in two families (Pyralidae and Noctuidae) have been found to attack maize in various parts of Africa. However, *Chilo partellus* (Swinhoe), *Chilo orichalcociliellus* Strand, *Busseola fusca* Fuller, *Sesamia calamistis* Hampson, and *Eldana saccharina* Walker are of great

* E-mail: ishtiyaq_raina@rediffmail.com

importance. The yield losses caused by stem borers to maize vary widely in different regions and range from 25-40 per cent according to the pest population density and phenological stage of the crop at infestation (Khan *et al.*, 1997a). Among them maize stem borer, *C. partellus* is most dominant contributing 90-95 per cent of the total damage in *kharif* season (Jalali and Singh, 2002). This pest is principal limiting factors in achieving the increased production of maize in tropical and subtropical regions especially in the northern India. Work in intermediate upper reaches on maize insect pests is still lacking, hence an attempt was made to study the insect pest complex of maize which is subjected to be the major concern in limiting the maize yield.

Experiments were conducted in free choice conditions at higher reaches of Jammu region during *kharif* season 2006 and 2007 under rainfed conditions. Maize variety "Kanchan" was planted at 60 x 20 cm spacing in 3 x 4 m plot size. The crop was raised in natural conditions (*i.e.* without any application of insecticides) to allow build up of insect populations. The crop was observed at weekly intervals throughout the growing season. To estimate the population of different beetles in maize agro-ecosystem, light trap was installed in the field to monitor the adult moths (mostly of nocturnal habit). The observations of the insect pests were recorded in morning as well as in the dusk period to take the advantage of sedentary nature of the insects. Larvae feeding on foliage and stalks were quantified on plant basis while, other feeding on cobs were quantified on cob basis. Incidence of beetles on each plant/cob in the plot was keenly observed. Mean of three plots was calculated to make further statistical analysis. Larvae hiding in the soil debris *i.e.*, cut worm was quantified by taking three samples of larvae excavating in 8400 cm^2 area *i.e.*, (25 x 25) 10 cm deep soil in the field. Mean number of larvae present per three samples was recorded (Chandla, 1986). Populations of insect-pests recorded were correlated with weather factors.

During both *kharif* seasons 2006 and 2007, various insect populations *viz.* lepidopteran insects: stem borer (*C. partellus*), cutworm (*A. ipsilon*), corn earworm (*H. armigera*) and red hairy caterpillar (*A. moorei*); coleopteran insects: wire worm/click beetle (*M. communis*), white grub (*Holotrichia* sp), blister beetle (*M. pustulata*), and flower eating beetle (*P. alboqattata*); aphid (*R. maidis*), pyrilla (*P. perpesula*), jassid (*Z. manaliensis*), red cotton bug (*D. koenigi*) and stink bug (*N. virudula*) and order orthoptera: grass hopper (*Ceracris* sp.) were recorded. Tobacco caterpillar (*S. litura*) appeared during 2007, however this pest remained absent during 2006. Thus, there were 14 and 15 insects recorded during 2006 and 2007, respectively. Five insects from each lepidoptera and hemiptera orders were observed to infest maize while, four insects from coleoptera and one insect from orthoptera were studied to infest maize during 2006 and 2007. The roots were attacked by white grubs; seedlings by wire worm and cutworm; stalk by stem borer; foliage by aphid, pyrilla, jassid, red cotton bug, stink bug, grass hopper and adults of white grub and cob by aphid, red hairy caterpillar, corn earworm, tobacco caterpillar and flower eating beetle while, blister beetle was recorded to feed on silk of cobs (Table 39.1).

Although, pest situation of maize in India was reviewed by several workers in which important insect pests owing to the low productivity were stem borer (*C. partellus*), pink stem borer (*S. inferens*), armyworm (*M. separata*), aphid (*R. maidis*) and thrips (*A. sudanensis*) (Nayar *et al.*, 1976). Similarly, Sarup *et al.* (1987) reported 130 insect pests which further increased up to 139 (Siddiqi and Marwaha, 1993). Mauricio (2000) reported insects attacking roots (rootworms, wireworms, white grubs, and seed-corn maggots), leaves (aphids, armyworm, stem borers, thrips, spider mites, and grasshoppers), stalks (stem borers, termites) and ears and tassels (stem borers, earworms, adult rootworms, and armyworm), respectively. Panwar *et al.* (1997) studied the insect pest scenario in maize based cropping system in *kharif* and *rabi* seasons.

Table 39.1: Insect Pest Complex on Maize and their Succession

Sl.No.	Common Name	Scientific Name	Family	Order	Period of Prevalence	Peak Period	Crop Stage	Status
				Insect pests				
1.	Wireworm	*Melanotus communis* (Gyllenhal)	*Elateridae*	Coleoptera	Jun.	Ist F.N. Jun.	Emergence–6th leaf	Minor
2.	Stem borer	*Chilo partellus* (Swinhoe)	Crambidea	Lepidoptera	Jun.-Oct.	Jul.	Tasseling	Major
3.	Cutworm	*Agrotis ipsilon* (Hufnagal)	Noctuidea	Lepidoptera	Jun.	Ist F.N. Jun.	Emergence–6th leaf	Major
4.	White grub	*Holotrichia* sp.	Melolonthidae	Coleoptera	Jun.-Sept.	Ist F.N. Jul.	Tasseling	Major
5.	Grass hopper	*Ceracris* sp.	Acrididae	Orthoptera	Aug.- Oct.	Ist F.N. Sep.	Miliking	Minor
6.	Aphid	*Rhopalosiphum maidis*	Apididae	Hemiptera	Jul.- Oct.	Aug.	Siliking	Major
7.	Jassid	*Zyginidia manaliensis*	Cicadellidae	Hemiptera	Aug.-Oct.	IInd F.N. Sep.	Dent	Minor
8.	Blister beetle	*Mylabris pustulata*	Meliodae	Coleoptera	Aug.	Ist F.N. Aug.	Siliking	Minor
9.	Stink bug	*Nezara virudula* (Linn.)	Pentatomidae	Hemiptera	Sep.- Oct.	IInd F.N. Sep.	Dent	Minor
10.	Hairy caterpillar	*Amsacta moorei* (Butler)	Arctiidae	Lepidoptera	Sep.-Oct.	IInd F.N. Sep.	Dough–Dent	Minor
11.	Red cotton bug	*Dysdercus koenigi* (Fab.)	Pyrrhocoreidae	Hemiptera	Sep.- Oct.	Ist F.N. Sep.	Miliking	Minor
12.	Pyrilla	*Pyrilla perpesula* (Walker)	Fulgoridae	Hemiptera	Aug– Sep.	Aug.	Siliking	Minor
13.	Corn ear worm	*Heliothis armigera*	Noctuidea	Lepidoptera	Sep.	Sep.	Dent	Minor
14.	Tobacco caterpillar	*Spodoptera litura* (Fab.)	Noctuidea	Lepidoptera	Sep-oct	IInd F. N. Sep	Dough–Dent	Minor
15.	Flower eating beetle	*Protaetia alboquattata* Vig.	Cetonidae	Coleoptera	Aug- Sep	2nd F.N Aug	Miliking	Minor
				Non-insect pests				
1.	Snail	*Helix pomatia*		Stylomatophora	Jul.-Aug.	Ist F.N. Aug.	Siliking	Minor
2.	Jackal	*Canis aureus* Linnaeus	Canidae	Carnivora	Aug.-Sep.	Ist F.N. Sep.	Miliking	Minor

F.N.: Fort Night.

A. ipsilon larvae ranged from 2 to 5 and 2 to 6 during initial three weeks of the crop season during 2006 and 2007, respectively. No larval population was recorded during rest of the growing season. Both *H. armigera* and *A. moorei* populations fluctuated from 1st week of July to 2nd week of September in both years. Initially *H. armigera* population increased then decreased according to increment of time and ranged from 2–9 and 1–9 adults/trap during 2006 and 2007 *kharif* seasons, respectively. Similar trend was found in case of *A. moorei* where, population ranged from 1–7 and 2–7 adults/trap during 2006 and 2007. Tobacco caterpillar populations ranged between 2–5 adults/trap with its peak in 2nd week of August during *kharif* 2007 (Table 39.2). As the damage of *A. moorei, H. armigera* and *S. litura* is concerned, these pests infested cobs from milking to the dent stage, but *S. litura* population and infestation was recorded during 2007 only. This might be due to shifting of this pest from the other hosts in that particular region. Correlation studies revealed that stem borer, red hairy caterpillar and corn earworm populations were positively correlated with maximum temperature, minimum temperature and relative humidity. Where as, cutworm populations was negatively correlated with all three weather parameters (Table 39.3).

Among these pests, population of adult stem borer appeared throughout the growing season with their respective peaks of 49 and 47 adults/trap during 2006 and 2007. Population of this pest increased from 21st standard week and showed decline after reaching its peak in both seasons. Intensity of infestation of stem borer was recorded from 3rd week of crop season till harvest. Maximum infestation (6.66 per cent) and (4.00 per cent) was recorded in the first week of July 2006 and in the 3rd week of August and 2007 (Table 39.4), respectively. This pest retained major status as infestation reached higher levels (*i.e.* 36.94 per cent in 2006 and 29.06 per cent in 2007). These results are in agreement with Odio (1984) who reported maize stem borers as major pests of maize in the savannah area of Ivory Coast. Moreover, Abraham *et al.* (1993) recorded maize stem borers like, *B. fusca, C. partellus* and various termite species (*Macrotermes* and *Microtermes* spp.) as key pests. Insects like armyworm, cutworms, chafer grubs, grasshoppers, leafhoppers, pink stalk borer and maize aphid were found to be sporadically important. Khan *et al.* (1997a, 1997b) and Wilson (1998) reported that stem borers seriously limit potentially attainable maize yields by infesting the crop from seedling stage to maturity.

Several workers generate the information on *C. partellus* infestation like Songa *et al.* (2001) reported highest intensity of infestation due to *C. partellus*. Similarly, Jalali and Singh (2002) studied that mean per cent infestation (35.4 per cent) of maize stem borers during *kharif* in fodder maize, while Georges *et al.* (2006) recorded maximum larval population (54.5 per cent) of this pest. These results were contradictory to Litsinger *et al.* (2007) in Mindanao, Philippines, as they reported the Asian corn borer, *Ostrinia furnacalis* as the most important pest. This might be attributed due the variation in agro-climatic conditions and diversity of insect pests in different agro-climatic zones.

Peak periods of wire worm and cutworm were found from emergence to 6th leaf stage of maize crop. Moreover, peak periods of stem borer and white grub was recognized in July at tasseling stage, however, aphids, pyrilla and blister beetle reached their peaks in the month of August at silking stage. At the milking stage pest infestation was found due to grasshopper, red cotton bug and flower eating beetle. Rest of insects *viz.* jassid, stink bug, red hairy caterpillar and corn earworm reached their peak incidence at dent stage during *kharif* season 2006, Similar findings were recorded during *kharif* 2007 but addition of tobacco caterpillar was recorded to infest maize at dent stage in this season. Incidence and distribution was found to be maximum in case of stem borer and aphid as these pests prevailed throughout the growing season. These results are in close conformity with Khan *et al.* (1997a) reported that stem borers seriously limit potentially attainable maize yields by infesting the crop from seedling

Table 39.2: Population of Lepidopteran and Coleopteran Pests Infesting Maize Crop During 2006

Standard Week	2006				2007				
	Stem Borer	Cut worm	Red Hairy Caterpillar	Corn Ear Worm	Stem Borer	Cut Worm	Red Hairy Caterpillar	Corn Ear Worm	Tobacco Caterpillar
	Adult/Trap	Larva/6250 cm^3	Adult/Trap	Adult/Trap	Adult/Trap	Larva/6250 cm^3	Adult/Trap	Adult/Trap	Adult/Trap
21	05	5	0	0	07	4	0	0	0
22	08	4	0	0	06	6	0	0	0
23	11	2	0	0	01	2	0	0	0
24	09	0	0	0	05	0	0	0	0
25	17	0	0	0	09	0	0	0	0
26	26	0	0	0	17	0	0	3	0
27	24	0	3	5	29	0	0	2	0
28	36	0	2	7	37	0	2	8	0
29	49	0	3	6	45	0	3	7	0
30	45	0	4	8	47	0	5	7	0
31	40	0	6	9	39	0	6	9	2
32	37	0	7	5	39	0	7	6	5
33	43	0	6	4	41	0	6	5	4
34	32	0	4	3	29	0	4	2	3
35	25	0	2	3	31	0	2	2	0
36	16	0	1	2	17	0	3	3	0
37	12	0	2	2	15	0	2	1	0
38	14	0	0	0	11	0	0	0	0
39	09	0	0	0	07	0	0	0	0
40	07	0	0	0	07	0	0	0	0
41	03	0	0	0	03	0	0	0	0
Range	03–49	0–5	0–7	0–9	3–47	0–6	0–7	0–9	0–5

Table 39.3: Correlation of Lepidopteran Insect Populations with Abiotic Factors During 2006 and 2007

Sr. No.	*Insect Pest*	*2006*			*2007*		
		Maximum Temperature	*Minimum Temperature*	*Relative Humidity*	*Maximum Temperature*	*Minimum Temperature*	*Relative Humidity*
1	Stem borer	0.605	0.525	0.885	0.367	0.422	0.889
2	Cutworm	–0.209	–0.091	–0.407	–0.180	–0.125	–0.290
3	Red hairy caterpillar	0.284	0.265	0.845	0.119	0.178	0.900
4	Corn earworm	0.554	0.413	0.320	0.461	0.492	0.802
5	Tobacco caterpillar		Absent		–0.006	0.008	0.701

Table 39.4: Lepidopteran Insect Pest Complex Infesting Maize Crop during *kharif* Season 2006 and 2007

Standard Week	*2006*				*2007*				
	Stem Borer/ 100 Plants	*Cut Worm/ 100 Plants*	*Hairy Caterpillar/ 100 Cobs*	*Corn Ear Worm/ 100 cobs*	*Stem Borer/ 100 Plants*	*Cut Worm/ 100 Plants*	*Hairy Caterpillar/ 100 cobs*	*Corn Ear Worm/ 100 cobs*	*Tobacco Caterpillar/ 100 cobs*
21	0.00	0.00	0.00	0.00	0.00	0.00	0.00	0.00	0.00
22	0.00	9.33	0.00	0.00	0.00	6.66	0.00	0.00	0.00
23	0.00	4.33	0.00	0.00	0.00	7.00	0.00	0.00	0.00
24	1.66	0.66	0.00	0.00	0.00	0.00	0.00	0.00	0.00
25	2.66	0.00	0.00	0.00	1.00	0.00	0.00	0.00	0.00
26	4.33	0.00	0.00	0.00	3.00	0.00	0.00	0.00	0.00
27	6.66	0.00	0.00	0.00	0.33	0.00	0.00	0.00	0.00
28	5.66	0.00	0.00	0.00	2.33	0.00	0.00	0.00	0.00
29	4.33	0.00	0.00	0.00	1.66	0.00	0.00	0.00	0.00
30	2.33	0.00	0.00	0.00	2.33	0.00	0.00	0.00	0.00
31	1.33	0.00	0.00	0.00	1.33	0.00	0.00	3.00	0.00
32	0.33	0.00	0.00	5.33	1.33	0.00	0.00	5.00	0.00
33	0.33	0.00	0.00	8.00	4.00	0.00	0.00	6.00	0.00
34	1.00	0.00	0.00	5.00	2.33	0.00	4.00	7.66	0.00
35	1.33	0.00	3.00	4.66	1.00	0.00	5.33	4.33	0.00
36	0.00	0.00	4.33	4.00	1.66	0.00	6.66	4.00	0.31
37	2.00	0.00	5.33	0.00	1.33	0.00	4.33	3.00	1.33
38	1.66	0.00	6.00	0.00	1.44	0.00	3.66	1.33	1.00
39	0.33	0.00	2.00	0.00	2.33	0.00	1.00	0.00	0.00
40	0.00	0.00	0.00	0.00	1.00	0.00	0.00	0.00	0.00
41	1.00	0.00	0.00	0.00	0.66	0.00	0.00	0.00	0.00
Range	0.33–6.66	0.66–9.33	2.00–6.00	4.00–8.00	0.33–4.00	6.00–7.00	1.00–6.66	1.33–7.66	0.13–1.33

stage to maturity. Aphids could become a threat to the early as well as late maize plantations (Fareed, 1991).

Acknowledgements

The authors thank to members of Department of Entomology for their help with some of the experiments, Director Research for providing essential facilities mainly the experimental farm. Can did thanks due Dr. R. N. Ganguli of Indira Gandhi Agriculture University, India for assistance in providing necessary advices.

References

Abraham, T., Ferdu, A., Assefa, G.A. and Adhanom, N., 1993. Highlights on maize insect pests and their management in Ethiopia. In: Benti T. and J.K. Ransom (eds.). *Proceedings of the First National Maize Workshop of Ethiopia*, 5–7 May, Addis Ababa, Ethiopia. IAR/CIMMYT, Addis Ababa, pp. 34–42.

Alejandro Ortega, C., 1987. *Insect Pests of Maize: A Guide for Field Identification*. Mexico, D.F.: CIMMYT.

Anonymous, 2007. *Statistical Year Book*, Jammu and Kashmir Government Dir. of Statistics and Planning, Janipur, Jammu (J&K), India.

Chandla, V.K., 1986. Pest complex of potato crop in Shimla hills and their management. *Ph.D. Thesis*, Y.S.P. University of Horticulture and Forestry, Solan (H.P.)

Fareed., 1991. Insecticidal and varietal studies on corn leaf aphid *Rhopalosiphum maidis* (F.). *M Sc. (Hon) Thesis*, Dept. of Entomology AUP, Pakistan, pp. 75.

Georges, O.O., Bruno, P.L.R., Stephane, D. and Moyal, P., 2006. Distribution, pest status and agro-climatic preferences of lepidopteran stem borers of maize in Kenya. *Ann. Society of Ent., France*, 42(2): 171–177.

Jalali, S.K. and Singh, S.P., 2002. Seasonal activity of stem borers and their natural enemies on fodder maize. *Entomon.*, 27(2): 137–146.

James, C., 2003. Global review of commercialized transgenic crops: 2002 Feature: *Bt* Maize. ISAAA Briefs No. 29. ISAAA: Ithaca, NY. *http: //www.isaaa.org*.

Khan, Z.R., Chiliswa, P., Ampong-Nyarko, K., Smart, L.E., Polaszek, A., Wandera, J. and Mulaa, M.A., 1997a. Utilisation of wild gramineous plants for management of cereal stemborers in Africa. *Insect Sci. and its Appl.*, 17: 143–150.

Khan, Z.R., Overholt, W.A. and Hassana, A., 1997b. Utilization of agricultural biodiversity for management of cereal stemborers and striga weed in maize–based cropping systems in Africa a case study. *International Centre of Insect Physiology and Ecology*, PO Box 30772, Nairobi, Kenya.

Litsinger, J.A., Dela Cruz, C.G., Canapi, B.L. and Barrion, A.T., 2007. Maize planting time and arthropod abundance in southern Mindanao, Philippines. I. Population dynamics of insect pests. *Inter. J. of Pest Managm.* 53(2): 147–159.

Mauricio, R. B., 2000. Participatory methods in the development and dissemination of new maize technologies. *CIMMYT World Maize Facts and Trends*, pp. 3–9.

Nayar, K.K., Krishnan, A.T.N. and David, B.V., 1976. *General and Applied Entomology*. Tata McGraw-Hill Publication. Co. Ltd., pp. 589.

Odio, A., 1984. Contribution à l'étude de, la biologie et'de l'écologie de *Sesamia calamistis* Hampson (Lepidoptera Noctuidae) en Côte d'lvoire centrale. These de doctorat de troisieme cycle,Université de Paris–sud, Orsay.

Panwar, V.P.S., Siddiqui, K.H. and Singh, N.N., 1997. Insect pest scenario in maize based cropping system. *Ind. J. of Ent.*, 59(1): 1–7.

Sarup, P., Siddiqui, K.H. and Marwaha, K.K., 1987. Trends in maize pest management research in India together with bibliography. *J. of Ent. Res.*, 11(1): 19–68.

Siddiqui, K.H. and Marwaha, K.K., 1993. *The Vistas of Maize Entomology in India*. Kalyani Publishers, Ludhiana.

Songa, J.M., Overholt, W.A., Mueke, J.M. and Okello, R.O., 2002. Regional distribution of lepidopteran stem borers and their parasitoids among wild grasses in the semi-arid eastern Kenya. *Afri. Crop Sci. J.*, 10(2): 183–194.

Willson, H., Kyamanywa, S., Kalule, T., Birgirwa, G. and Kauma, T., 1998. Management of maize stalk borer in Uganda. *FIPM CRSP Symposium 98 Abstracts.htm.*

Plant Protection, Diversity and Conservation (2011) *Pages* **525–531**
Editors: **D. Prasad & Satya Kumar**
Published by: **BIOTECH BOOKS, NEW DELHI**

Chapter 40
Management Strategies for Phytonematodes Associated with Oilseed Crops

D. Prasad

Division of Nematology, IARI, New Delhi – 110 012

Oilseeds occupy an important position in the agricultural economy of India and has largest area under cultivation but low productivity as these energy rich crops are grown in energy depleted conditions, In India, six major oilseed crops namely, groundnut, sunflower, soybean, mustard-rapeseed sesamum and linseed are cultivated to a lesser or greater extent in almost all states. Among all, groundnut occupies an important place occupying an area of about 12.0 million hectare with a total production of 19.5 mt. Thus, India has the largest area under this crop in the World but production and productivity of the crops is quite low (Av. 847 kg/ha). One of the major constraints in oilseed production is wide prevalence of root-knot and reniform nematodes on ground nut, soybean, castor and sunflower. However, among plant-parasitic nematodes, endoparasites cause 22–27 per cent loss in groundnut by *Meloidogyne arenaria* (Prasad, 1989) whereas ectoparasites cause 11–22 per cent and *Rotylenchulus reniformis* and *M. incognita* together cause 30–40 per cent. Kalahasti Malady disease was reported to cause as high as 40.1 per cent per cent loss in case of groundnut due to *Tylenchorhynchus brevilineatus* (Prasad and Rangappa, 1994).

Management of Plant Parasitic Nematodes on Oilseed Crops

The production of oilseed crops fluctuate widely from year to year due to outbreak of pests and pathogens and also due to climatic conditions. But, the main reason for the low production of oilseed crops in that this energy rich crop is cultivated largely under conditions of energy starvation. However, nematodes cause great loss both in early stages of plant growth and also at the stages of maturity. It is therefore, imperative to manage root-knot nematode which causes heavy damage to the crops. A long medium term strategy used by harmonising low input components (Fallowing, summer ploughing, solarization, growing of resistant cultivars and judicious use of nematicides) to bring down the nematode population below threshold level, should be the main aim for successful cultivation of oilseed crops. Efficient strategies are therefore, required for the management of nematode by judicious use of nematicides so that there is maximisation of advantages and minimisation of disadvantages.

Fallowing

It is the practice of keeping the land fallow, *i.e..*, without host crops. This practice is widely used in Northern India, after harvest of rabi crops. The main principle in this practice is starvation of nematodes due to non-availability of host crop and dessication due to heat during summer. But if fallowing is followed with summer ploughing. The decrease of endoparasite and semi endoparasites especially root-knot and reniform nematodes were recorded to be as much as 40 per cent.

Summer Ploughing

Three or more ploughings during summer months exposes the nematodes to solar energy which are being killed due to dessication. More than 60 per cent nematodes were killed due to summer ploughing. Three ploughings (15 cm deep) in summer months at the interval of 15 days decreased population of *Tylenchorhynchus vulgaris, Hoplolaimus indicus, Helicotylenchus indicus* and *Pratylenchus zeae* (Haque and Prasad, 1983) (Table 40.1).

Table 40.1a: Effect of Ploughings on Nematode Populations in Wheat per cent Decrease Over Control

Treatments	*Hoplolamus indicus*	*Helicotylenchus indicus*	*Tylenlhorhynchus vulgaris*	*Pratycenchus zeae*	*Yield (kg)*
One ploughing	78.56	180.00	261.90	57.08	7.70
Two ploughings	104.43	211.77	331.09	200.75	7.70
Three ploughings	157.97	224.12	369.30	300.00	7.80

Table 40.1b: Effect of Ploughings on Nematode Populations in Maize Per cent Decrease Over Control

Treatments	*Hoplolamus indicus*	*Helicotylenchus indicus*	*Tylenlhorhynchus vulgaris*	*Pratycenchus zeae*	*Yield (kg)*
One ploughing	85.11	151.4	302.37	6.00	9.09
Two ploughings	127.28	142.87	277.46	100.00	9.11
Three ploughings	220.00	388.83	376.19	220.12	10.04

Solarisation

This practice is applicable only in limited areas. In this management tactics, the solar energy is preserved with the help of transparent polyethylene sheet to increase soil temperature enough to kill the nematodes present in a particular site, Kumar *et al.* (1993) reported significant decrease of weed seed emergence by over 90 per cent and reduced the population of plant parasitic nematodes by about 80 per cent and increased seed yield by upto 70 per cent and growth of soybean was improved following solarisation for 40 days during May-June months. Prasad *et al.* (1995) found about 90 per cent kill of root-knot and reniform nematodes where solarised for 30 days during summer months in Delhi State. However, under field conditions, solarisation when coupled with triazophos application @ 1.5 t/ha completely controlled the plant parasitic nematodes in fields.

This method provides an economical alternative for multiple pest disease weed control in small areas like nursery beds, orchards, forestry plantation and high value crops in regions with hot summers.

Intercropping

Mixed cropping either with non-host or antagonistic crops such as Tagetes, sesamum, mustard, etc. have also been reported to reduce root-knot and reniform nematodes. However, the use of trap crops as green manure, not only reduce the nematode population but improves soil fertility in many other ways and promotes antagonistic microbial activity. Gaur (1995) reported that Sunhemp or Dhaincha which allow invasion by *Meloidogyne* spp. but does not support development, can serve as good trap crop.

Resistant Varieties

Resistance is an effective, economical and environmentally safe means of reducing losses from diseases caused by these pests. Use of resistant plants against plant-parasitic nematodes enables the grower to manage these pests without increasing extra inputs associated with the purchase of expensive chemicals, applicators and in the numerous mechanical operations that go into the production of crop. The main aim of breeding for nematode disease resistance is to produce varieties which suffer less damage from nematode attack. Prasad (1989) screened about 500 varieties/lines of groundnut crop against *M. arenaria* and exhibited two lines as resistant namely C 41 (NRCG-31) and NCAC-2196 (NRCG 1060). Sakhuja and Sethi (1985) also reported local Ambali, G-201, Local-256 and Japtin-220-15 as resistant against *M. arenaria* (Table 40.2).

Table 40.2: Reaction of Selected Cultivars of Groundnut Against *Meloidogyne arenaria* (Out of 500 varieties)

Cultivars	*Number of Galls or Eggsacs per Root System*	*Av. Eggs/Larvae per Eggsacs*
C-41 (NRCG-31)	3.3	0
NCAC-2196 (NRCG-1060)	10.3	0
Local Ambali	4.0	22.0
G-201	1.8	33.6
Local-256	0.0	0.0
Japtui-220-15	0.0	0.0
J-11 (susceptible)	20.0	89.0

Bhat and Krishnappa (1989) screened 21 cultivars/lines of groundnut against *M. arenaria* in pot experiment, all recorded a high gall index. Prasad and Chawla (1990) correlated yield of soybean with degree of resistance. Varieties like JS-79-291 and DS seemed to withstand moderate *M. incognita* infestation and at the same time gave highest yield per plant (Table 40.3). Raut and Sethi (1980) found soybean cultivars like forest, JS-2, Selection, DS 74-20-1 were found resistant against *M. incognita.*

Prasad and Rangappa (1994) reported that two varieties namely Tirupati-1 and Tirupati-2 have been found resistant against Kalahasti Malady disease caused by phytonematodes in Tirupati Region of A.P.

Table 40.3: Performance of Soybean Cultivars Against *Meloidogyne incognita*

Cultivars	*Root-knot Index (1-5)*	*Yield/Plant (g)*
DS-8	1.0	7.5
MACS-212	0	5.0
MACS-124	0	2.8
MACS-153	4.0	2.6
MACS-125	2.0	5.2
JS-79-81	2.0	4.9
JS-79-277	0	4.9
JS-79-291	4.0	12.6
PK-695	4.0	3.6
PK-890	1.0	3.3

Amarantha and Krishnappa (1989) screened 10 sunflower cultivars against *M. incognita*, an found that all were susceptible to varying degrees. After 50 days, EC 68415 had fewest galis and eg masses per root system. Kaul (1977) indicated inter varietal differences in sunflower to the infection c *M. incognita* however, variety EC 69874 and sunrise selection exhibited lower damage.

Chemical Method

The use of nematicides are quite effective and cannot be replaced by any other method of nematod control. But, the continuous use of nematicides pose the residual toxicities in plant soil and caus environmental pollution. Seed pelleting, seed soaking or foliar sprays have greater scope for managin nematodes and maximising yield without affecting environment and human health. Therefore, efficier strategies are required for the management of nematode by judicious use of nematicides so that ther is maximisation of advantages and minimization of disadvantages. Nematicides can be used befor the infection taken place (Prophylactic measures) or the infection is suspected to cause damage (curativ measures). Prophylactic measures include seed soaking, seed pelleting and foliar sprays.

Prasad and Gill (1986) soaked seeds of groundnut in aqueous solutions of metasysto monocrotophos, zolone, carbosulfan, triazophos, oxamy, benfuracarb and phenamiphos @ 125, 25 and 500 ppm for 12 hours. Among the different concentrations of the chemicals 500 ppm was found t be most effective in reducing the juveniles penetration by *M. arenaria*. Phenemphos at lower dose wa most effective in reducing gall index and enhancing groundnut growth. Sakhuja (1984) reported th seed soaking in phenamiphos, aldicarb, oxayml or carbofuran at 1.5 a.i./ha gave reasonably goo control of *M. javanica* and increased the growth of groundnut plant significantly over inoculate check. Seed dressing with isophenophos and carbofuran @ 2 and 4 per cent w/w exhibite phytotoxicity. Both the chemical at lower dose provided sufficient initial protection of growing seedling of groundnut.

Prasad and Chawla (1992) reported that pre-sowing application of benfuracarb and basudin o soybean @ 1, 2, 3, 4 and 5 kg a.i./ha resulted better growth in lower doses of both the chemicals. Out c these two chemicals, basudin was found to be quite effective at 2 kg which was at par with 3 kg c benfuracrab application.

Amarnatha and Krishnappa (1992) used carbofuran, ethophrop or phorate @ 2 to 6 a.i./ha i sunflower against *M. incognita* and found that ethoprop @ 5 kg a.i. applied 7 days before nematod ıoculation gave greatest reduction in nematode development and reproduction. However, growth of ınflower plants was highest with phorate @ 6 kg a.i./ha.

ıtegrated Nematode Management

An integrated strategy for crop pest management includes use of resistant varieties, modifying gronomic practices to evade and reduce pest incidence, biological control and other noval approaches or pest suppression and only need based and judicious use of chemical pesticides. Integrated nematode ıanagement emphasizes the need for simpler and ecologically safer measures for pest control to educe environmental pollution and other problems caused by excessive and indiscriminate use of esticides.

Integration of aldicarb, neem cake and soobabul leaves under field conditions resulted better rowth and higher yield of soybean (Prasad and Khan, 1990). All three combinations resulted in aximum yield of soybean besides maintaining low equilibrum nematode population, maximum eduction of nematodes being at 15 days of sampling.

Dwivedi and Upadhyaya (1992) evaluated carbofuran as seed treatment together with soil eatment of carbofuran, phorate or ethoprophos for the control of *M. javanica* on groundnut in field ial in Uttar Pradesh. Carbofuran as seed treatment combined with soil treatment of phorate at kg/ha gave best control of *M. javanica* and increase in yield.

Combinations of chopped madar leaves + 2 per cent carbosulfan seed treatment + triazophos gnificantly lowered the populations of *Meloidogyne arenaria and Rotylenchulus reniformis.* Data on ield of ground nut significantly increased however, maximum yield was observed in only madar aves treatment. Moreover, when madar leaves was coupled with trinzophos, the yield was lowest tributing to antagonistic effect of madar with triazophos. (Prasad *et al.*, 1997) (Table 40.4).

Table 40.4: Effect of Madar Leaves (*Calotropis procera*) Seed Treatment with Carbosulfan and Triazophos Spray on Yield of Groundnut

Treatment	*Yield/5 Plant (g)*	*Per cent Increase Over Control*
Chopped Madar leaves 109/ha (T_1)	47.0	59.0
Seed Tr. with Carbosulfan 2 per cent w/w T_2	32.0	7.8
Triazophos spray 500 ppm (twice) (T_3)	33.0	11.3
$T_4 = T_1 + T_2$	33.0	13.6
$T_5 = T_1 + T_3$	24.0	–19.4
$T_6 = T_2 + T_3$	40.0	36.5
$T_7 = T_1 + T_2 + T_3$	40.0	37.5
T_8 = Control	29.0	–

uturistic Approach for Management Strategies

India which is a major grower of oilseeds in the world has yet low production in the front. Due to gh content of oil and protein, these crops are more prove to attack by insect pests and diseases cluding nematodes. One among the various factors responsible for low production is the non adoption suitable plant protection measures and slow transfer of technology to the fields. About 10–15 per cent of loss is inflicted by pests in oilseed crops in our country (Jayaraj, 1993). However, there are cases where 40–50 per cent loss have been reported by the root-knot and reniform nematodes (Prasad, 1994). Oilseeds along with pulse crops have a cropped area share of 58 per cent and are cultivated under rainfed conditions. In view of the risks involved in their cultivation the pesticides share of these crops is only 6–7 per cent. These crops do not receive adequate plant protection umbrella as they are mostly cultivated by resource poor farmers. This also led to poor adoption of plant protection measures. Application of nematicides at the right time and at proper season would minimise the infection by the nematodes.

There are few management practices if used at suitable time will certainly enhance the production and productivity of oilseed crops by managing nematode pests.

1. Available cultivars/varieties resistant to the nematodes may be given priority for cultivation in the infested area.
2. Cultural practices like ploughings during summer months or suitable crop rotation have tremendous impact on the nematode population thereby increasing yield.
3. Use of judicious use of nematicides to combat the initial invasion and attack by the endoparasites. Nematicides may be used as seed treatment or seed soaking in aqueous solutionn of nematicides.
4. Integrated nematode management combining biocontrol, resistant varieties and cultural control measures is the most appropriate solution to ensure high oilseed production and pollution free environment. Biotechnological approaches can be highly rewarding in developing biocides and pest and disease resistant.
5. Transfer of improved technology is an integrated part of the research and development programmes. In oilseeds, the levels of adoption of management strategies by farmers has been rather vary low. For instance, in Maharashtra, even the basic input of seed of improved resistant varieties, the coverage was only 3 per cent in ground nut, 35 per cent in sunflower and 40 per cent in sesame. Hence, a high priority has to be given to transfer of technology programme.

References

Amarnatha, B.S. and Krishnappa, K., 1989. Effect of different inoculum levels of *Meloidogyne incognita* on sunflower. *Int. Nematode Network Newsl.*, 6: 9–10.

Amarnatha, B.S. and Krishnappa, K., 1992. Evaluation of certain pesticides for the management of *Meloidogyne incognita* infesting sunflower. *Indian J. Nematol.*, 22: 125–133.

Bhat, A.I. and Krishnappa, K., 1990. Chemical control of *Meloidogyne javanica* infecting groundnut. *Indian J. Nematol.*, 20: 95–100.

Chawla, M.L., Prasad, D. and Singh, S., 1990. Performance of soybean cultivars under concomitant infestation of three nematode and an insect species. *Current Nematol.*, 1: 43–46.

Dwivedi, Kusum and Upadhayay, K.D., 1992. Management of root knot nematode infesting groundnut with some promising nematicides. *Indian J. Nematol.*, 22: 9–10.

Gaur, H.S., 1995. Some ecological considerations in integrated nematode management. In: *Nematode Pest Management: An Appraisal of Ecofriendly Approaches*, (Eds.) Gopal Swamy, D.R. Dasgupta and J.S. Gill. NSI, New Delhi, p. 28–36.

Haque, M.M. and Prasad, D., 1983. Effect of agronomical practices and D.D. application on nematode population and yield of wheat and maize. *Indian J. Nematol.*, 13: 226–229.

Jayaraj, S., 1993. Biopesticides and IPM systems for sustainable oilseed production. (Abstr.) *Nat. Seminar Oilseeds Res. and Development in India: Status and Strategies*, 25 August, Hyderabad, p. 97.

Kaul, V.K., 1977. Pathogenicity and control of root-knot nematode on sunflower by some systemic nematicides. *M.Sc. Thesis*, IARI, New Delhi.

Kumar, B., Yaduraju, N.T. Ahuja, K.N. and Prasad D., 1993. Effect of soil solarization on weeds and nematodes under tropical Indian conditions. *Weed Res.*, 33: 423–429.

Prasad, D., 1989) Assessment of yield loss in groundnut due to ectoparasitic nematodes. *Adv. Biosci.*, 8: 175–179.

Prasad, D., 1989. Pathogenicity and screening of groundnut cultivars against *Meloidogyne arenaria. Pak J. Nematol.*, 7: 97–101.

Prasad, D. and Chawla, M.L., 1992. Presowing nematicidal treatment for the management of root-knot nematode on soybean. *Current Nematol.*, 3: 201–202.

Prasad, D. and Gill, J.S., 1986. Seed soaking–chemotherapy against root-knot nematode infesting groundnut. In: *(Abstr) Nat. Cont. Short Term-Long Term Hazards of Pesticides and Strategies for their Safe Use*, 24–26 February, INSA, New Delhi, p. 82.

Prasad, D., Nagia, D.K., Kumar, S. and Saini, M.L., 1994. Integrated approach for management of plant parasitic nematodes in brinjal. *Pl. Protec. Bull.*, 46: 31–33.

Prasad, D., Mittal, A. and Yaduraju, N.T., 1995. Effect of deep ploughing, solarisation and triazophos singly and in combination to the field soil for the management of plant parasitic nematodes. In: *(Abstr) Nat. Symp. Perspective* in *Ecofriendly Approaches to Pl. Protec.*, 7–8 September, SPPS, New Delhi, p. 75.

Prasad, D. and Rangappa, T., 1994. Distribution and estimation of avoidable loss due to nematodes associated with Kalahasti Malady of groundnut. *Ann. Pl. Protec. Sci.*, 2: 37–42.

Prasad, D., Dubey, K.N. and Mittal, A., 1997. An integrated approach for management of plant parasitic nematodes in groundnut. *Ann. Pl. Protec. Sci.*, 5: 217–219.

Raut, S.P. and Sethi, C.L., 1980. Studies on the pathogenicity of *Meloidogyne incognita* on soybean. *Indian J. Nematol.*, 10: 166–174.

Sakhuja, P.K., 1984. Studies on the inter relationship between root-knot nematode and root rot fungi on groundnut. *Ph.D. Thesis*, IARI, New Delhi.

Plant Protection, Diversity and Conservation (2011) *Pages* **532–548**
Editors: **D. Prasad & Satya Kumar**
Published by: **BIOTECH BOOKS, NEW DELHI**

Chapter 41
Pests of Jute and their Management

T. Ramasubramanian* and S. Satpathy
Division of Crop Protection,
Central Research Institute for Jute and Allied Fibres (ICAR),
Barrackpore, Kolkata – 700 120, W.B.

Jute, "the golden fibre of India" though occupies only 0.47 per cent of the gross cropped area, provides livelihood to more than 4 million farm families. The production of raw jute has increased from mere 3.3 million bales during 1950-51 to 11.3 million bales during 2006-07. The productivity of raw jute has become doubled from nearly 10 q/ha during 1950-51 to more than 22 q/ha during 2006-07. However, the area under jute has been dwindling around 0.8 million hectares over the six decades. It may be difficult to break this plateau, but concerted efforts may be taken to enhance the productivity. There are several constraints in increasing the productivity of jute, of which the loss due to insect pests is of major concern and hence, the Jute Entomology deserves more attention for economic production of the crop (Ramasubramanian and Ghosh, 2008).

Pest Management in Jute

After partition of the country in 1947, a large area was brought under the cultivation of jute and other bast fibre crops to meet the demand of large number of mills that would have remained idle otherwise (Tripathi and Ram, 1971a). However, the intensive cultivation of high yielding, fertilizer responsive cultivars of jute brought forth the problem of insect pests and mites. A few relatively less important pests turned to be major ones. Bihar hairy caterpillar, *Spilosoma obliqua* Walker that was once considered as a sporadic pest on jute (Dutt, 1958) is now a major threat to jute crop during later part of the season every year. Semilooper, *Anomis sabulifera* Guenee, stem-weevil, *Apion corchori* Marshall, yellow mite, *Polyphagotarsonemus latus* (Banks), and the indigo caterpillar, *Spodoptera exigua* Hübner are the major pests of jute (Das *et al*., 1999). The jute girdler, *Nupserha bicolor postbrunnea* Dutt, scale insect, *Pinnaspis* sp., thrips, *Ayyaria chaetophora* Karny, mealy bugs, *Ferrisia virgata* Ckll. and *Pseudococcus*

---

* E-mail: tramasubbu@rediffmail.com

filamentosus var. *corymbatus* are the minor pests of jute (Tripathi and Ram, 1971a). Grey weevil *Myllocerus discolor* Boheman was first time recorded as a pest on *tossa* (*Corchorus olitorius* L.) jute in 1973 (Das and Ghosh, 1973). *Tossa* jute occupies 80 per cent of the jute growing area as against 20 percent by the *white* jute, *Corchorus capsularis* (Saha, 2000), but unfortunately the incidence of major pests except *A. corchori* is more on *C. olitorius* than on *C. capsularis*.

Major Pests

Jute Semilooper, *Anomis sabulifera* Guenee (Noctuidae: Lepidoptera)

Biology

Lefroy (1907) recorded *A. sabulifera* (Figure 41.1) as a major pest of jute. It is one of the most important foliage pests of the crop (Tripathi, 1967a; Tripathi and Ghosh, 1964; Das and Singh, 1976; Singh and Das, 1979; Das *et al.*, 1995) and occurs in all the jute growing tracts of India. The crop loss due to this pest was estimated at up to 50 per cent (Dutt, 1958). The eggs are laid on both the surfaces of leaf and preferably on the lower surface. About 51 percent of the eggs are laid between the growing point and the fourth leaf, and 32 percent between fifth and eighth leaf (Tripathi, 1963). The larval period ranges from 9-16 days and the third instar larvae were observed to be the most destructive (Tripathi and Ram, 1971a).

Nature of Damage

High forenoon relative humidity for 15 days followed by bright sunshine during daytime and drizzling during night hours were observed to be more conducive for the outbreak of the semilooper. Damage starts in all cases from unopened leaves or buds, which represent the most susceptible portion. It was observed that, 81 per cent of the damage was limited to seven fully opened leaves from top and up to 95 per cent down to the ninth leaf (Dutt, 1958). Repeated damage by this pest checks crop growth and induces profuse branching, resulting in reduction of fibre yield (Tripathi and Bhattacharya, 1963). Pods and unripe seeds are also being damaged by *A. sabulifera*. The *C. olitorius* pods are more susceptible than the *C. capsularis* pods. (Singh and Das, 1979). Although the damage caused by *A. sabulifera* to the jute crop was recognized as early as 1954 in Hooghly district of West Bengal (Dutt, 1958), cultivators are realizing the magnitude of loss due to such ravages only recently. Hence, prevention of huge crop loss as often occurs due to semilooper attack deserves the utmost consideration.

Management

The higher dose of fertilizers makes the jute plant more susceptible to semilooper attack both in *white* (JRC 212, JRC 321, JRC 7447) and *tossa* jute varieties (JRO 7835, JRO 524 and JRO 878) (Das and Singh, 1985c, 1986b). Among the limited number of *white* jute cultivars screened for resistance to semilooper, NDC-8812 and NDC-9801 were reported to be moderately resistant (Prasad *et al.*, 2004).

Sisiropa formosa, Trichologa sorbilans, Litomastix sp., Mesochorus sp. were recorded as parasitoids on semilooper (Das *et al.*, 1999). As the mass multiplication technique has not yet been perfected, these parasitoids are not being utilized for applied biological control to suppress the semilooper population. *Bacillus thuringiensis* Berliner was reported to be effective against the semilooper, *A. sabulifera*. Complete dissolution of midgut epithelial cells was observed 12 hours after the ingestion of *B. thuringiensis* spores (Das and Singh, 1976).

The use of plant products for the control of crop pests is highly encouraged to reduce the environmental contamination of chemical insecticides. In this endeavour, insecticidal property of herniarin, a constituent of *Herniaria glabra* (Family: Illecebraceae), was tested against the last instar

Figure 41.1

larvae of *A. sabulifera*. Heavy knockdown effect was observed immediately after spraying (1.0 per cent - 0.1 per cent) and cent percent mortality of larvae was observed within 24 hours of treatment (Mallick and Banerji, 1989b). The compounds of plant origin such as didymocarpin, isodidymocarpin, pedicellin (extracted from *Didymocarpus pedicellata*), and aurentiacin (obtained from *Didymocarpus aurentiaca*) were found to exhibit insecticidal property against the semilooper larvae. The LC_{50} values of aurentiacin, didymocarpin, isodidymocarpin and pedicellin were 0.16, 0.20, 0.17 and 0.18, respectively (Chatterji *et al.*, 1980). The methanol extract of *Ocimum sanctum* Linn. showed promising antifeedant effect (5 per cent and 10 per cent). Though the methanol extract of the *O. sanctum* did not exhibit any insecticidal property, it showed excellent repellent action against the larvae of semilooper (Mallick and Banerji, 1989a). Ethyl acetate extract of *Swertia chirata* (10 per cent) showed promising antifeedant effect against the semilooper (Mallick *et al.*, 1985). Sisalina oil exhibited heavy knockdown effect (at 0.1 and 1.0 per cent) as well as insecticidal property (100 per cent mortality at 0.07 per cent concentration in 48 hours) against the jute semilooper (Mallick, 1989b).

Insecticidal interference may be avoided up to 15 percent of infestation by the semilooper on JRO 524 (Chatterjee *et al.*, 1988). However, control measures need to be initiated as soon as the semilooper infestation reaches 15 percent level. Determination of suitable size of sampling unit is a prerequisite for proper estimation of semilooper population. Row length of 1.5-2m was determined as the suitable size for getting the most reliable estimates (Das and Basak, 1995). The most suitable direction for sampling was observed to be south for semilooper (Das, 1995b).

Endosulfan (0.07 per cent) has been recommended for the management of the semilooper since 1969 (Mookerjee and Chatterjee, 1969). Carbaryl and fenvalerate were found to be superior in managing the semilooper on the basis of their median lethal concentration (Chatterji and Das, 1983; Das, 1985). Later cypermethrin (0.03 per cent) was reported to be the most effective insecticide against the semilooper as compared with the standard check endosulfan (Das *et al.*, 1995).

Jute Stem Weevil, *Apion corchori* Marshall (Curculianidae: Coleoptera)

Biology

The jute stem weevil, *Apion corchori* Marshall occurs in all the jute growing tracts of India. The adult female lays eggs singly in holes gnawed by it on the stem at the bases of apical petioles. Up to 675 eggs are laid by a female during the oviposition period of around 124 days. The rate of oviposition is reduced towards the end of the cropping season and ceases in December. Egg, larval and pupal periods last for 3, 8-18 and 4 days, respectively. The grub makes tunnel within the stem, feeds on the internal tissues and pupates within. Adults survive up to 209 days (Das, 1944).

Nature of Damage

Though the stem weevil is a pest of both the species of *Corchorus, C. capsularis* suffers more than *C. olitorius*. The insect exists throughout the cropping season and it damages the early sown crop sometimes to such an extent that hardly any fibre can be obtained. *C. capsularis* is more susceptible to stem weevil infestation starting from seedling stage to harvest. The *C. olitorius* plant suffers little due to apion possibly because of its higher tannin content (Tripathy and Ram, 1971a). The yield loss due to stem weevil was estimated as about 18 per cent on JRC 212 (Tripathi and Bhattacharya, 1968). There may be one or more punctures by the female at the nodes, and there are corresponding numbers of grubs in the plant. Nodal region below leaf base is the spot most susceptible to damage (Tripathi and Ram, 1971a). The feeding of tissues by grubs affects the growth of the plant, and where the growing point is lost, branching is induced (Figure 41.2). Loss of apical growing point checks the vertical growth and thus affects the fibre both in terms of quality and quantity. It affects more the quality of fibre than it does with the yield. In older plants mucilaginous substances produced around the tissues damaged by the larva bind the fibres together; the fibres break at these points during processing and result in 'knotty fibre'. Of the wild species of jute, *Corchorus acutangulus* was found to be very much susceptible to attack and *Corchorus trilocularis* and *Corchorus fascicularis* were also susceptible, but to a lesser extent.

Management

No germplasm was found to be immune to stem weevil (Pradhan, 1995), but the *tossa* jute variety JRO 878 was found to be less susceptible (Das and Singh, 1985c, 1986b). Among the *white* jute cultivars, NDC-8812 and NDC-9801 were reported to be resistant to *A. corchori* (Prasad *et al.*, 2004). Both *tossa* and *white* jute varieties sown in the late April had shown remarkably less incidence of stem weevil as compared to those sown in late March or early April (Das and Singh, 1985c). Application of phosphatic and potassic fertilizers reduces the attack, whereas nitrogenous fertilizers increase the incidence of

Figure 41.2

the pest. It is recommended to remove and destroy the stubbles and self sown plants in order to avoid the carryover of the pest. Quick steeping of harvested jute and stacking it for not more than four days will also reduce the incidence of this pest.

Entedon sp. (Eulophidae: Hymenoptera), *Eupelmus urozonus* Dalm. (Eupelmidae: Hymenoptera), *Bracon greeni* Ashm. (Braconidae: Hymenoptera), *Lymaenon* sp. (Mymaridae: Hymenoptera) and *Psilopa* sp. (Ephydridae: Diptera) were recoded as parasitoids on *A. corchori* grub. *Entedon manilensis* Crawford was reported to be the most efficient among the parasitoids and the per cent parasitization often exceeded 50 per cent during the cropping season (Dutta and Ghosh, 1964-65).

The adults feed very little on the apical leaves and rest of the life stages are passed well hidden inside the stem; as a result the *A. corchori* always escapes from the direct contact of the insecticides. Endosulfan (0.075 per cent) followed by carbofuran (1 kg a.i ha^{-1}) were effective in managing the stem weevil on *white* jute (Das and Singh, 1986a). The synthetic pyrethroid, cypermethrin was found superior to endosulfan in managing the stem weevil (Das *et al.*, 1995). Generally, the infestation is more in the periphery of the field than in the core of the plot and hence the insecticidal spray may be concentrated in the periphery of the field.

Yellow Mite, *Polyphagotarsonemus latus* Banks (Tarsonemidae: Acarina)

Biology

Yellow mite, *Polyphagotarsonemus latus* Banks has established itself as a serious pest of jute since the forties. It is now widely distributed in all the jute growing tracts of India (Nair, 1986). The newly hatched adults of both the sexes are *white* in colour, which gradually changes to yellow after the expiry of 24 hours. With the ageing of the mites the colour becomes deeper. The females are larger than males and their average life-span is about one week. The average life span of males is 6 days. Just after hatching male yellow mite moves very slowly for 2-3 hours and sucks juice from the ventral surface of leaf. After that the male starts moving over the entire leaf lamina in search of quiescent female. As soon as male gets a quiescent female, it catches that female with the sucker and lifts the female vertically up and moves very quickly for safer places. One male can carry one quiescent female at a time. When the adult female comes out from the quiescent stage, the male still carries the adult for some more time. Then the male mates with the young female and the mating period lasts for 20 seconds. Just after mating the female rests for sometime before it lays eggs in a suitable place. For this purpose the females prefer the young leaves which remain more or less like buds. The female lays eggs singly on the ventral surface of leaves in between veinlets. Approximately 23 eggs are laid throughout the life period. The egg stage lasts for 30 hours. The active nymph of yellow mite is crystal white in colour and three pairs of legs mark the nymphal stage, but during adult stage, the fourth pair of legs is developed. The nymphal period in case of female is on an average 24 hours, whereas in case of male the same is 18 hours. In the inactive quiescent stage, the nymphs are boat-shaped and crystal white in colour. This quiescent stage has also been termed as 'pupa'. The life cycle is usually completed in about 5 days. The mite has 14 generations in a season on jute crop and sometimes one overlaps the other. Due to number of generations the population build-up is faster; consequently the incidence is also noticeable to a greater extent (Das and Singh, 1988).

Nature of Damage

P. latus is one of the important destructive pests of jute (Das and Roychaudhuri, 1979; Das and Singh, 1985a, 1985b; Nair, 1986; Pradhan and Saha, 1997). Both nymphs and adults suck the sap from younger leaves. Due to sucking of sap from the ventral surface of the young leaves, leaf curls ventrally (Figure 41.3) (Das and Singh, 1988). As a result, the leaves do not grow to their full size; they turn coppery-brown and drop prematurely. The vertical vegetative growth of the crop is arrested, internodes become shortened and significant yield loss occurs regularly (Das *et al.*, 1987). The *tossa* jute suffers more due to yellow mite infestation as compared with *white* jute (Das, 1989a). Under severe damage the internodal length is reduced by two folds. In healthy plants the average internodal length was recorded as 6 cm whereas in the yellow mite infested plants the average internodal length was 3.4 cm only. The yield loss due to this pest has been estimated between 10 and 42 per cent depending upon the level of infestation between 20 and 90 per cent, respectively. The quality of the fibre was also found to be deteriorated due to the attack of yellow mite. The fibre strength was reduced to 12.13 g tex^{-1} in the infested plants as against 15.43 g tex^{-1} in the healthy plants. Though it attacks the crop during mid-June to mid-August, cloudy weather with high humidity and high morning temperature were found congenial for multiplication of yellow mite in the field (Pandit *et al.*, 2002).

Management

JRO 524, JRO 7835 and JRC 212 were reported to be least susceptible to this pest (Das *et al.*, 1999). JRO 878 was reported to be highly susceptible to yellow mite (Das *et al.*, 1987). Crops sown during March and April are liable to more damage than the late sown crops. Whenever the population crosses the ETL of 110 mites per leaf (preferably in the second unfolded leaf) (Das *et al.*, 2002), insecticidal interference is required. Two rounds of spray with endosulfan (0.04 per cent) at fortnightly intervals were reported to be most effective and economical in managing the yellow mite infestation on jute (Das and Singh, 1985a, 1985b).

Grey Weevil, *Myllocerus discolor* Boheman (Curculianidae: Coleoptera)

Biology

Myllocerus discolor Boheman causes considerable damage to *tossa* jute by defoliating the crop. This was first recorded as a pest of *C. olitorius* by Das and Ghosh (1973). Fecundity of *M. discolor* is 200

Figure 41.3

eggs during the oviposition period of 20-25 days. Egg, larval and pupal periods varied from 3-4, 20-55 and 6-8 days, respectively. The adult's longevity is about two months. The weevil was found to hibernate in the larval stage during winter (Mallick, 1985).

Nature of Damage

Unopened tender apical leaves are the first choice of attack by *M. discolor* (Mallick, 1986). Irregular defoliation is the typical symptom of damage due to grey weevil attack (Figure 41.4) One or two pre-monsoon showers, temperature range of 35-39°C and relative humidity of 85-94 per cent were observed to be conducive for the maximum incidence of the grey weevil (Mallick, 1989a).

Management

No variety was found resistant to grey weevil. JRO 7835 was observed to be the most susceptible variety (Mallick, 1995a). Sample size comprising 10-20 percent of the population can give reliable estimates for the infestation by major pests of jute. For grey weevil, however, a larger sample size of 25-30 per cent population was observed to be the most suitable (Das, 1995a). South was found to give most reliable estimates for all the pests except grey weevil for which, east was found the most reliable (Das, 1995a, 1995b). Two rounds of cypermethrin (0.03 per cent) at 15 days interval was found the most effective followed by oxydemeton methyl (0.05 per cent), carbaryl (0.1 per cent) and endosulfan (0.075 per cent) (Mallick, 1989a, 1995b).

Bihar Hairy Caterpillar, *Spilosoma obliqua* Walker (Arctiidae: Lepidoptera)

Biology and Nature of Damage

Spilosoma obliqua Walker is a polyphagous pest attacking several crops including the bast fibre crops jute and mesta. However, jute was found to be the preferred host than mesta; further the *tossa* jute was somewhat tolerant to the hairy caterpillar as compared to *white* jute (Pandit, 1985). It was once considered as a sporadic and irregular pest on jute, but now-a-days it is a regular and becomes a major pest on jute causing significant loss to the jute crop. The female moths lay eggs in groups and a single female may lay up to 1000 eggs. Upon hatching the larvae are gregarious (Figure 41.5) and scrap the chlorophyll content of the leaves. Later they disperse to entire field and prefer to defoliate the older leaves. The larval period ranges from two to three weeks and pupation occurs in the soil.

Management

In the early stage, when the caterpillars remain gregarious on leaf, it is easy to destroy them after plucking such infested leaves and then dipping them in kerosenised water or by burning them (Tripathi and Ram, 1971a). When caterpillars dispersed, their control is achieved by insecticidal spraying. Cypermethrin has been proved to be most toxic to the hairy caterpillar (Das, 1989b). Endosulfan was also reported to be effective in reducing the incidence of hairy caterpillar under field condition (Hath and Chakraborty, 2004). Ploughing the field after the harvest of the crop may expose the larvae to the hot sun. *Apanteles obliquae, Protapanteles obliquae* and *Eurytoma* sp. were recorded to parasitise the larvae of *S. obliqua* (Dutta, 1964; Ramasubramanian, 2008).

Indigo Caterpillar, *Spodoptera exigua* Hübner (Noctuidae: Lepidoptera)

Spodoptera exigua Hübner that was once considered as a minor pest on jute with a maximum yield loss of 20 percent (Dutt, 1958) is now become a major pest responsible for entire crop failure. Resowing is hence, warranted very often to maintain the desired plant population in the field. It is an important pest of seedling jute and occurs in all the jute growing tracts of India. The eggs are laid on the under surface of the leaves and covered with buff-coloured hairs. A single female lays more than 1000 eggs

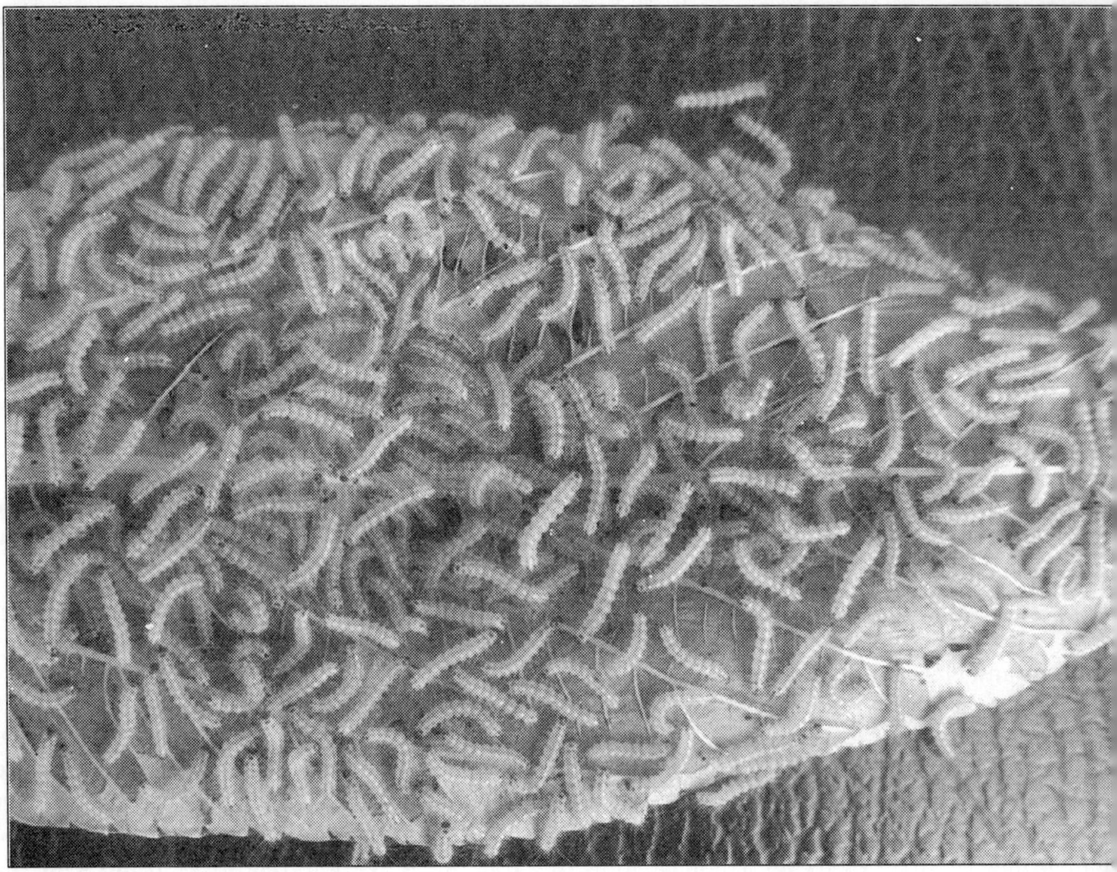

Figure 41.4

in 2-11 days. The first instar caterpillars are gregarious in habit. Egg, larval and pupal stages last 2 13-16 and 5-7 days, respectively. The *C. capsularis* jute grown early is more liable to damage by this pe than the late sown *C. olitorius* plant is. This pest can be controlled with insecticides like endosulfa (0.075 per cent).

Minor Pests

Red Mite, *Oligonychus coffeae* Nietner (Tetranychidae: Acarina)

Das (1948) and Tripathi (1967b) have described this mite as a pest of jute. It is more common o *white* jute. The mite lives under web and suck the sap from upper surface of the leaves. Its incidence i sporadic in nature, but occasionally it causes serious damage to the crop. The affected leaves becom leathery, turn yellow and drop off prematurely. Intermittent shower followed by dry spell with hig humidity are quite favourable for the incidence of the mite. Crop sown during March and April ar likely to be more affected because during the following two months, conditions are generally conduciv for multiplication of the mite (Tripathi and Ram, 1971a). Dicofol (0.04 per cent) is effective in managin the red spider mite on jute (Tripathi and Ram, 1970; Das *et al.*, 1999; Das, 2000). *Physalis minima* i serving as an alternate host to red mite (Das, 1980) in the jute ecosystem and hence, weed free conditio

Figure 41.5

needs to be maintained to avoid carryover of the pest. Among the *white* jute varieties, Tripura *capsularis* was found to be least susceptible to red mite (Das *et al.*, 1999).

Jute Girdler, *Nupserha bicolor postbrunnea* Dutta (Cerambycidae: Coleoptera)

This pest was first recorded in 1949 on *tossa* jute (Nair, 1986). This pest was observed only on *tossa* jute, but *white* jute varieties are resistant to the jute girdler. Among the *tossa* jute varieties, 'Chinsurah green' showed highest degree of susceptibility. Sufficient rainfall during early crop season is conducive for early emergence of the pest. Unlike most other Cerambycidae, the larva practically causes no damage to the host plant. In *olitorius* jute pith disintegrates early, resulting in the formation of hollow central region in the stem. On hatching, the borer larva moves down along the central hollow and feeds upon the remnant of the pith tissue. Such feeding has no apparent adverse effect on the growth of the plant. The larva neither travels to the side branches, if any, nor makes any tunnels to cause damage to the fibre. The female cuts the cortex of the stem at two points by its mandibles, making two circular rings (girdles) about one cm apart around the stem and thrust the egg in the pith of the stem between the girdles. The girdling is restricted to that portion of the stem, which falls within the stem diameter of 0.2 - 0.4cm. The girdled plant withers, droops and finally dies above the level of lower girdle. The loss of stem varies from 12.51- 17.11cm on an average in different height groups, while 101-200 cm height group seems to suffer most by this pest. Oviposition period extends about a month during which a female lays about 60 eggs singly. Larval period varied between 30 and 50 days and the mature larva enters in to diapause. The larva constructs a chamber within the stem and makes cuts on either side of it. This larval chamber falls down to the soil to undergo the diapause. It has been observed that the larva can survive under soil for more than a year (Dutta, 1952b, 1952c, 1954, 1956, 1961, 1964). High humidity is necessary to break the diapause, and at 97 percent humidity 70 per cent of the diapaused larvae emerge as adult (Dutta, 1963). *Neocatolaccus nupserhae* Dutta and Ferr. (Pteromalidae), *Goryphus* sp. (Ichneumonidae), *Ipobracon* sp. (Braconidae), *Spathius* sp. (Braconidae), *Epibracon* (Braconidae) and *Norbanus acuminatus* Dutta and Ferr. (Pteromalidae) are the hymenopteran parasitoids associated with this pest (Dutta and Ferriere, 1961). Among these parasitoids, *N. nupserhae* and *N. acuminatus* are the most important. Nowadays the incidence and the extent of damage are very insignificant. Collection and destruction of drooping shoots and casings containing the diapausing larvae may reduce the further spread of the pest. Any contact insecticide may take care of the pest population.

Mealy Bugs

Ferrisia virgata Ckll (Pseudococcidae: Hemiptera)

This species is widely distributed in tropical countries including most parts of India. The damage is caused mostly by the nymphs, which suck the sap from leaf, petiole, stem and pod. Repeated attacks on the stem cause the development of crust due to which fibre bundles resist separation at the time of retting, resulting in the formation of barky fibre. More damage is caused to seed crop, which does not put forth more number of pods (Kundu *et al.*, 1959; Tripathi and Ram, 1971a).

Pseudococcus filamentosus var. *corymbatus* (Pseudococcidae: Hemiptera)

It is another species of mealy bug causing serious damage to *capsularis* and *olitorius* jute. The infestation by this pest arrests the vertical growth of the stem and the attacked region is characterized by its deep green colour, swelling and shortened internodes. The leaves arising from the closely opposed nodes give the attacked region a bushy appearance. The first instar and early second instar nymphs mainly induce this kind of malformation. Apical meristem is the most susceptible portion of the crop (Kundu *et al.*, 1959; Tripathi and Ram, 1971a).

Thrips, *Ayyaria chaetophora* Karny (Thripidae: Thysanoptera)

Tossa jute particularly in seedling stage was found to be infested by thrips, whereas *white* jute is resistant to thrips attack. This pest in general avoids light and remains within the folds of apical buds. They macerate the leaf tissues in between veins and when an infested bud begins to unfold, the pest moves up to the next bud and thus macerates all the buds in regular succession. Soon the macerated tissues die and become prominent as white streaks in unfolded leaves (Tripathi and Ram, 1971a).

Integrated Pest Management for Jute

The following strategies may be considered for inclusion in the IPM programme after detailed investigation in the research institutes (Ramasubramanian and Ghosh, 2008):

- ☆ The tolerant/least susceptible varieties such as JRO 524 (Navin) and JRC 212 (Basudev) may be recommended for cultivation in the yellow mite endemic areas (Das, 2000). Cultivation of JRO 7835 may be discouraged in grey weevil endemic areas.
- ☆ Studies on the seasonal incidence and population dynamics of major pests of jute revealed that the age of the crop has very little influence on the incidence pattern of pests with the exception of grey weevil. The intensity of attack however, was observed to be influenced by the date of sowing. The intensity of damage by stem weevil, grey weevil and yellow mite were more on early sown crops, but the reverse was true with respect to semilooper incidence (Das, 1995a). The optimal date of sowing is the 4th week of March for JRO 524, whereas 2nd week of April was reported to be optimum for other varieties in order to minimize the pest incidence (Das *et al.*, 1999).
- ☆ *Physalis minima* was reported as an alternate host for yellow mite (Das and Roychaudhuri, 1979) and hence, the field should be maintained weed free to avoid the carryover of the pest.
- ☆ Negative impact on the gut physiology of the semilooper larvae was observed due to the ingestion of *Bacillus thuringiensis* Berliner spores (Das and Singh, 1976). Hence *Bt* commercial formulations may be incorporated as a component of Jute IPM after detailed investigations in the field. This may also promote the organic farming of jute.
- ☆ Neem oil @ 4ml l^{-1} was reported to be effective against the jute semilooper, *A. sabulifera* (Das, 2000). This may be recommended as one of the important components in the IPM practices.
- ☆ Prior to spray insecticides, it is suggested to pluck the leaves with early stage gregarious larvae of Bihar hairy caterpillar and kill them by dipping in the insecticide solution.
- ☆ Carbaryl @ 0.1 per cent (1kg a.i. ha^{-1}) may be recommended to manage the semilooper, *A. sabulifera* (Source: Central Insecticide Board and Registration Committee; http://cibrc.nic.in/).
- ☆ Endosulfan @ 0.075 per cent may be applied to manage the major pests of jute with the exception of yellow mite for which, endosulfan @0.035 per cent may be recommended as soon as the infestation reaches 10 per cent level (Das, 2000). Cypermethrin (0.03 per cent) was found on par with endosulfan (0.075 per cent) against the jute stem weevil and semilooper (Das *et al.*, 1995). Moreover, residues of cypermethrin (0.03 per cent) reduced over 90 percent within 10 days after spray and no residues could be detected after 20 days (Das and Pradhan, 1995).
- ☆ The incidence gradually declines from border to inner rows in case of yellow mite and grey weevil with reverse trend in case of semilooper and no such trend with respect to jute stem weevil (Das, 1995b). Hence, care should be taken accordingly to get reliable population estimates as well as spraying insecticides.

Future Strategies

Need for Resistance Risk Assessment

Endosulfan was reported to be more effective against all the major pests of jute (Das and Singh, 1986a). It has been in use for more than four decades in the jute growing tracts of India. The possibility of development of resistance to endosulfan due to several decades of exposure cannot be ruled out. A detailed investigation on status of susceptibility of the jute pests to endosulfan from different parts of the India is required to prevent/delay the development of resistance, if any, to endosulfan. Moreover, the exposure of the jute pests to the new chemistry in jute growing belts of West Bengal as well as other parts of the Indian subcontinent is nil or at least minimal, and hence, it is the ideal time to get the susceptible population for generating baseline susceptibility to insecticides with novel modes of action. For effective pest management, detection and continuous monitoring of insecticide resistance are essential. This will also help to phase out the insecticides for which resistance has been built up, and thus the expenditure on ineffective insecticides by the resource poor jute farmers may be minimized. (Ramasubramanian, 2007).

Need for Transgenic Jute

Lack of resistant varieties as well as hidden nature of the stem weevil, which precludes the contact from insecticides, necessitates genetic engineering of jute with coleopteran specific *cry* genes. Moreover, the emerging threat from *A. sabulifera* and *S. exigua* also warrants genetic engineering with lepidopteran specific *cry* genes. Hence, it may be better to go for pyramiding both lepidopteran and coleopteran specific *cry* genes to contain the colossal loss from major pests (Ramasubramanian and Ghosh, 2008).

Need to Standardize the Artificial Diet

Mass rearing of jute pests over several generations is a pre-requisite for the studies related to insecticide toxicology. Since the field populations are heterogeneous in nature, the F_1 generation is usually preferred for the bioassay of insecticides. Moreover, the field population of *A. sabulifera, S. exigua* and *S. obliqua* might be naturally infected with entomopathogens and hence the rearing of lepidopteran pests on natural host may lead to complete wipe out of the population in the middle of the experiments. Hence it is inevitable to standardize the artificial diet for *A. sabulifera, S. exigua* and *S. obliqua*. Mass culturing of *A. sabulifera* on artificial diet had already been attempted by some workers (Pandit, 201). However, it is yet to be perfected for continuous culturing of the population over several generations (Ramasubramanian, 2007).

References

Chatterji, S.M., and Das, L.K., 1983. Relative toxicity of some insecticides to the larvae of semilooper, *Anomois sabulifera* Guen. *J. Entomol. Res.*, 7: 80–81.

Chatterji, S.M., Das, L.K. and Singh, B., 1988. Effect of different degrees of infestation by jute semilooper, *Anomis sabulifera* Guen. on fibre yield of jute crop. *Sci. Cult.*, 52: 136–137.

Chatterji, S.M., Das, L.K., Mallick, R.N., Aditya Chowdhuri, N. and Deb Kirtaniya, S., 1980. Relative contact toxicity of some botanical insecticides to the last instar larvae of jute semilooper, *Anomois sabulifera* Guen. *J. Entomol. Res.*, 3: 223–224.

Das, G.M., 1944. Studies on the jute stem weevil *Apion corchori* Marshall I: Bionomics and life history. *Indian J. Agric. Sci.*, 14: 295–303.

Das, G.M., 1948. Insects and mite pests of jute. *Sci. Cult.*, 14: 186–190.

Das, L.K., 1985. Relative toxicity of some insecticides to the larvae of jute semilooper *Anomis sabulifera* Guen. *Sci. Cult.*, 51: 423–424.

Das, B.B., 1989a. Studies on the seasonal incidence and population dynamics of major insect pests of jute. In: *Annual Report: 1987–1988 and 1988–1989*. Central Research Institute for Jute and Allied Fibres, Barrackpore, p. 93–94.

Das, L.K., 1989b. Relative toxicity of newer insecticides against jute pests under laboratory condition. In: *Annual Report: 1987–1988 and 1988–1989*. Central Research Institute for Jute and Allied Fibres, Barrackpore, p .95.

Das, B.B., 1995a. Studies on the seasonal incidence and population dynamics of major insect pests of jute. In: *Annual Report: 1990–1991 and 1991–1992*. Central Research Institute for Jute and Allied Fibres, Barrackpore, p. 66.

Das, B.B., 1995b. Studies on the seasonal incidence and population dynamics of major insect pests of jute. In: *Annual Report: 1993–1994 and 1994–1995*. Central Research Institute for Jute and Allied Fibres, Barrackpore, p. 47.

Das, L.K., 2000. Integrated management of jute pests. In: *Workshop-cum-Training on Adaptive Research on Improved Varieties of Jute and Allied Fibres and their Utilization for Enhanced Income Generation of farmers*" (Eds.) S. Pathak. Central Research Institute for Jute and Allied Fibres, Barrackpore, pp. 15–16.

Das, B.B. and Basak, S.L., 1995. Size of sampling units for the estimation of jute semilooper incidence. *Indian Biol.*, 27: 59–61.

Das, L.K. and Ghosh, T., 1973. *Myllocerus discolor* Boheman on *Corchorus olitorius*: A new pest. *Jute Bull.*, 36: 199.

Das, L.K. and Pradhan, S.K., 1995. Studies on dissipation of important effective pesticides in/on jute crop. In: *Annual Report: 1990–1991 and 1991–1992*. Central Research Institute for Jute and Allied Fibres, Barrackpore, p. 66–67.

Das, L.K. and Roychaudhuri, D.N., 1979. *Physalis minima* (Solanaceae): A new host plant of yellow mite, *Polyphagotarsonemus latus* (Banks). *Sci. Cult.*, 45: 169–170.

Das, L.K. and Singh, B., 1976. The effect of *Bacillus thuringiensis* Berliner on the gut of the jute semilooper, *Anomis sabulifera* Guen. *Sci. Cult.*, 42: 567–569.

Das, L.K. and Singh, B., 1985a. Effective control measures against yellow mite, *Polyphagotarsonemus latus* (Banks) of jute. *Jute Dev. J.*, p. 28–29.

Das, L.K. and Singh, B., 1985b. Number of sprays suitable against yellow mite, *Polyphagotarsonemus latus* (Banks) of jute. *Sci. Cult.*, 51: 376–377.

Das, L.K. and Singh, B., 1985c. Effect of varieties, sowing dates and fertilizer levels on the incidence of pest complex of *capsularis* and *olitorius* jute. In: *Annual Report: 1985*. Central Research Institute for Jute and Allied Fibres, Barrackpore, pp. 68–69.

Das, L.K. and Singh, B., 1986a. Effective control measures against the pest complex of *capsularis* jute *Corchorus capsularis* L. *Jute Dev. J.*, p. 25–27.

Das, L. K. and Singh, B., 1986b. Effect of varieties, sowing dates and fertilizer levels on the incidence of pest complex of *capsularis* and *olitorius* jute. In: *Annual Report: 1986*. Central Research Institute for Jute and Allied Fibres, Barrackpore, pp. 59–60.

Das, L.K. and Singh, B., 1988. Life history of yellow mite, *Polyphagotarsonemus latus* (Banks) on jute crop. *Ann. Agric. Res.*, 9: 20–25.

Das, L.K., Laha, S.K. and Pandit, N.C., 1999. Entomology. In: *Fifty Years of Research on Jute and Allied Fibres Agriculture"*, Central Research Institute for Jute and Allied Fibres, Barrackpore, pp. 142–164.

Das, L.K., Mitra, G.C. and Roychaudhuri, D.N., 1987. Anatomical changes of jute stem by yellow mite. *Sci. Cult.*, 53: 252–253.

Das, L.K., Rao, P.V., Pradhan, S.K., and Chakraborty, A.K., 2002. Studies on determination of economic threshold level and economic injury level of major pests of jute. In" *Annual Report: 2000–2001 and 2001–2002*. Central Research Institute for Jute and Allied Fibres, Barrackpore, p. 69.

Das, L. K., Singh, B., and Pradhan, S. K., 1995. Efficacy of different synthetic pyrethroids insecticides against pest complex of jute. *Sci. Cult.*, 61: 203–204.

Dutta, N., 1952b. *Nupserha bicolor* Thoms ssp. *postbrunnea* Breun., a new pest on jute (*Corchorus olitorius* Linn. *Nature*, 170: 287–288.

Dutta, N., 1952c. *Pinnaspis* sp. a new coccid pest on *Corchorus capsularis* Linn.. *Curr. Sci.*, 21: 73–74.

Dutta, N., 1954. Diapause in *Nupserha bicolor* Thoms ssp. *postbrunnea* Breun and its bearing on control. *Jute Bull.*, 17: 286–287.

Dutta, N., 1956. Studies on *Nupserha bicolor* Thoms ssp. *postbrunnea* Breun. (Coleoptera: Lamiidae). IV. Preliminary observations on immature stages and elimination of larval diapause. *Jute Bull.*, 18: 254–256.

Dutta, N., 1958. *Anomis sabulifera* Guen and *Apion corchori* Marsh – Incidences and control. *Jute Bull.*, p. 121–128.

Dutta, N., 1961. Studies on bionomics of jute girdler, *Nupserha bicolor postbrunnea* Dutt. *Sci. Cult.*, 29: 302.

Dutta, N., 1963. Water uptake in the diapause mechanism of jute stem girdler, *Nupserha bicolor postbrunnea* (Col., Lamiidae). *Bull. Entomol. Res.*, 51: 765–779.

Dutta, N., 1964. Jute pests. In: *Entomology in India*. Entomological Society of India, New Delhi, pp. 59–71.

Dutta, N., and Ferriere, C.H., 1961. On the Chalcidoid parasites of jute stem girdler from West Bengal. *Indian J. Agric. Sci.*, 31: 139–142.

Dutta, N., and Ghosh, S.K., 1964–65. Observations on the parasites of jute stem weevil, *Apion corchori* Marsh. and their relative efficacy under field condition. *Proc. 51st–52nd Indian Sci. Cong. III* (Abstracts), pp. 561.

Hath, T.K. and Chakraborty, A., 2004. Towards development of IPM strategy against insect pests, mites and stem rot of jute under North Bengal conditions. *J. Entomol. Res.*, 28: 1–5.

Kundu, B.C., Basak, K.C. and Sircar, P.B., 1959. *Jute in India: A Monograph*. Indian Central Jute Committee, Calcatta, pp. 159–170.

Lefroy, H.M., 1907. Insect pests of jute. *Agric. J. India*, 2: 100–115.

Mallick, R.N., 1985. Studies on the biology, bionomics and control of grey weevil (*Myllocerus discolor* Boheman) on *olitorius* jute. In: *Annual Report: 1985*. Central Research Institute for Jute and Allied Fibres, Barrackpore, p. 65.

Mallick, R.N., 1986. Studies on the biology, bionomics and control of grey weevil (*Myllocerus discolor* Boheman) on *olitorius* jute. In: *Annual Report: 1986*. Central Research Institute for Jute and Allied Fibres, Barrackpore, p. 56.

Mallick, R.N., 1989a. Studies on the biology, bionomics and control of grey weevil (*Myllocerus discolor* Boheman) on *olitorius* jute. In: *Annual Report: 1987–1988 and 1988–1989*. Central Research Institute for Jute and Allied Fibres, Barrackpore, p. 94.

Mallick, R.N., 1989b. Studies on the efficacy of newer compounds of plant origin against major pests of jute. In: *Annual Report: 1987–1988 and 1988–1989*. Central Research Institute for Jute and Allied Fibres, Barrackpore, p. 95–96.

Mallick, R.N., 1995a. Studies on the biology, bionomics and control of grey weevil (*Myllocerus discolor* Boheman) on *olitorius* jute. In: *Annual Report: 1990–1991 and 1991–1992*. Central Research Institute for Jute and Allied Fibres, Barrackpore, p. 66.

Mallick, R.N., 1995b. Studies on the biology, bionomics and control of grey weevil (*Myllocerus discolor* Boheman) on *olitorius* jute. In: *Annual Report: 1993–1994 and 1994–1995*. Central Research Institute for Jute and Allied Fibres, Barrackpore, p. 48.

Mallick, R.N. and Banerji, A., 1989a. Effects of methanol extracts of *Ocimum sanctum* Linn. on jute semilooper, *Anomis sabulifera* Guen. *Indian J. Entomol.*, 51: 84–89.

Mallick, R.N. and Banerji, A., 1989b. Insecticidal effect herniarin, a constituent of *Herniaria glabra* against the jute semilooper, *Anomis sabulifera* Guen. *Sci. Cult.*, 55: 211–213.

Mallick, R.N., Das, P.C. and Chatterji, S.M., 1985. Antifeeding properties of *Swertia chirata* against jute semilooper, *Anomis sabulifera* Guen. *Curr. Sci.*, 54: 1110–1114.

Mookerjee, A.L. and Chaterjee, D.K., 1969. Studies on the control of jute semilooper (*Anomis sabulifera* Guen.) with new chemicals. *Pesticides (Bombay)*, 3: 19–20.

Nair, M.R.G.K., 1986. *Insects and Mites of Crops in India*. Indian Council of Agricultural Research, New Delhi, pp. 122–126.

Pandit, N.C., 1985. Studies on the bioecology of major pests of mesta. In: *Annual Report: 1985*. Central Research Institute for Jute and Allied Fibres, Barrackpore, p. 65.

Pandit, N.C., 2001. Studies on the microbes and preparation of biocides for the control of insect pests of jute and allied fibre crops. Project completion report. Central Research Institute for Jute and Allied Fibres, Barrackpore.

Pandit, N.C., Rao, P.V. and Chakraborty, A.K., 2002. Studies on the biotic and abiotic factors on the incidence of yellow mite of jute. In: *Annual Report: 2000–2001 and 2001–2002*. Central Research Institute for Jute and Allied Fibres, Barrackpore, p. 71.

Pradhan, S.K., 1995. Screening of jute germplasms against major pests of jute. In: *Annual Report: 1990–1991 and 1991–1992*. Central Research Institute for Jute and Allied Fibres, Barrackpore, p. 68.

Pradhan, S.K. and Saha, M.N., 1997. Effect of yellow mite (*Polyphagotarsonemus latus* Bank) infestation on the major nutrient contents of *tossa* jute (*Corchorus olitorius* L.) varieties. *J. Entomol. Res.*, 21: 123–127.

Prasad, S.S., Yadav, U.S., Srivastava, R.K., Prakash, N. and Pandey, B.P., 2004. Performance of certain *capsularis* jute cultivars against jute pests. *Proc. Nat. Acad. Sci. India. (Biol. Sci.)*, 74: 189–193.

Ramasubramanian, T., 2007. Need for resistance risk assessment in major pests of jute to insecticides. *Resist. Pest Manage. Newsl.*, 17: 2–4.

Ramasubramanian, T., 2008. Baseline toxicity and field efficacy of newer insecticides/acaricides against major pests of jute. In: *Annual Report: 2007–2008*, (Eds.) B.S. Mahapatra, S.K. Hazra, T. Ramasubramanian, S.K. Sarkar and A.R. Saha. Central Research Institute for Jute and Allied Fibres, Barrackpore, p. 65.

Ramasubramanian, T. and Ghosh, S.K., 2008. Pest management in jute and allied fibres: present status and future strategies. In: *Jute and Allied Fibre Updates: Production and Technology*, (Eds.) P.G. Karmakar, S.K. Hazra, T. Ramasubramanian, R.K. Mandal, M.K. Sinha and H.S. Sen, pp. 194–216.

Saha, A., 2000. Improved varieties of jute for maximization of fibre yield. In: *Workshop-cum-Training on Adaptive Research on Improved Varieties of Jute and Allied Fibres and their Utilization for Enhanced Income Generation of Farmers*, (Ed.) S. Pathak. Central Research Institute for Jute and Allied Fibres, Barrackpore, pp. 5–6.

Singh, B. and Das, L.K., 1979. Semilooper (*Anomis sabulifera* Guen.) escalating on jute pods. *Sci. Cult.*, 45: 121–123.

Tripathi, R.L., 1963. Selection of oviposition site on jute leaves by *Anomis sabulifera* Guen. (Lepidoptera: Noctuidae). *Jute Bull.*, 26: 187–188.

Tripathi, R.L., 1967a. Relative contact toxicity of some insecticides to the larvae of jute semilooper, *Anomis sabulifera* (Guen.) (Lepidoptera: Noctuidae). *Indian J. Entomol.*, 29: 234–236.

Tripathi, R.L., 1967b. Insects and other pests of jute and their control measures. *Jute Bull.*, 30: 90–98.

Tripathi, R.L. and Bhattacharya, S.P., 1963. Studies on the comparative efficacy of some modern insecticides in the control of leaf injuries in jute by *Anomis sabulifera* Guen (Lepidoptera: Noctuidae). *Jute Bull.*, p. 164–165.

Tripathi, R.L. and Bhattacharya, S.P., 1968. Estimation of loss in jute fibre yield due to jute stem-weevil, *Apion corchori* Marshall (Curculianidae: Coleoptera). *Indian J. Agric. Sci.*, 38: 870–873.

Tripathi, R.L. and Ghosh, S.K., 1964. Studies on prophylactic and curative measures against jute semilooper, *Anomis sabulifera* Guen (Lepidoptera: Noctuidae). *Jute Bull.*, p. 75–77.

Tripathi, R.L. and Ram, S., 1970. Acaricidal effect of some pesticides against red spider mite, *Oligonychus coffeae* Nietner (Tetranychidae). *PANS (London)*, 16: 162–164.

Tripathi, R.L. and Ram, S., 1971a. *A Review of Entomological Researches on Jute, Mesta, Sunhemp and Allied Fibres*. ICAR Technical Bulletin No. 36.

Plant Protection, Diversity and Conservation (2011) *Pages* **549–564**
Editors: **D. Prasad & Satya Kumar**
Published by: **BIOTECH BOOKS, NEW DELHI**

Chapter 42

Pest Complex and their Succession in Cotton (*Gossypium* spp.) in Madhya Pradesh

Yogesh Patel[1], H.B. Sharma[2] and S.B. Das[3]

[1]*Department of Entomology, College of Agriculture, J.N. Krishi Vishwa Vidhyalaya, Ganjbasoda Distt. Vidisha, M.P.*
[2]*BSA College of Mathura, U.P.*
[3]*J.N. Krishi Vishwa Vidhyalaya, Jabalpur – 482 004, M.P.*

India is one of the major producers of cotton in the world with the largest acreage, which is almost one fourth of world area. The production share is however, only 13.5 per cent ranking third in the world. Amongst various cardinal factors responsible for poor yield of cotton in India, the damage caused by large number of insect pests during different stages of crop growth are of prime importance. Rainwater (1952) remarked that cotton plant is specially tailored for insect. About 1326 insects and mites from all over the world were reported (Hargreaves 1948). The pest spectrum of a crop varies from place to place as par variation in the agro-climatic condition of the locality. Visualization of real pest problem of an area is the pre-requisite for formulating a sustainable IPM strategy. Keeping these points in view investigation was undertaken to study the pest complex and their succession in cotton tract of the Madhya Pradesh.

In order to explore information on the insect pests problems associated with cotton in Nimar region of south-west MP, were conducted at the J.N. Krishi Vishwa Vidhyalaya, Agricultural Research Station, Khargone and Khandwa, M.P. during 2004-05 and 2005-06. The cotton genotype JK-4 was sown on 29th June and 25th June during 2004 and 2005 respectively at a spacing of 60X60 cm. Normal agronomic practices recommended for the region were followed for raising the crop. No plant protection measure was taken throughout the crop season. The observations were taken twice in a week starting from the initial appearance to final disappearance on randomly tagged plants. The population of major sucking pests *viz.* jassid, aphid, whitefly and thrips recorded on five leaves, two each from lower, middle and one from upper canopy of the plants. The per cent damage caused by bollworms was recorded by counting total number of damaged and healthy fruiting body *viz.* squares, green boll, open boll and locule per plant. Shaded fruiting bodies were also collected from these plants and dissected to record bollworms damage. The observation on the population of predator of cotton pests recorded on per plant bases. The collected insect specimens from the field were identified by comparing with the available identified specimen and appropriate taxonomic litreature. Insect pests were categorized as per the method given by Reddy *et al.* (1998), as major, minor and stray pests on the basis of their mean level of infestation throughout the season. To find out the succession of insect pests and their natural enemies during the each year, the periodical mean population and incidence were worked out. Overall information explored from the investigation were compiled and depicted in Tables 42.1–42.3.

Accounts of insect pest fauna associated with cotton ecosystem and their natural enemies obtained from field have been presented in Tables 42.1–42.3.

Insect Pest Complex of Cotton

The results revealed that the total number of nineteen insect pests and their nine natural enemies were recorded in cotton crop in the Nimar region of Madhya Pradesh. Among these, highest eight insect-pests belong to order Lepidoptera, six to Hemiptera and two to Coleoptera were observed. These finding were in accordance with the findings of Choudhary and Garg, (2002) who reported twenty species of insect pests on upland cotton crop in Madhya Pradesh. Sohi (1964) gave a list of 160 insects and mites on cotton in India. Most of these insects belong to four orders namely Lepidoptera, Orthroptera, Coleoptera, and Hemiptera.

Out of these nineteen insect pests, on basis of their incidence level, nature of damage and duration of activity, seven insect pests categorized as major, nine minor and three stray, pests of cotton. The major pests were Aphid, *Aphis gossypii* (Glover) (Aphididae; Hemiptera), Jassid, *Amrasca biguttula bigutttula* (Ishida) (Cicadelliae; Hemiptera), Thrips, *Scirtothrips dorsalis* (Hood) and *Thrips tabaci* (Lindeman)(Thripidae; Thysanoptera), Whitefly, *Bemisia tabaci* (Gennadius)(Aleyrodidae; Hemiptera), American or green bollworm, *Helicoverpa armigera* (Hubner) (Noctuidae; Lepidoptera),Spotted bollworm *Earias vittella* (Fabricius) and *E. insulana*,(Boisduval) (Noctuidae; Lepidoptera) and Pink bollworm *Pectinophora gossypiella* (Saunders) (Gelechiidae; Lepidoptera). These insect pests infested the crop continuously and remained active almost throughout the season or appeared for the few months but assume severe form, there by considered as 'major pests'. Nine insect pests, *viz.*, Leaf miner, *Leriomyza trifoli* Burgess; Grey weevil, *Myllocerus undecimpustulatus, maculosus* Desbrocher; Red cotton bug, *Dysdercus cingulatus* Fabricius; Dusky cotton bug, *Oxycarenus laetus* Kirby; Cotton semilooper (Green), *Anomis flava* Fabricius; Tobacco caterpillar, *Spodoptera litura* Fabricius; Cotton leaf roller, *Syleptu derogata* Fabricius; Green sting bug, *Nezara viridula* (Linn.); Termite (White ant), *Odontotermes obesus* (Rambur) were categorized as minor pests, because of they infested crop continuously for almost throughout the cropping season but their incidence level was very low and did not cause economic losses.

Major pests, further may be classified in two groups based on their feeding habits, first sucking pests comprises with four insect namely Aphid, *Aphis gossypii* (Glover) (Aphididae; Hemiptera), Jassid, *Amrasca biguttula bigutttula* (Ishida) (Cicadelliae; Hemiptera), Thrips, *Scirtothrips dorsalis* (Hood)

Table 42.1: Insect Pest Associated with Cotton Ecosystem in Madhya Pradesh

Sl.No.	Common Name	Scientific name	Order–Family	Period of Activity		Damaging Stage of the Insect Pest
				Prevalence	Peak Incidence	
1.	Leaf miner	*Leriomyza trifoli* Burgess	Diptera–Agromyzidae	June–July	Early July	Maggot
2.	Grey weevil	*Myllocerus undecimpustulatus maculosus* Desbrocher	Coleoptera–Curculionidae	July–December	Mid October	Grub, Adult
3.	Jassid (Leaf hopper)	*Amrasca biguttula biguttula* Ishida	Hemiptera–Cicadellidae	July–November	Mid August	Nymph, Adult
4.	Aphid	*Aphis gossypii* Glover	Hemiptera–Aphidae	July–November	Mid August	Nymph, Adult
5.	Thrips	*Thrips tabaci* Lindemann	Thysenoptera–Thripidae	August–November	Early-mid September	Nymph, Adult
6.	White fly	*Bemisia tabaci* (Gennadius) Fabricius	Hemiptera–Aleyrodidae	August–December	Late September	Nymph, Adult
7.	Spotted Bollworm	*Earias insulana* (Boisduval) *Earias vitiella* (Fb.)	Lepidoptera–Noctuidae	July–November	Late August	Caterpillar
8.	American Bollworm	*Helicoverpa armigera* (Hubner)	Lepidoptera–Noctuidae	July–January	Late Aug. to September	Caterpillar
9.	Pink Bollworm	*Pectinophora gossypiella* (Saunders)	Lepidoptera–Gelechiidae	September–January	Late Nov. to December	Caterpillar
10.	Red cotton bug	*Dysdercus cingulatus* Fabricius	Hemiptera–Pyrrhocoridae	Late October–January	Early December	Nymph, Adult
11.	Dusky cotton bug	*Oxycarenus laetus* Kirby	Hemiptera–Lygaeidae	September–December	Late October	Nymph, Adult
12.	Cotton semilooper (Green)	*Tarache notabilis* (Walker) *Anomis flava* Fabricius	Lepidoptera–Noctuidae	August–November	September	Larval
13.	Tobacco caterpillar	*Spodoptera litura* Fabricius	Lepidoptera–Noctuidae	Early August–Late Sep.	Early September	Caterpillar
14.	Cotton leaf roller	*Sylepta derogata* Fabricius	Lepidoptera–Pyraustidae	Late August–Late Sep.	Mid September	Larval
15.	Green sting bug	*Nezara viridula* (Linn.)	Hemiptera–Pentatomidae	Mid Sept.–Early Nov.	Late October	Nymph, Adult
16.	Soybean green semilooper	*Plusia acuta* Walker	Lepidoptera–Noctuidae	Mid August–Early October	Early September	Larval
17.	Soybean black Semilooper	*Plusia orichalcia* Fabricius	Lepidoptera–Noctuidae	Mid August–Early Oct.	Early September	Larval
18.	Termite (White ant)	*Odontotermes obesus* (Rambur), *Microtermis* obesi Holmgren	Isoptera–Termitidae	Mid July–Early October	August and Late September	Nymph, Adult
19.	Cotton stem weevil	*Pempherulus affinis* Faust.	Coleoptera–Curculionidae	Mid Sept.–Early Nov.	Early October	Grub

Table 42.2: Succession, Nature of Damage and Status of Pest Complex in Cotton Ecosystem in Madhya Pradesh

Sl.No.	Common Name	Stage of the Crop Attacked		Nature of Damage of the Insect Pest to the Crop	Pest Status
		Growth Phase	Crop Age (Days)		
1.	Leaf miner	Cotyledionaous	1–35	Damaged is caused by maggot in leaf tissues by making wandering tracks, which reduces the photosynthetic activity, leaves dry, and fall off prematurely. In severe cases plant ultimately dies due to wilting.	Minor
2.	Grey weevil	Vegetative	1–30	Damaged is caused by both grub and adults, which feed on roots of cotton seedlings, leaves, squares and bolls causing irregular outer boundaries of these parts, each grubs can destroys upto ten seedlings.	Minor
3.	Jassid (Leaf hopper)	Vegetative	1–50	Nymphs and adults of were found to suck the cell sap from lower surface of leaves. Severe infestation resulted in curling of leaf margins downward, reddening of leaf margins, stunted growth of the plants and sickly appearance of the crop.	Major
4.	Aphid	Vegetative	1–50	Nymphs and adults were found sucking the sap from under surface of leaves, twigs and apical portion of the stem. Severe infestation resulted in curling of leaf margins upward. Aphid also secreted honeydew, which attracts ants and sooty mold, aiding to the development of pathogenic bacteria.	Major
5.	Thrips	Vegetative/ Square formation	1–70	The nymphs and adults suck sap from the lower surface of the leaves lacerating the leaf tissues. Thrips feed on the young leaves and the buds and stunt the crop growth. A common sign of a heavy thrips infestation is the distorted leaves that have turned brownish around the edges and cup upward.	Major
6.	White fly	Square/ Boll formation	35–110	The nymphs and adults feed on the cell sap, reduce the vitality of the plant interfering with normal photosynthesis due to the excretion of honeydew and formation of sooty mold all over the surface of the leaf and lint of the opened bolls, resulting in process of blackening. Chlorotic spots develop on leaves and in severe cases, the veins become translucent. Sooty mould contaminates the lint. The insect helps in transmitting leaf curl virus (CLCV).	Major
7.	Spotted Bollworm	Square/Flowering/ Boll formation	35–110	Caterpillars cause damage by boring into the growing shoots, buds, flowers and bolls. The attacked shoots wither, droop and ultimately die, and flowers and buds drop off. Attacked squares flared and fell down on the ground. Small punctures were seen on squares plugged by the faecal matter. Thread like faecal matter was found on bolls. The infested bolls when dissected showed cream–coloured caterpillars.	Major
8.	American Bollworm	Square/Flowering/ Boll formation	45–110	Caterpillar damaged squares, flowers and bolls. Small punctures were seen near the base of squares and bolls. Attacked squares flared up and dropped on ground. Pallets like faecal matter were seen between the bracts and bolls and also on the soil under the plants. The puncture was large and circular.	Major

Contd...

Table 42.2–Contd...

Sl.No.	Common Name	Stage of the Crop Attacked		Nature of Damage of the Insect Pest to the Crop	Pest Status
		Growth Phase	Crop Age (Days)		
9.	Pink Bollworm	Fowering/ Boll formation	65–110	Caterpillar attacked squares, flowers and bolls. They damaged squares by feeding on underdeveloped anthers and style and were feeding down the column into the ovary in opened flowers. The affected flowers were not open fully and showed characteristic resetting. Such flowers buds were shed at boll stage. The larvae enter the boll as an early instar and remain inside feeding on immature seed and developing lint. Bolls open prematurely. The full–grown larvae were found feeding inside the bolls and also cut down the window holes in two adjoining seeds and joined them together to form double seeds.	Major
10.	Red cotton bug	Boll formation	50–110	The adult and nymph suck sap from the leaves and green bolls, stain the lint with their excreta or body juices that give way for the growth of bacteria and fungi.	Minor
11.	Dusky cotton bug	Boll formation	45–100	The nymphs and adults sucks sap from immature seeds which do not ripe, remain light weight. Adults found in the lint get crushed during ginning and stain the lint emitting bad odour.	Minor
12.	Cotton semilooper (Green)	Vegetative	35–110	The young larvae congregate in group and move actively, feed on the leaf lamina making a small punctures. The caterpillar feed on the tender, shoot, bud and bolls. In case of severe infestation, the plant became totally denuded, leaving the midrib and side veins of leave intact.	Minor
13.	Tobacco caterpillar	Vegetative	45–125	The larvae feed gregariously on the under surface of the leaves. In severe infestation leaves becomes skeletonize, later on larvae attack on flower, buds and squares causing considerable losses.	Minor
14.	Cotton leaf roller	Vegetative	35–110	Larva folds the leaf and feed in side the roll starting with leaf margins. Rolled leaves are found drooping on the plant with their leaf petiols. In severe infestation, plants are completely defoliated.	Minor
15.	Green sting bug	Vegetative	35–110	The bugs feeding on buds, green bolls and shoots of cotton plants.	Minor
16.	Soybean green semilooper	Vegetative	35–110	The young larvae congregate in groups and move actively, feed on the leaf lamina making small puncture. The grown up larvae feed voraciously leaving only the midribs and veins. They feed by chewing the leaves from margin towards the leaf veins. The caterpillar feed on tender shoots and buds.	Stray
17.	Soybean black semilooper	Vegetative	35–110	The larvae were found to feed on the green foliage. They feed the leaves voraciously and leaving only the midribs and veins. They start feeding the leaves from margin towards the leaf veins.	Stray
18.	Termite (White ant)	Vegetative	25–65	Nymph and adults of worker termites feeds on root and rootlets. Infested plants are easily uprooted from the soil.	Minor
19.	Cotton stem weevil	Vegetative and Reproductive	30–80	The grub tunneling within the stems, the attacked points swell up and such swelling are seen generally at the base of the stems. Thirty to fourty days old plants when attacked succumb; older plants may survive but suffer in vigour.	Stray

Table 42.3: Natural Enemies Recorded in Cotton Ecosystem in Madhya Pradesh

Sl.No.	Common Name	Scientific Name	Order–Family	Activity	Host	Population Status	Status
1.	Lady Bird beetle	*Coccinella septumpunctata* L. *Menochilus sexamaculatus* (Fab.)	Coleoptera–Coccinellideae	July–December	Aphid, Jassid, Whitefly	High	Predator
2.	Green lace wing	*Chrysoprela cornea* Stephens	Neuroptera–Chrysopdae	July–December	Aphid, Jassid, Whitefly Boll worm egg	High	Predator
3.	Syrphid fly	*Syrphus serarius* (Wiedmann)	Diptera–Syrphidae	August–Early Oct.	Aphid	Medium	Predator
4.	Dragonfly	*Platygomphus dolobratus* Selys	Odonata–Gomphidae	June–Dec.	Boll worms	Medium	Predator
5.	Paintatomid bug	*Cantheconida furcellata* Wolff	Hemiptera–Pentatomidae	Late July–Nov.		Low	Parasitoid
6.	Trichogramma	*Trichogramma* chilonis *Trichogramma ishi*	Hymenoptera–Trichogramma tidae	August–late Nov.	Boll worms	Low	Parasitoid
7.	Ichneumonid Wasp	*Ecphorosis perditincutus*	Hymenoptera–Ichneumonidae	August–November		Low	Parasite
8.	Braconids	*Bracon gelechiae* Ashmead	Hymenoptera–Braconidae	August–November	Bollworm	Low	Parasite
9.	Apanteles	*Apanteles flavipes* (Cameron) *Apanteles colomani*	Hymenoptera–Braconidae	August–November	Boll worm	Low	Parasite
10.	Spider	*Oxyopes shweta* Tikader *Oxyopes pandae* Tikader	Araneae–Oxyopidae	July–December	Boll worm	High	Predator

or *Thrips tabaci* (Lindeman) (Thripidae; Thysenoptera) and Whtefly, *Bemisia tabaci* (Gennadius),(Aleyrodidae; Hemiptera) and the other group included Bollworm pests complex comprises with three insects namely, American or green bollworm, *Helicoverpa armigera* (Hubner) (Noctuidae; Lepidoptera), Spotted bollworm, *Earias vittella* (Fabricius) or *E. insulana*,(Boisduval) (Noctuidae; Lepidoptera) and Pink bollworm, *Pectinophora gossypiella* (Saunders) (Gelechiidae; Lepidoptera).

Natural Enemies of Cotton Pests

The present study reveled that total number of nine natural enemies was found to feed on the insect pests attacking the cotton crop *viz.* Lady bird beetle, *Coccinella septumpunctata and Menochilus sexamaculatus* (Coleoptera: Coccinellideae), Green lacewing, *Chrysoprela cornea* (Neuroptera: Chrysopdae), Syrphid fly, *Syrphus serarius* (Wiedmann) (Diptera: Syrphidae), Dragonfly, *Platygomphus dolobratu* Selys (Odonata; Gomphidae), Paintatomid bug, *Cantheconida furcellata* Wolff (Hemiptera; Pentatomidae) *Trichogramma chilonis* Hymenoptera; *Trichogramma tidae*, Ichneumonid Wasp, *Ecphorosis perditincutus* (Hymenoptera; Ichneumonidae), *Bracon gelechiae* (Hymenoptera:Braconidae), *Apanteles flavipes (Cameron)*and *Apanteles colomani* (Hymenoptera; Braconidae) and spider, *Oxyopes shweta Tikader*, (Araneae; Oxyopidae).These finding are in confirmty with those of Choudhary and Garg (2002).

Succession of Major Insect Pests of Cotton and their Potent Natural Enemies

The succession of the insects pests showed that the population of different species occurred in an overlapping manner and that the crop was continuously under the presence of the harmful pests and beneficial insects. Among the various insect pests infesting the cotton crop, the activity of serpentine leaf miner (*Leriomyza trifoli* Burgess) was found in the most early crop age, *i.e.* cotyledonary stage. Early season sucking pests *viz.* jassid and aphid were active between July to November. The activity of thrips was observed during August to November whereas white fly invade the crop between August and December. These insects though appeared for a short time on crop they showed their impact on crop growth. The sucking pests devitalize the plant in earlier period of the crop growth stage, thereby, affecting the yield. Among the bollworm, infestation of spotted bollworm was appeared at vegetative stage (by damaging apical twigs) and remained till third flush of reproductive stage (*i.e.* July to November). The american/green boll worm remained active from vegetative stage till last picking *i.e.* late July to January, whereas, the infestation of pink boll worm initiated during September and ended by January (late reproductive stage). Activity of bollworms were maximum during September to November coincides with the flowering and fruiting stage. Most of the foliage feeders *i.e.* grey weevil, cotton green semilooper, tobacco caterpillar, soybean semilooper and cotton leaf roller infested the crop plants during vegetative stage. Population of natural enemies *i.e.* directly generated with the availability of food materials *i.e.* pests population on which these insect feed. The crop was attacked by one or more insect species from the time it was planted to harvest. All the parts of cotton plant were vulnerable to pests attack. termites attacked the germinating seeds. Later, dusky cotton bugs, jassids, aphids, thrips, whiteflies, leaf rollers, semilooper and tobacco caterpillars attacked the foliage. During reproductive phase the crop was attacked by spotted bollworm, American bollworm, Pink Bollworm, Red cotton bug and sucking pests. These finding were in accordance with those of Gupta *et.al.*; (1971), Aston and Winfield, (1972); Geddes and Iles; (1991), Dhawan; (2001) and Choudhary and Garg; (2002).

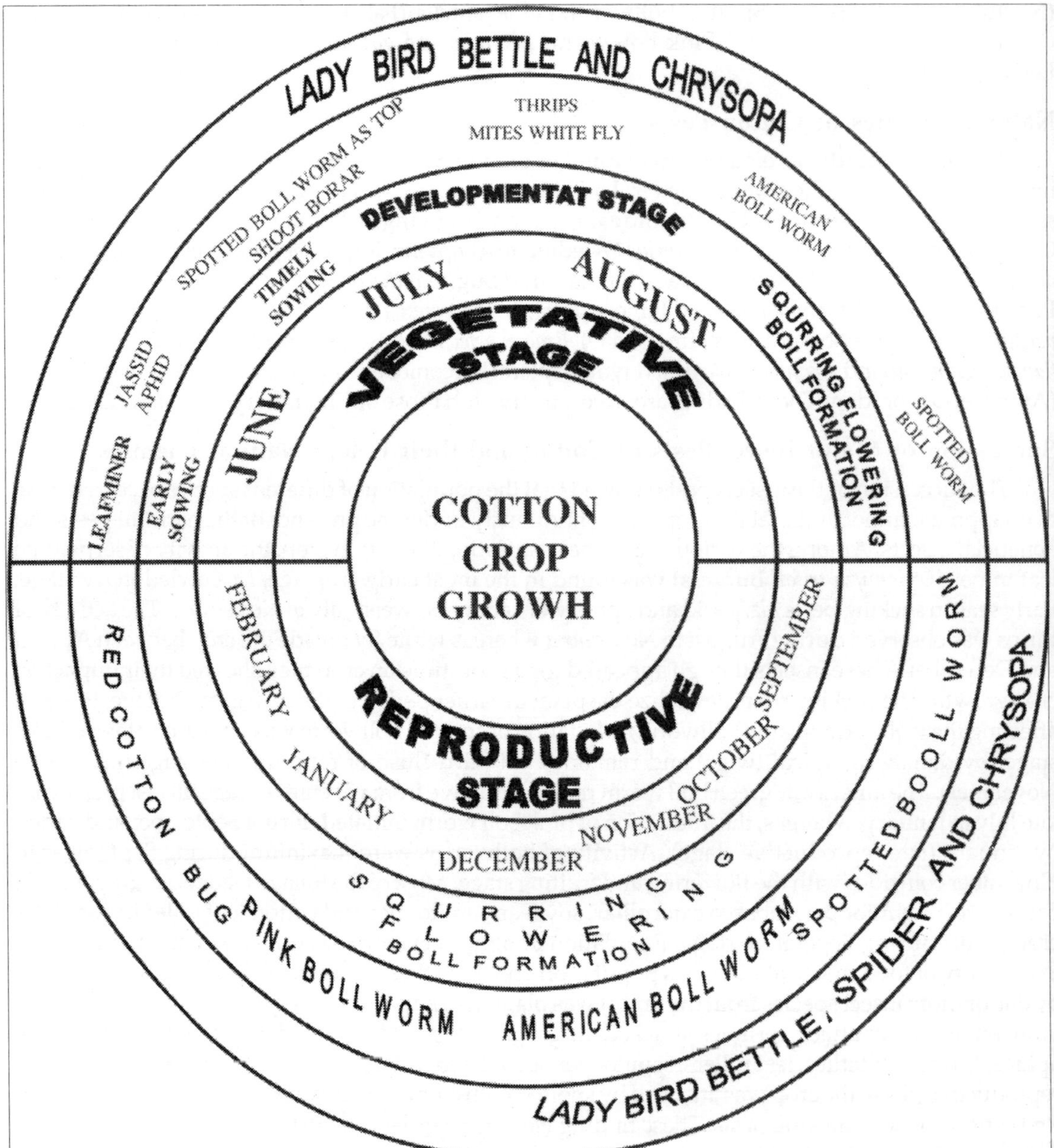

Figure 42.1: Succession of Insect Pests of Cotton

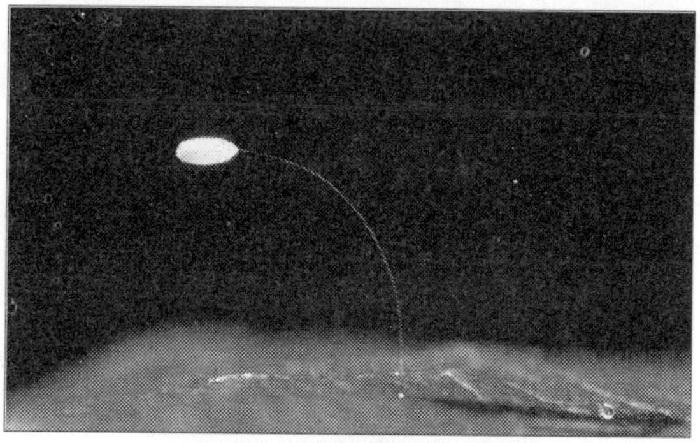

Egg of *Chrysoperla carnea*

Pupa of *Chrysoperla carnea*

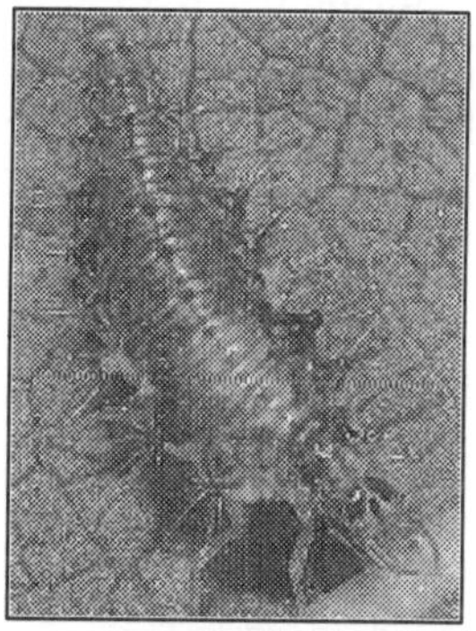

Larvae of *Chrysoperla carnea*

Adult of *Chrysoperla carnea*

Plate 42.1: Different Stages of Wing Lace Beetle, *Chrysoperla carnea*

Plate 42.2: Different Stages of Spotted Bollworm, *Earias* spp.

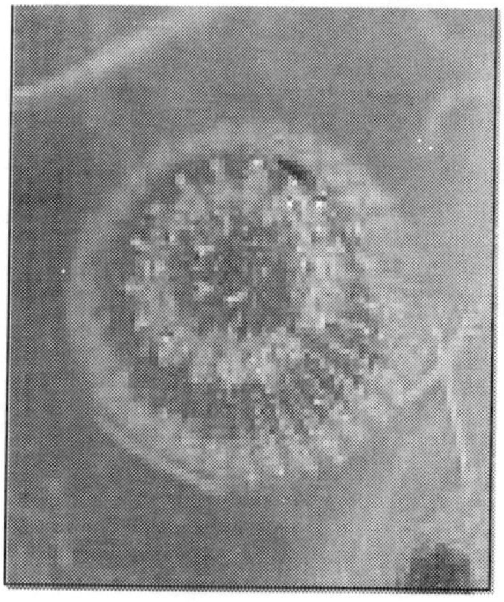

Egg of *Earias* spp.

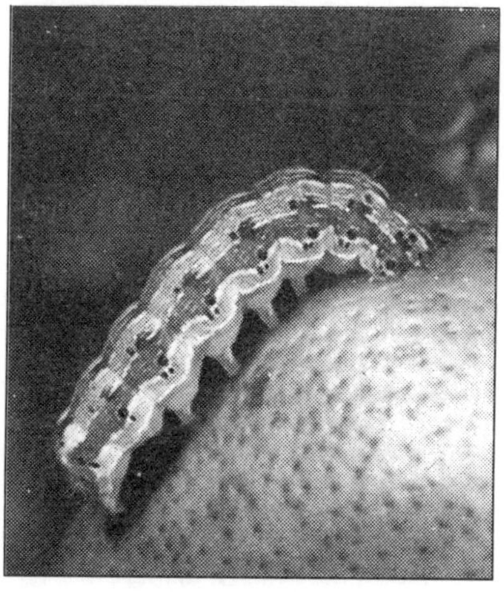

Larvae of *Earias* spp.

Plate 42.2–Contd...

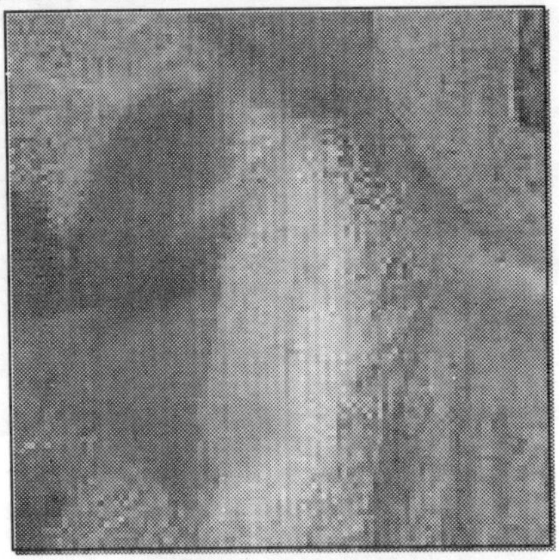

Pupae of *Earias* spp.

Adult moth of *Earias* spp.

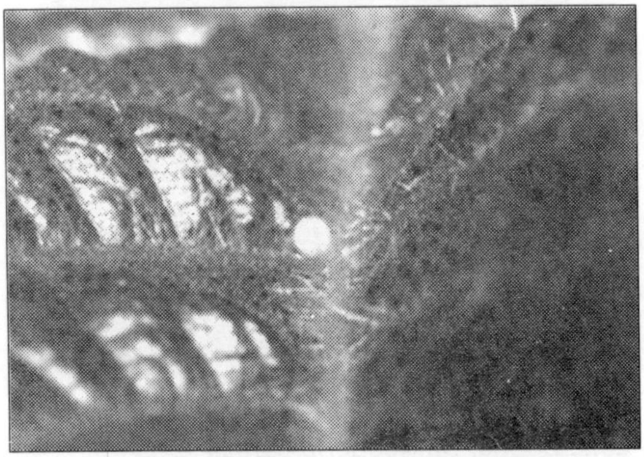

Egg of *H. armigera* on Leaf of Cotton

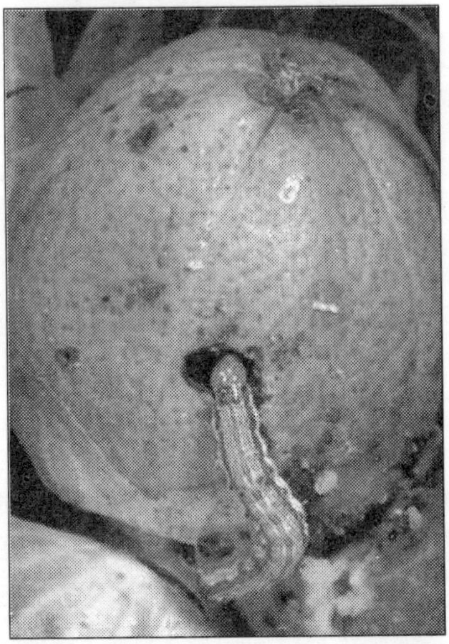

Larva of *H. armigera* Damaging Boll of Cotton

Adult Moth of *H. armigera*

Pupae of *H. armigera* in Soil

Plate 42.3: Different Stages of American bollworm, *H. armigera*

Eggs of *P. gossypiella*

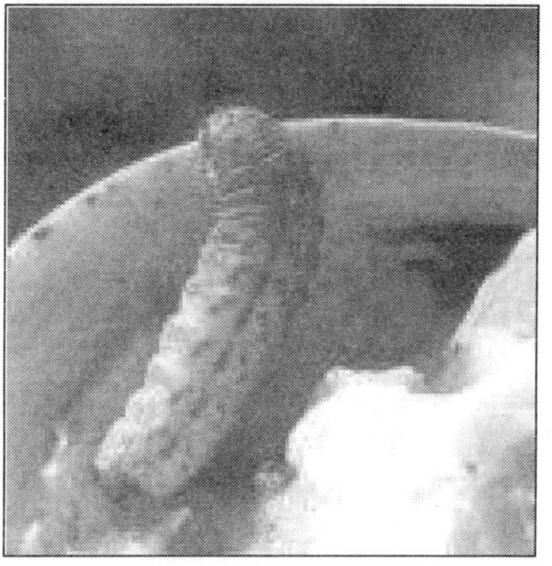

Larva of *P. gossypiella*

Pupa of *P. gossypiella*

Adult Moth of *P. gossypiella*

Plate 42.4: Different Stages of Pink Bollworm, *P. gossypiella*

Nymph of *B. tabaci*

Damage of *A. biguttula*

Adults of *B. tabaci*

Adult of *A. biguttula*

Adult of *B. tabaci*

Plate 42.5: Different Stages of White Fly, *Bemisia tabaci*

Nymph of *A. biguttula*

Plate 42.6: Cotton Jassid, *Amrasca biguttula*

Cotton Aphid Damage to Leaves

Nymph of *A. gossypii*

Adult of *A. gossypii*

Plate 42.7: Cotton Aphid, *Aphis gossypii*

Thrips Injuring to Cotton

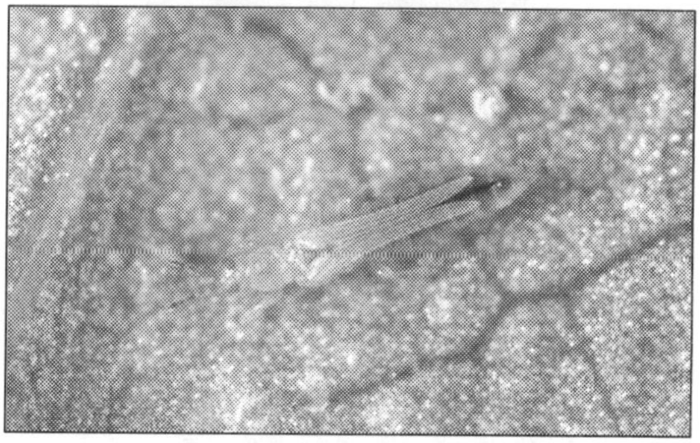

Nymph of *T. tabaci*

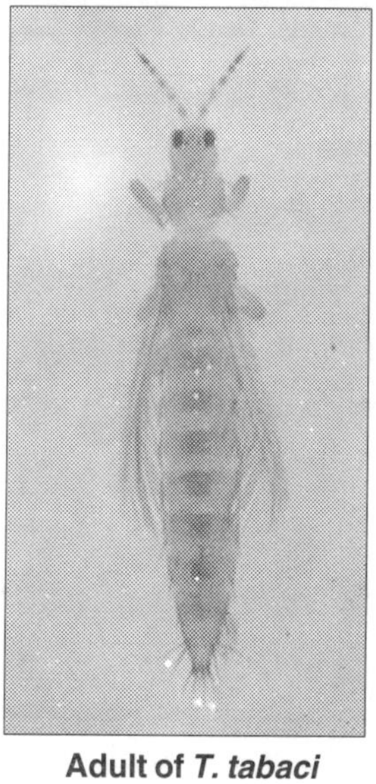

Adult of *T. tabaci*

Plate 42.8: Thrips, *Thrips tabaci*

Eggs of *C. septumpunctata*

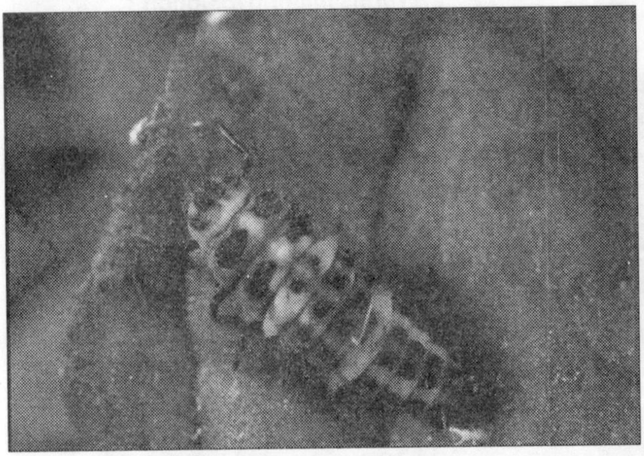

Larva of *C. septumpunctata*

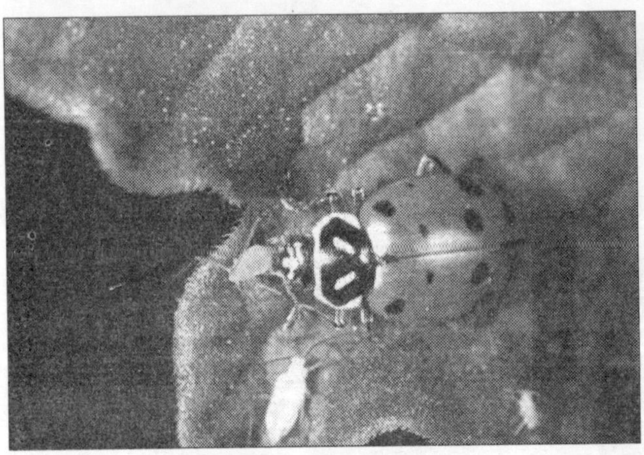

Adult of *C. septumpunctata*

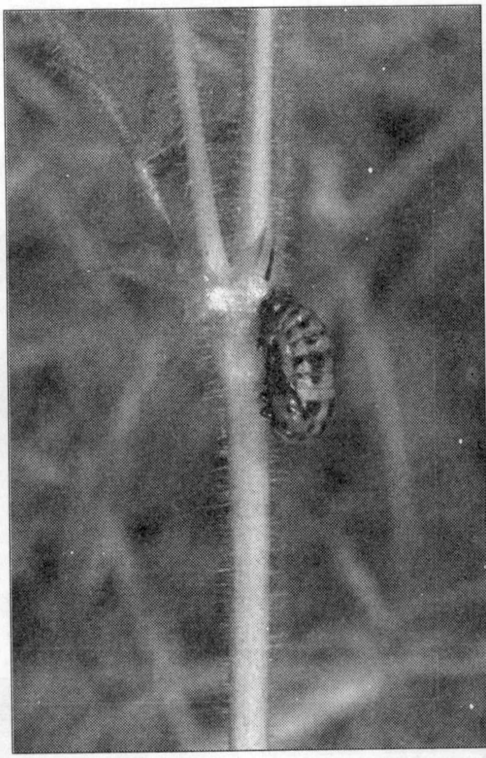

Pupa of *C. septumpunctata*

Plate 42.9: Lady Bird Beetle, *Coccinella septumpunctata*

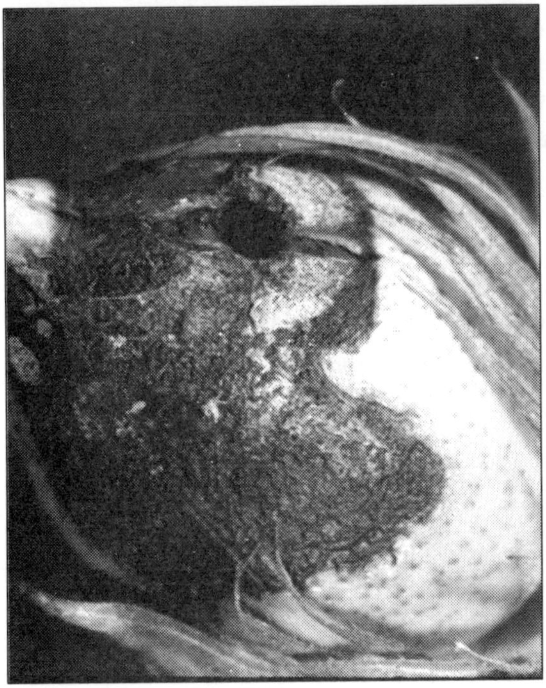

Plate 42.10: Typical Damage to Boll by Cotton Bollworms

Plate 42.11: Lady Bird Beetle Larvae Feed Cotton Aphid

Plate 42.11: A Minor Pest of Cotton *Soptodoptera litura*

References

Aston, J.L. and Winfield, B., 1972. Insect sections: Introduction cotton. *Publ. Ciba-Geigy Agrochemicals* pp. 44–45.

Chodhary, R.K. and Garg,V.K., 2002. Succession of pest complex and their natural enemies on cotton (*Gossypium spp.*) in Madhya Pradesh. *J. Cotton Res. Dev.*, 17(2): 180–185.

Dhawan, A.K., 2001. Impact of some new insecticides on natural enemy complex of cotton ecosystem. *Pestology*, 24(5): 8–15.

Gupta, R.L., Purohit, M.L., Rathod, V.S. and Sood, N.K., 1971. Pest of cotton in MP. *Tech. No 13.* Department of Entomology Zoology Directorate of Research, JNKVV, Jabalpur, pp. 20.

Hargreaves, H., 1948. *List of Recorded Cotton Insects of the World*. Commonwealth Inst. of Ent., Londonm pp. 1–50.

Rainwater, C.F., 1952. Progress in research on cotton insects. *U.S.D.A. Yearb Agric.*, pp. 497–500.

Reddy, C., Singh, Narendra, Singh, Yeshbir and Vijai, S., 1998. Pest complex and their succession on pigeon pea variety. *Indian J. Ent.*, 60(4): 334–338.

Sohi, G., 1964. Pests of cotton. In: *Entomology in India.* Entomological Society of India, New Delhi, p. 111–148.

Plant Protection, Diversity and Conservation (2011) *Pages* **565–577**
Editors: **D. Prasad & Satya Kumar**
Published by: **BIOTECH BOOKS, NEW DELHI**

Chapter 43

Coconut Diseases in the Bay Islands and their Management

Krishna Kumar, Ajanta Birah, Someshwar Bhagat, Deepa Balakrishnan and R.C. Srivastava

Central Agricultural Research Institute, Port Blair – 744 101, Andaman and Nicobar Islands

A Prologue on Coconut in Bay Islands

The coconut palm, *Cocos nucifera* L., is one of the important and useful palm cultivated through out the world. It provides a wide variety of useful products like food, fuel and timber. Since every part of the tree is being utilized in one or other way and hence it is called the "tree of heaven"-the tree that provide all the necessities of life. Coconut is grown in more than 93 countries in the world. Asia and Pacific countries share is about 80 per cent of the area and production. India ranks third in the world, with an area of about 1,935 hectares with a production of 12,832.9 million nuts. The productivity is 6,632 nuts per ha. (2004-05). The Andaman and Nicobar Islands comes under the humid tropics with average rainfall of about 3000 mm annually. About 86 per cent of the total geographical area is under forest cover. Out of the total geographical area of 82, 49,000 ha, around 50,000 ha at present is under agriculture practices. Coconut is a unique plantation crop of Andaman and Nicobar Islands and has been associated with socio-cultural facts of people of these Islands. Topography of these Islands is characterized by hilly and mountain terrain coupled with heavy rainfall and undulating nature of soil, which are ideally suitable for growing coconut. Though the agro-biodiversity of coconut is unique in these islands, the production and productivity is very poor and stagnating since last two decades. In Andaman and Nicobar Islands coconut is grown in an area of about 21,689 ha with a production of 81.90 million nuts annually. The productivity is 3776 nuts per hectare. The average yield of the palm in the Islands are very low (20 nuts/palm/year) as compared to the national average of 44 nuts/palm/year. Low yield is due to poor genetic makeup and poor management of existing plantation along with points enumerated below:

1. Planting on hill slopes having shallow/eroded soil.
2. Inadequate plant protection measures.
3. Inadequate soil and nutrient management.
4. Inadequate knowledge and poor adoption of technologies.

Planting of high yielding varieties and adopting modern production technologies can increase the productivity. In this review, plant protection perspective is taken in to account and significant management outcomes are summarized.

Scenario of Coconut after *Tsunami*

The most destructive and devastating earth quake followed by *Tsunami*, which hit these unique Islands on 26.12.04, not only took the lives of several thousand people, left home less for many others, but also caused intensive damages to live stocks and agricultural crops. The preliminary survey revealed that out of 25,000 ha under coconut about 6660 ha area have been badly damaged due to the inundation of sea-water during the *Tsunami* and for replanting about 13 to 14 lakhs coconut seedlings are required. For producing the required quantity of the planting material, seed nuts from the recommended and high yielding varieties can be used.

The prolonged submergence of coconut palms due to stagnant water has already started exhibiting wilt systems. In order to safe-guard the trees from the adverse effect of accumulated salts the following measures have to be under taken immediately in the existing plantation.

1. Scrapping of surface salt followed by leaching with rainwater in the affected area is recommended.
2. The land should be immediately flooded with water during rainy season and drained. Making of channels in between the coconut rows for free flow of water in the areas subject to seawater inundation during high tides was largely recommended.
3. Gypsum @ 2kg/tree should be applied at the basin of the tree.
4. One-foot deep trenches can be dug out around the tree with in a radius of 1.8 to 2.0 m distance from tree and incorporation of large amount of organic manures should be done.

On priority basis Central Agricultural Research Institute has raised nursery with 5000 coconut seedlings from the identified elite palms for distribution to the *Tsunami*-affected areas of A&N Islands.

Coconut Diseases in the Bay Islands

Coconut palm is affected by number of diseases in countries where coconut is cultivated. Of these few are lethal while other diseases are debilitating plant vigor and adversely affecting yield potential. The diseases which are known to cause severe losses in these Islands are Grey Blight, *Ganoderma* wilt, Stem Bleeding and Bud rot. Leaf rot being caused by fungus could be managed by adopting integrated management practices. Since coconut wilt disease caused by *Phytoplasma spp* are very difficult to control by adopting the general plant protection measure. Therefore, adoption of integrated management strategies along with certification of planting material is the need of days to come. Once the transplanting is over systematic roughing of plants could be prevent further spread of viral, phytoplasmal and other vector borne diseases. Major constraints of the farmer are non-availability of quality planting material. Hence a program for production of disease free elite planting material has been initiated to fulfill the need of quality planting material in these Islands. Development of root wilt disease symptom is very slow and takes several months or even years between infection and full symptom expression. Moreover abiotic factors like seed and stress conditions leads the similar symptoms like root wilt. Hence it is very important to select disease free mother palms for further multiplication for use as a elite planting material. For this reliable and robust diagnostic test is needed for detection of viral and phytoplasmal has to be implemented. Latent infection in viral and phytoplasmal diseases are not visible and it can only be detected after full symptom expression or by deploying molecular tools like spot hybridization, PCR detection and other serological techniques.

Ganoderma Wilt

In India, BSR (Basal Stem Rot disease), caused by *Ganoderma lucidum* (Leyss.) Karst., and *Ganoderma applanatum* is a major limiting factor in coconut production. The disease is also referred to as Thanjavur wilt, bole rot, *Ganoderma* disease and Anabe (Vijayan and Natarajan 1972; Nambiar and Rethinam 1986; Bhaskaran *et al.*, 1990). The disease was first reported in Thanjavur district, Tamil Nadu in 1952 after the cyclone and was confined to the coastal districts of Tamil Nadu till 1966. Gradually the disease spread and by 1980 it was present in all coconut growing areas.

Symptoms

The visual symptoms of *Ganoderma* wilt disease are generally found in adult coconut palms of more than 10 years old. The initial symptom is characterized oozing out of reddish brown viscous fluid from the basal portions of stem (bleeding patches) near the ground level. Occasionally, some infected palms do not show bleeding symptom. Bleeding patches gradually extend upwards in progress with the disease intensity. Extensive root rotting and discoloration of roots are caused due to the disease incidence. The roots become watery with a distinct alcoholic smell, red below the hypodermis and brownish towards stele. Production of new roots is also progressively reduced. In affected palms, leaves drop down and become dried. As the disease advances, the remaining leaves also drop down in quick succession and the spindle alone remains. Subsequently heavy button shedding, upward extension of stem decay and finally death of the palm are observed. Sporophore or fruiting body of *G. lucidum* appear at the base of the trunk in some palms prior to wilting or just after the death of the palm. Sporophores are pictorially depicted in Figure 43.1. Normally the time taken from the initial appearance of bleeding patches on the stem to death of the palm is 6 to 54 months, the average being 24 months.

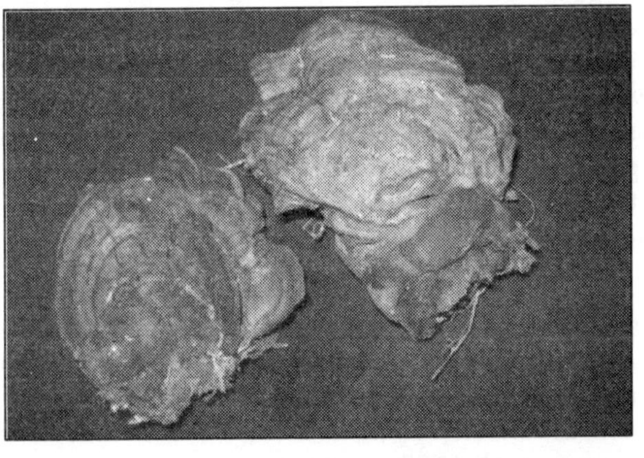

Figure 431: Sporophores of *Ganoderma* (Lethal Stage)

Pathogen

Ganoderma lucidum was first described under name *Fomes lucidus* (Leys) Fr. (Butler, 1909). According to Butler and Bisby (1931), the synonym for *G. lucidum* is *F. lucidus*. The mycelium is hyaline, aerial, thin walled, branched with frequent clamp connections and 1.4 to 2.9 µm in diameter, chlamydospores are abundant, ellipsiod, slightly thick walled both terminal and intercalary, sometimes in chains and 8.8 to 11.8 × 3.7 to 5.9µm in size; cuticular cells from crustose layer are hyaline to light brown, round to irregular in shape and closely packed (Govindu *et al.*, 1983). The naturally produced basidiocarps are perennial, stipitate, usually lateral sometimes sessile, corky, becoming woody later, usually 10 to 12 × 3 × 4 cm, but may grow up to 30 cm or more and upper surface is shiny, laccate crust, ox-blood in color and smooth. Hymenial surface is whitish or creamish, turning brown later, pores are small, round and measures 90 to 250 µm in diameter. Basidiospores are brown, thick walled, minutely verrucose, truncate at one end and 8.3 to 10 × 5.8 to 6.7 µm in size (Govindu 1983; Adaskaveg and Gilbertson, 1988). Hseu *et al.* (1996) differentiated and grouped several isolates of *Ganoderma* using techniques *viz.*, RAPD-PCR on the basis of internal transcribed space regions. The colony is white to pale yellow and even, felty to floccose at the optimum temperature on potato dextrose agar medium (Seo, 1987. Adaskaveg and Gilbertson, 1989). The colony becomes more yellowish under exposure to light. A few reports have described the formation of aberrant fruit bodies of *G. lucidum in vitro* (Adaskaveg and Gilbertson, 1989). The best medium for growth of *G. lucidum* was Waksman's medium. Basidiocarp production was observed in saw dust medium (moist saw dust 300 g, 10 per cent malt extract and 15 ml of biotin 5 ppm) two months after inoculation. However, the fungus was observed to grow in a wider range of pH, maximum growth of the fungus occurred at pH 5.5 in culture. Glucose was the best carbon source for the fungus, though it can utilize a wide range of carbohydrates including mono, di- and polysaccharides. Peptone served as the best nitrogen source for *G. lucidum* (Nambiar and Radhakrishnan Nair, 1973). Molecular characterization by using, ITS1 (– 5' – TCCGTAGGTGAACCTGCGG – 3') and ITS4 (– 5' – TCCTCCGCTTATTGATATGC – 3') should be employed, for further identification.

Epidemiology

The disease incidence is observed in all soil types but it is predominant in sandy and sandy loam soils. Soil moisture stress during summer, presence of old infected stumps, irrigation through infected palms or fields and non adoption of regular cultural practices increase the disease spread. Coconut palms of 10-50 years of age are generally susceptible to the disease. Coconut hybrids are highly susceptible and are affected by the disease even at the age of five years. Though the disease incidence was found through out the year, disease intensity was high during February – August in Tamil Nadu. The disease incidence was positively correlated with soil temperature and negatively correlated with soil moisture. The disease spreads through root to root contact and transportation of infected soil and plant debris, other host plants like subabul, glyricidia, neem, tamarind, citrus, mango, eucalyptus etc., are also responsible for the disease spread. Irrigated coconut gardens with high water table had more sporophores and the production is more during the rainy seasons (Karunanithi *et al.*, 2005).

Management

Adoption of photosanitary measures like removal of dead palms, burial of affected roots bole in the pit. Application of neem cake 5 kg or 200 kg of FYM or 50 kg of green manure per tree. Irrigation should be done in summer once in 15-20 days. Isolation of affected palm from the healthy ones by digging of trenches around the affected palm. Drenching of the root system with 40.1 of 1 per cent Bordeaux mixture of 0.1 per cent calixin 251/palm. Application of 100 ml calixin 5 per cent through root feeding at quarterly intervals for one year.

Biological Control

Application of *Trichoderma harzanium* and *T. viride* (Gunasekaran *et al.* (1986) and Bhaskaran *et al.* (1988) were found to be antagonistic to *G. lucidum* and application of neem cake encouraged the saprophytic soil microflora especially *Trichoderma spp.* in the control of *Ganoderma* disease. Bhaskaran *et al.* (1988) reported that the soil application of *T. harzanium* at 500 g inoculum along with various organic manure *viz.*, green leaves (50 kg/palm) or FYM (50 kg/palm) or neem cake (5 kg/palm) significantly reduces the disease. Application of *T. viride* and *Pseudomonas fluorescens* talc formulation at the rate of 200 g each to palm in combination with 50 kg FYM was found effective against the disease (Karthikeyan *et al.*, 2005c).

Stem Bleeding

The disease was reported in Andaman and Nicobar Islands by Mitra in 1929. The fungus is reported to be a wound parasite entering the host through wounds or growth cracks on the stem (Briton-Jones, 1941, Menon, 1948, Sundararaman, 1922). The characteristic stem bleeding on coconut is depicted pictorially in Figure 43.2. The disease is caused by stem wounding in the presence of the soil borne plant-pathogenic fungus, *Chalara paradoxa* (syn. *Thielaviopsis paradoxa and Ceratocystis paradoxa*) in a moist environment. *Chalara* sp. is pathogenic to various palm species, causing a range of bud rots and stem rots. The fungus is mainly an opportunist that survives on plant debris in soils. The disease is severe on palms because as monocots they lack the woody dicot's ability to seal off or heal wounds before opportunistic pathogen, such as *Chalara* sp. can cause disease (Nelson, 2005).

Symptoms

Stem bleeding is characterized by the exudation of a dark reddish-brown liquid from growth cracks and wounds on the stem trickling down for a distance of several inches to several feet. Beneath the bleeding lesions the tissue is yellow to brown and finally black in advanced stages, whereupon liquid exudate ceases to flow and turns black. Affected tree cease to bear nuts, the crown dwindles, and the tree finally dies. Young palms having soft stem tissue are more susceptible to infection than older palms. The necrotic breakdown progresses in both directions in the stem, moving very rapidly in

Figure 43.2: Severe Stem Bleeding Symptoms in South Andaman

young palms and producing a soft, watery and tan-coloured rot which yields to slight pressure when applied to the rind (Mitra, 1929).

Pathogen

The causative organism is *Thielaviopsis paradoxa*. The Sphaeronemoid fruit body of *T. paradoxa* has been shown to be a perithecium which bears very characteristic appendages and produces very fugitive asci. The identity of the perithecial form with the conidial forms of *T. paradoxa* has been proved in culture. Work with vegetations originating from single ascospores has shown that the fungus is heterothallic. Molecular characterization by using ITS1 (– 5′ – TCCGTAGGTGAACCTGCGG – 3′) and ITS4 (– 5′ – TCCTCCGCTTATTGATATGC – 3′) should be employed for further identification.

Epidemiology

The disease is usually soil borne and can also proceed after Rhinocerous beetle attack. The disease is most common where wounds occur to palm, stems in highly irrigated or moist landscapes. Wounds exposed to *Chalara*-infested soil are prone to infection, especially where nozzle irrigation impacts the wound. Infections can start from *Chalara* spores or mycelium that survives on decaying plant litter. The pathogen can be spread from soils to coconut stems by splashing rain or irrigation water. *C. paradoxa* infested most soils in Hawaii, especially where sugarcane, pineapple, banana are grown. Growth cracks on the coconut trunk, severe downpours, water stagnation, imbalance in nutrition, excess salinity and plant stress can act as predisposing and aggravating factors (Nelson, 2005).

Management

The disease can be managed by improving the drainage system and avoiding excess stagnation of water during rainy season. Providing irrigation during summer, by scooping out the diseased tissue and sterilizing open wound with heat of a burning torch of coconut leaves followed by the application of hot tar; after which cavity is filled with saw dust and tar. Burn the infected tissue removed by scooping before hot tar treatment. Root feeding of 100 ml of calixin 5 per cent thrice a year during June, October and January prevents further spread of lesions.

Biological Control

Trichoderma viride exerted maximum inhibition of *Thielaviopsis paradoxa* (90 per cent) followed by *T. harzanium* (86.6 per cent) and *Aspergillus niger* (50 per cent) causing stem bleeding in coconut palms.

Bud Rot

The disease is caused by a fungus *Phytophthora palmivora* and *P. katsurae.* Palms of all age groups are susceptible to bud rot disease. A serious fatal disease of coconut occurs commonly in all growing countries (Chandramohanan and Basheer, 2007).

Symptoms

The first visible symptom is withering of the spindle marked by pale colour. The spear leaf or spindle turns brown and bends down. Farmers often experience difficulties in detection of the disease at this stage. Though it is essential to diagnose the disease at the initial stage to take curative measures the affected spindle can easily be pulled out as the basal portion of the spindle is completely rotten emitting a foul smell. Symptoms are later observed in younger leaves next to the spindle. Later the inner leaves also fall away one by one leaving only outer whorl of matured leaves in the crown. Ultimately the palm succumbs to the disease with the death of the growing bud (Chandramohanan and Basheer, 2007) depicted in Figure 43.3.

Figure 43.3: Wilted Crown Showing Severe Symptom of Bud Rot

Pathogen

Malt medium is used for cultivation of *Phytophthora katsurae* and obtaining high mycelium density (Ruth *et al.*, 2010). Oospores are made visible through Microscopy studies. PCR amplification of ITS1 (5'TCCGTAGGTGAACCTGCGG3') and ITS4 (5'TCCTCCGCTTATTGATATGC3') and Amplified Fragment Length Polymorphism (AFLP) employing restriction enzymes like *HinfI*, *MspI*, *AluI*, *HaeIII*, *RsaI* and *EcoRV* to detect polymorphisms (Chowdappa, 2003).

Epidemiology

The pathogens are soil borne and the disease persuade after a rainy season. The microclimatic parameters are found to influence the incidence of bud rot disease. Minimum temperature of 21-24°C and high relative humidity of 97-100 per cent were found optimum for the incidence of the disease. Disease development was found to be determined by the duration of such favourable days. In young palms of 5 to 10 years, the occurrence of such favorable days was more frequent and hence there was more disease incidence in such young palms (Chandramohanan and Basheer, 2007).

Management

When spindle leaves start withering, cut and remove all affected tissues of the crown and apply Bordeaux paste and protect it from rain by providing polythene covering till normal shoots emerge. Burn the disease affected tissues removed from the palm, spray 1 per cent Bordeaux mixture on spindle leaves and crowns of affected palms as well as neighbouring palms before the onset of monsoon.

Biological Control

Numerous studies have examined biological control of *P. palmivora*, using microbial antagonists such as *Bacillus* spp, *Aspergillus tamarii*, *A. gigentus*, *Penicillium purpurescens* and *Pseudomonas fluorescens*, with some success (Galindo, 1992). Two species of the soil dwelling genus *Myrothecium* were found to provide a possible option for biocontrol of bud rot in coconut.

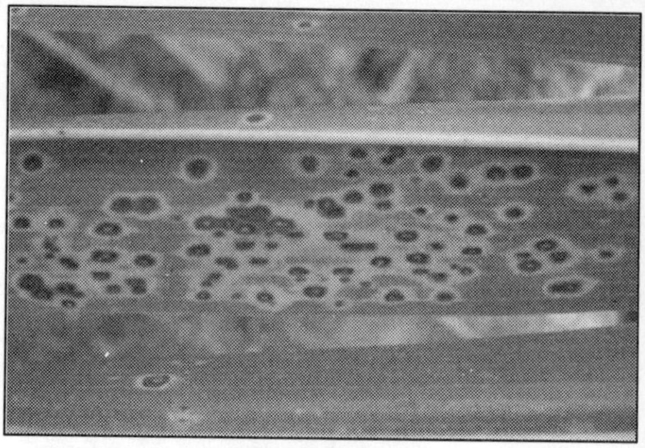

Figure 43.4: Initial Symptoms of Severe Grey Blight

Grey Leaf Blight

This pathogen has previously been reported on coconut from Andaman and Nicobar Islands (Rao, 1988). It is caused by *Bipolaris incurvata*, a fungal pathogen. Figure 43.5 depicts the presence of grey blight in CARI Farm. First report of *Bipolaris* leaf blight of coconut (*Cocos nucifera*) caused by *B. incurvata* in mainland India was reported by Kamalakannan *et al.* (2005). Nevertheless the pathogen is widely prevalent in Andaman and Nicobar Islands, having no lethal damage but more severe. It is otherwise called as Grey Leaf Spot and *Drechslera* leaf spot.

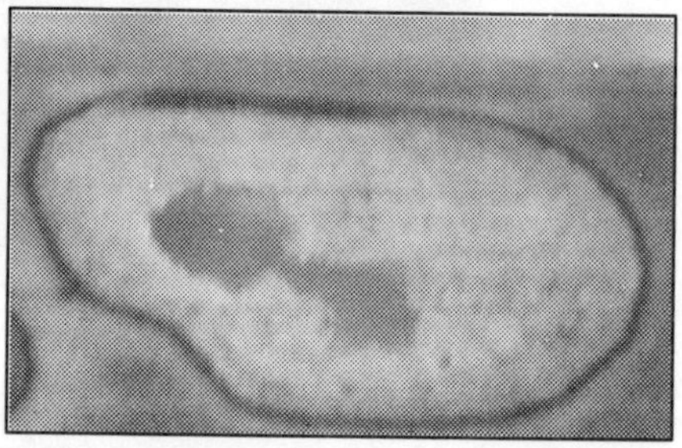

Figure 43.5: Characteristic Grey Blight

Symptoms

The leaf spots appear at first as small, oval, dark brown spots, usually towards the leaf tip, later enlarging and becoming pale buff in the center with a broad, dark brown margin. With multiple infections the lesions coalesce and the leaf becomes blighted from tip to backwards. The incidence of the disease was recorded upto 0.12 per cent (South Andaman), 0.1 per cent (Little Andaman) and 1 per cent (Campbell bay) of the palms covered under preliminary survey (Rao, 1988).

Morphology and Cultural Characteristics of Pathogen

Bipolaris incurvata produces spores that are long, curved, multicellular, and gray to pale olive-brown in color. Pure cultures of *B. incurvata* are relatively easy to establish from diseased plants but the sporulative phase is more difficult to maintain, as compared to other *Bipolaris* species. Spores of this fungus germinate by producing a germ tube (fungal thread) from both ends of the spore, that is germination is bipolar. Germ tubes grow to establish a fungal colony or mass of fungal threads called mycelium. These colonies produce specialized fungal threads called conidiophores, which produce more spores or conidia. The fungus can consistently be isolated from diseased leaves onto potato dextrose agar medium at 20 ± 2°C under near UV light. The mycelium appears septate, branched and brown in colour. Straight or flexuous conidiophores mainly arise in small groups and are 500 µm long, 7–12 µm thick and often swollen to 16-20 µm at the base. The conidia are typically curved, fusiform, septate with 8-13 pseudo-septa, pale straw coloured with a size ranging from 100-150 × 19-22 µm. Based on the symptoms, mycelia and conidial characters, the fungus *B. incurvata* (Ellis, 1971) can be identified.

As the disease advances, the remaining leaves also drop down in quick succession and the spindle alone remains. Subsequently heavy button shedding, upward extension of stem decay and finally death of the palm are observed.

Epidemiology

Moisture favors the production of hundreds of conidiophores that emerge from the epidermis of the coconut leaves. Lower temperatures (*20° C or 68 F*) favor spore formation while higher temperatures (31° C or 88 F) are not conducive. *Bipolaris* spores are spread by wind and rain. Insects are also potential carriers of this pathogen (Uchida and Aragaki, 1991). It gets worse when young coconuts are overcrowded or in heavy shade or have poor soil fertility. Heavy nitrogen fertiliser increases the amount of disease. It gets worse when there is dew on the leaves (Holliday, 1980).

Management

Removal of the older 2-3 disease affected leaves and spraying the foliage with 1 per cent Bordeaux mixture will check the spread of the disease. Sea salt application or K fertilizers are often effective.

Biological Control

Fungal antagonists such as *Trichoderma viride, T. harzanium* and among the bacterial antagonist's *B. subtilis* and *P. fluorescens* were found to inhibit the pathogen. In the field trial *T. harzanium* sprayed coconut palms showed lowest disease index.

Integrated Diasease Management of Coconut

Sanitation and Replanting

Removal and destruction of dead palms and the palms in an advanced stage of the diseases will be useful to prevent spread of the diseases as a whole. Providing regular basin irrigation (once in 7 to 10 days depending on soil type) to diseased palms during summer is recommended. Avoiding irrigation through diseased palms/field is also essential to control the spread of the diseases. The earliest recommendation includes the removal of affected bark tissues and applying hot coal tar on the chiseled surface. Hot coal tar doesn't kill the pathogen completely since it cannot penetrate the affected inner tissues. This treatment is not advisable to the palms below the age of ten years, as it is phytotoxic. For early infections, chisel or scrape out the rotten portion of the disease affected palm. High cost of input for plant protection measures have been often pointed out as the constraints in proper implementation of recommended integrated management practices against diseases of coconut.

Cut and Removal of Severely Affected Palms or Dead Palms

Cut and removal of palms which are in the advanced stage palms dead due to the diseases are very important operations to be carried out in all the affected gardens for better control of the diseases. This will help reduce the inoculums load in the garden as well as secondary spread within the garden and to the neighboring gardens. Rotten position of the crown should be destroyed by burning.

Nutrient Management

Besides integrated disease management techniques, nutrient management practices for the affected palms as well as healthy palms are also important for improving the health and vigor of the palms for higher yield. The fertilizer dose generally recommended for an adult palm is 500 g N, 320 g P_2O_5 and 1200 g K2O per palm per year. Under rain fed condition fertilizer may be applied in two split doses. After the receipt of summer showers, apply one third of the recommended dose of fertilizer around the palm within the radius of 1.8. Dig circular basins of 1.8 cm radius and 25 cm depth around the palm in September – October and spread organic manure in the form of compost or green leaves @ 50 kg per palm in the basin. Then apply two third of the recommended dose of fertilizer over the organic manure and cover the basin with soil. Along with 50 kg organic manure apply 5 kg neem cake containing the antagonistic fungi *Trichoderma* culture to the basin during September. Apply recommended doses of organic manures and chemical fertilizers. Application of recommended dose of NPK fertilizers *i.e.* 1.3 kg Urea, 2 kg super phosphate and 2 kg potash along with 50 kg of FYM is desirable. Soil application of 5 kg neem cake/palm/year will be useful to check the disease. Growing sunhemp, daincha and Kolingi and ploughing *in situ* is also helpful to contain the disease. Soil application of neem or pungam or prosopsis leaves @ 50 kg/palm, will control the disease spread in the palms.

Prophylactic Spraying

Removal of the older 2-3 disease affected leaves and spraying the foliage with 1 per cent Bordeaux mixture will check the spread of the disease. Chemical methods like root feeding of 2 ml of *Tridemorph* in 100 ml of water thrice at quarterly interval combined with soil drenching of 40 litres of 1 percent Bordeaux mixture is effective for managing the disease. Soil application of 500 g each of calcium sulphate and magnesium sulphate/palm/year reduce the disease severity. Coconut stem boring insects like *Xyleborus, Diocalandra* should be controlled by applying Carbaryl 50 per cent WP on the trunk @ 3g per litre water. Chiseling affected tissue and dressing the wound with 5 per cent Calixin effectively manages diseases of coconut in general. The healthy palms which are in the vicinity of diseased and treated ones should be sprayed with 1 per cent Bordeaux mixture as a prophylactic measure to prevent the disease. The spraying has to be directed to the spindle and to the base of 2–3 inner leaves. It is also important to give prophylactic spraying to all palms before the onset of monsoon, in gardens with a previous history of bud rot incidence. In certain dwarf palms, phytotoxic symptoms like brown sunken spots followed by nut shedding have been observed when Bordeaux mixture is sprayed. Keeping perforated sachets containing Mancozeb (Indofil M-45) in the innermost leaf axils during rainy season is found to be useful.

Curative Treatment

Avoid any mechanical injury to the stem. To avoid spread of disease on to upper portion of trunk and root feeding with 5 per cent calixin may be adopted 3 times a year (April-May, September-October and January-February) when the diseases are diagnosed in the early stages, curative measures can save the palms. The appropriate period would be the stage when the spindle leaf has just started showing the symptoms of withering remove the spindle leaf by pulling it out and cut and remove the infected tissues completely. Two or three healthy leaves adjacent to the spindle may have to be removed of all rotten portions and through cleaning. Apply 10 per cent Bordeaux paste to the wound and cover with polythene sheet to prevent entry of rain water. Increasing the spacing of plants, improve the soil fertility, make the nursery less shaded, fertilise young plants with potassium and phosphorus fertilisers and using shade cloth will effectively combat Grey Blight, improving drainage conditions and application of potassic fertilizers will also effectively arrest the progressiveness of the disease.

Biological Management

Biological control summarily involves the application of biofertilizers 200 g *Azotobacter* or Phosphobacteria along with FYM is advised to check the disease spread. Quarterly application of biocontrol agents *viz.* 200 g each of *T. viride* along with *P. flourescens*/palm is effective in controlling the soil borne diseases. Application of *T. viride* and *P. fluorescens* talc formulation at the rate of 200 g each/palm in combination with 50 kg FYM was found effective against *Ganoderma* disease (Karthikeyan *et al.*, 2005c).

Problems and Research Needs

Specific primers and probes for indexing the mother palms are in scarce. Moreover, microscopy like the electron, epi-flourescence microscopes are present minimally for apt identification of pathogens like *Phytoplasma* and characterization of viroids like Cadang- Cadang. Shortage of biomass may surpass the protection scenario as the farmer doesn't know whether he has provided enough nutrition and plant protection. The thrust mainly lies in the above said criteria as the Bay Island is by and large dependant on Organic Coconut Production.

Future Perspectives

Characterization of prevalent diseases in an efficient manner through application of appropriate probes and primers is obvious. Proper learning programs initiated to teach the farmer of the possibilities of the infection and adoption of combating measures. Programs like *Kisan Mela* organized by CARI, Port Blair aims at education through experimentation.

References

Adaskaveg, J.E. and Gilbertson, R.L., 1988. Basiodiospores, pilocystidia and other basidiocarp characters in several species of the *Ganoderma lucidum* complex. *Mycologia*, 80: 493–507.

Adaskaveg, J.E. and Gilbertson, R.L., 1989. Cultural studies of four North American species in the *Ganoderma lucidum* complex with comparisons to *G. lucidum* and *G. tsugae*. *Mycol. Res.*, 92: 182–191.

Andre, Drenth and David, Guest I., 2004. Managing *Phythopthora* diseases: Diversity and management of *Phythopthora* in South East Asia. *ACIAR Monograph*, 114: 156–157.

Anonymous, 1979. Nematodes, fungi, insects and mites associated with the Coconut palm. *Tech. Bull.*, CPCRI, Kasargod, Kerala, India, 2: 236.

Anonymous, 2000. Treatment of stem bleeding in Andaman and Nicobar Islands. *Ind. Coco. J.*, p. 18.

Arunachalam, V., Jerard, B.A., Elangovan, M., Ratnambal, M.J., Dhanapal, R., Rizal, S.K., Damodharan, V., 2001. Unexploited diversity in Coconut palm (*Cocos nucifera* L.). *PGR Newsletter*, p. 121–124.

Batugal, P., Ramanatha Rao, V. and Oliver, J. (Eds.) 2005. *Coconut Genetic Resources*. International Plant Genetic Resources Institute – Regional Office for Asia, the Pacific and Oceania (IPGRI–APO), Serdang, Selangor DE, Malaysia.

Bhaskaran, R., 1990. Biological control of Thanjavur wilt disease of coconut. In: *National Symposium on Biocontrol of Root Disease*, Annamalai University, Annamalainagar, p. 7–8.

Bhaskaran, R., Karthikeyan, A., Marimuthu, R., Ramanathan, A. and Sundaran, T., 2007. Leaf blight disease of coconut palm. *Ind. Coco. J.*, p. 4–6.

Briton-Jones, H. R., 1941. *The Diseases of the Coconut Palm*. Williams and Wilkins Co., Baltimore, 176 p.

Butler, E.J., 1906. Some diseases of palms. *Agril. J. India*, 1: 297–310.

Carter, W., Latta, R.K. and Suah, J.R.R., 1965. The symptoms of lethal yellowing disease of coconut palms. *FAO Plant Prot. Bull.*, 13: 49–55.

Chandramohanan, R. and Mohammad Basheer, B.M., 2007. Integrated management of bud rot disease of coconut. *Indian Coconut J.*, 38(5): 2–3.

Chowdappa, P., Brayford, D., Smith, J. and Flood, J., 2003. Molecular discrimination of *Phytophthora* isolates on Cocoa and their relationship with Coconut, black pepper and bell pepper isolates based on r DNA repeat and AFLP fingerprints. *Curr. Sci.*, 84(9): 1235–1238.

Clerk, G.C., 1971. Germination of Sporangia of *Phytophthora palmivora* (Butl.) Butl. *Annals of Botany*, 36 801–807.

Deng, S. and Hiruki, C., 1991. Amplification of 16S rRNA genes from culturable and non culturabl mollicutes. *J. Microbiol. Methods*, 14: 53–61.

Dulce, R.N. Warwick and Edson, E.M. Passos, 2009. Outbreak of stem bleeding in coconuts caused b *Thielaviopsis paradoxa* in Sergipe, Brazil. *Tropical Plant Pathology*, 34(3): 175–177.

Ellis, M.B., 1971. *Dematiaceous Hyphomycetes*. Commonwealth Mycological Institute, Kew, England.

Govindu, H.C., Rao, A.N.S. and Keshavamurthy, K.V., 1983. Biology of *Ganoderma lucidum* (Leys. Karst. And control of Anabe Roga of coconut. In: *Coconut Research and development*, (Ed.) N.M Nayar. Wiley Eastern Limited, pp. 325–332.

Gundersen, D.E. and Lee, I.-M., 1996. Ultrasensitive detection of *phytoplasmas* by nested-PCR assay using two universal primer pairs. *Phytopathologia Mediterranea*, 35: 144–151.

Harrison, N.A., Narvaez, M., Almeyda, H., Cordova, I., Carpio, M.L. and Oropeza, C., 2002. First repor of Group 16SrIV *Phytoplasmas* infecting Coconut palms with Leaf Yellowing Symptoms on th pacific Coast of Mexico. *BSPP.*

Hodgetts, 2007. Taxonomic groupings based on the analysis on the 16s/23s spacer regions whicl shows greater variation than the normally used 16srRNA gene results in classification similar t that derived from 16s rRNA data but with more detailed subdivisions. *Plant Pathology*, 56: 357- 365.

Holliday, P., 1980. *Fungus Diseases of Tropical Crops*. Cambridge Editions, pp. 143.

Hseu, R.S., Wang, H.H., Wang, H.F. and Moncalco, J.M., 1996. Differentiation and grouping of isolate of the *Ganoderma lucidum* complex by random amplified polymorphic DNA – PCR compared wit grouping on the basis of internal transcribed spacer sequences. *Applied and Environmenta Microbiology*, 64: 1354–1363.

Kamalakannan, A., Valluvaparidasan V., Rabindran, R., Khabbaz, S.E. and Ladhalakshmi, L., 2005 First report of Bipolaris leaf blight of coconut (*Cocos nucifera*) caused by *Bipolaris incurvata* i mainland India. *BSPP*. ID = 012018.

Karunanithi, K., Sarala, L., Manickam, G., Rajarathinam, S. and Khan, H.H., 2005. Management o Basal stem rot of Coconut. *Ind. Coco. J.*, p. 10–11.

Koshy, 2000. Leaf rot disease of coconut. *Ind. Coco. J.*, 31: 4–10.

Koshy, P.K., Kumar, Yogesh, Jayasree, D., Joseph, Udaya and Sosamma, V.K., 2002. Integrate management of leaf rot disease and insect pests on coconut. *Ind. Phytopath.*, 55: 45–50.

Leonard, K.J. and Suggs, E.G., 1974. *Sclosphaeria prolata*, the ascigerous stale of *Exscrohihim prolalum* Mycologia, 66: 281–297.

Menon, K.P.V. and Pandalai, K.M., 1948. *The Coconut Palm: A Monograph*. Times of India Press, Bombay 384 p.

Mitra, M., 1929. Some diseases of crops in the Andaman Islands. *Pusa Agr. Res. Inst. Bull.*, 195, 14 p.

Musetti, R. and Favali, M.A., 2004. Microscopy Techniques Applied to the Study of *Phytoplasm* Diseases: Traditional and Innovative Methods. Current Issues on Multidisciplinary Microscop Research and Education, p. 72–80.

Naik, R.G., Palanimuthu, V., Hanumanthappa, M. and Indiresh, K.M., 2000. Prevalence and intensity of basal stem rot disease of Coconut in Arsikere taluk of Karnataka. *Ind. Coco. J.*, 31: 8–11.

Nair, M.K., 2001. Disease and pest management. In: *Handbook of Horticulture*, (Ed.) K.L. Chadha. Indian Council of Agricultural Research Publications, p. 854–857.

Nambiar, K.K.N. and Radhakrishnan Nair, R., 1973. Investigations on Anabe disease of Arecanut caused by *Ganoderma lucidum* (Leys.) Karst. *J. Plant Crops.*, 1 (suppl.): 119–123.

Nambiar, K.K.N. and Rethinam, P., 1986. *Thanjavur Wilt/Ganoderma Disease of Coconut*. Pamphlet No. 30. Central Plantation Crops Research Institute, Kasaragod.

Nelson, S., 2005. Stem bleeding of Coconut palm. *Plant Disease*, 30: 1–2.

'lavsic-Banjac, B., Hunt, P. and Maramorosch, K., 1972. *Mycoplasma* like bodies associated with lethal yellowing disease of coconut palms. *Phytopathology*, 62: 298–299.

.ao, S.C., 1988. Leaf spot disease of coconut incited by *Bipolaris incurvata* (Bernad) Alcorn. *Current Science*, 57: 674–675.

.uth Yoa, N.A., Goran, B.N., Allou, K., Konan, K.J. and Kouassi, P., 2010. Growth characteristics of *Phytophthora katsurae* isolates from four coconut tree growing areas in Ivory coast. *Eur. J. Scientific Res.*, 41(1): 162–171.

alam, M.A., Srivastava, R.C., Pandey, G.S., Singh, D.R., Sharma, T.V.R.S., Kumar, Ramesh, Hassan, M.M., Rekha, S.C. and Singh, Arti, 2010. Status of coconut cultivation in A&N Islands. In: *Proceedings of National Workshop on Coconut Based Farming System in A&N Islands*, p. 31–34.

amiyappan, R., Karthikeyan, G., Kandan, M., Raguchander, T. and Rajendran, T., 2006. Basal stem rot disease in coconut: Recent developments. *Ind. Coco. J.*, p. 2–12.

anchez, V., Rebolledo, O., Picaso, R.M., Cardenas, E., Cordova, J., Gonzalez, O. and Samuels, G.J., 2007. *In vitro* antagonism of *Thielaviopsis paradoxa* by *Trichoderma longibrachiatum*. *Mycopathologia*, 163: 49–58.

eo, G.S., 1987. Studies on cultural characteristics of *Ganoderma lucidum* (Fr.) Karst. *M.Sc. Thesis* (in Korea).

inivasan, N. and Gunasekaran, M., 1996. Incidence of fungal species associated with leaf rot disease of coconut palms in relation to weather and stage of lesion development. *Annals. of. App. Biol.*, 129: 433–449.

inivasan, N., 1991. Occurrences of coconut leaf rot in relation to Root Wilt disease. *Ind. Coco. J.*, 21(10): 14–18

ındararaman, S., 1922. The coconut bleeding disease. *Pusa Agr. Res. Inst. Bull.*, 127, 8 p.

chida, J.Y. and Aragaki, M., 1991. *Bipolaris* and *Exserohilum* leaf spots. In: *Diseases and Disorders of Ornamental Palms*, (Eds.) Chase and Broschat. APS Press, St. Paul Minnesota, p. 55.

tomo, C. and Niepold, F., 2000. Development of diagnostic methods for detecting *Ganoderma*–infected oil palms. *J. Phytopath.*, 148: 507–514.

jayan, K.M. and Natarajan, S., 1972. Some observations on the coconut wilt disease of Tamil Nadu. *Coconut Bulletin*, 2(12): 2–4.

arokka, J .S., Jones, P. and Dickinson, M.J., 2006. Detection of *Phytoplasmas* associated with Kalimantan wilt disease of coconut by the polymerase chain reaction. *Journal Littri.*, 12(4): 154–160.

Plant Protection, Diversity and Conservation (2011) *Pages* **578–599**
Editors: **D. Prasad & Satya Kumar**
Published by: **BIOTECH BOOKS, NEW DELHI**

Chapter 44

Present Status of Insect-Pests of Vegetables in Bay Islands and their Future Strategies for Pest Management

Ajanta Birah*, Krishna Kumar, Someshwar Bhagat and R. C. Srivastava
Central Agricultural Research Institute,
Port Blair – 744 101, Andaman and Nicobar Islands

The Andaman and Nicobar group of Islands forms an arched string of 572 Islands and islets stretching from Burma in the north to Sumatra in the south. Andaman and Nicobar group of Islands are located in the Bay of Bengal and falls under the island ecosystems. The climate is unique with humid tropical (temp 28-32°C and RH 75-95 per cent), endowed with the occurrence of both South-West and North-East monsoon and the rainy season is of more than eight months in a year with an average rainfall of 3100mm. In Andaman and Nicobar Islands the total available land for Agriculture is limited to 50,000 ha. *i.e.* 6 per cent of the total geographical area only. In the total cultivated area around 22 per cent is being utilized for rice cultivation, about 50 per cent of the area is covered by coconut, 9 per cent each by arecanut and fruit crops and the remaining area is covered by spices and vegetables.

The agro climatic conditions of Andaman and Nicobar islands are extremely suitable for the cultivation of variety of crops. Intensive agronomic practices and indiscriminate use of insecticides has disrupted the delicate balance between the insect pests and their natural enemies. Biorational or biointensive pest management is an ecofriendly pest management technique which devoid of harmful chemical pesticides and includes biopesticides based on bacteria, fungi and viruses and naturally occurring bio chemical pheromones, insect growth regulators, plant derivatives and neem based pesticides etc. Almost all types of vegetables are being grown in the Bay Islands. Earlier the vegetable

* E-mail: ajantabirah@rediffmail.com

crops were grown as inter or subsidiary crop rather than sole crop. But due the intensive and participatory effort of CARI, Port Blair, farmers started to grow and cultivate vegetables as sole crops. But still only 1/3rd of total demand of vegetables can be produced right now and the productivity of vegetables in the bay Island is far below than the national average. Vegetables cover as a major crop in these Islands in about 4598 ha every year, vegetables are cultivated in Kharif season in the limited area of South Andaman, Neil Island, Havelock and little Andaman where the soils are mostly porous, well drained and suitable for vegetable cultivation. Generally indigenous vegetables like okra, cowpea, brinjal, cucurbits, amaranthus, radish etc. are cultivated in Kharif season. Cultivation of vegetable is taken up throughout the Islands in the Rabi season in the fallow land available after harvest of paddy. The major chunk falls in Diglipur, Mayabunder, Neil island and south Andaman. Though vegetables are cultivated in summer season their cultivation is confined to selected areas only where there are facilities for irrigation either from sunken well, ponds etc. Some of the polyphagous pests *viz.*, fruit borers, cut worms, pod borers, jassids, whiteflies etc. attacks majority of the vegetables. In this paper we will emphasize on pests of common vegetable crops grown in Bay islands.

A. Solanaceous Vegetables

Brinjal (*Solanum melongena*)

Shoot and Fruit Borer *Leucinodes orbonalis* Guen. (Noctuidae: Lepidoptera)

Distribution

The fruit and shoot borer is the major insect pest of brinjal, (*Solanum melongena*) in Andaman & Nicobar Islands.

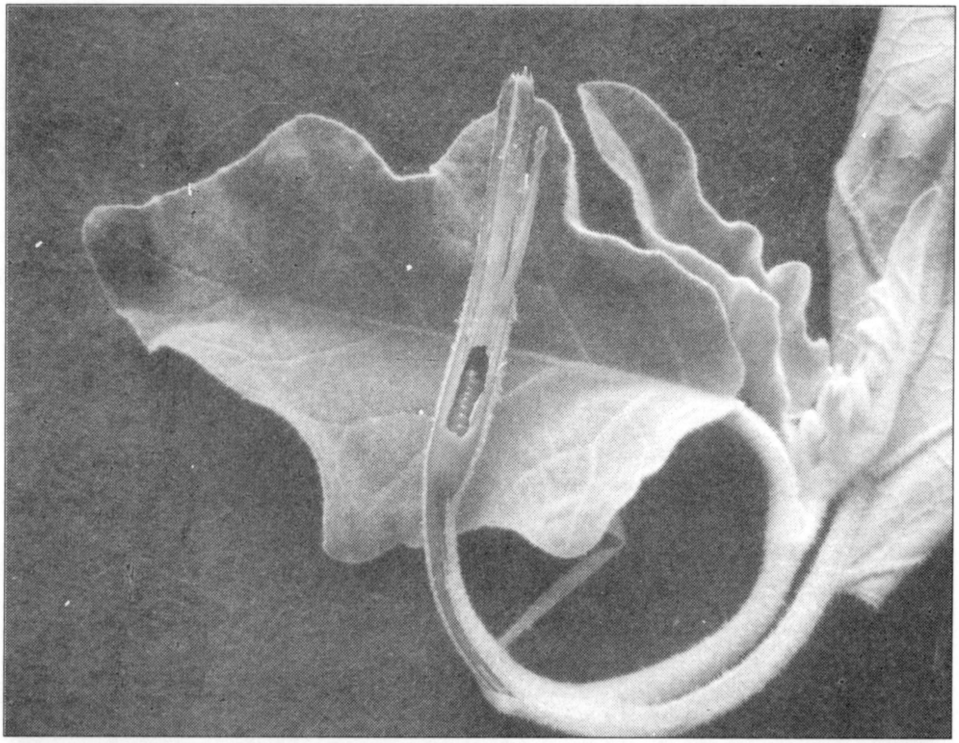

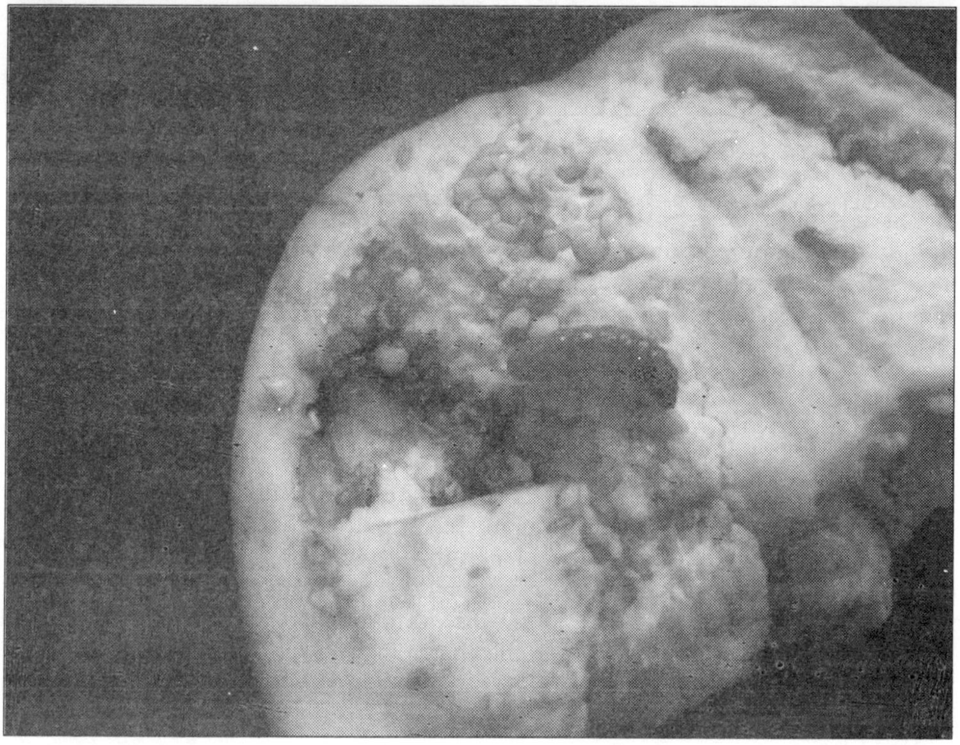

Nature of Damage

The larva bores into tender shoots in the early stage and causes "dead hearts". It also bores into flower buds and developing fruits causing shedding of buds and making the fruits unfit for consumption and marketing. The infestation may go as high as 50-70 per cent on brinjal. Larvae bore into shoots during the vegetative growth stage and later in flowers and fruits rendering fruit unfit for human consumption. In Neil Island insecticide applications frequently exceed three times in a week.

Life History

The caterpillars are creamy white turning pink colored with maturity and measure 18–23 mm in length with brownish head. The moth is white with pale brown spots on dorsal side. The wings are white with a pinkish or bluish tinge and are ringed with small hairs along the apical and anal margins. The forewings are ornamented with many black, pale brown spots. The female lays about 250 eggs singly on tender shoots and developing fruits of brinjal. The pinkish larva with sparsely distributed hairs on warts on the body and a brownish head measures 16–20 mm long. It pupates in a tough grayish cocoon on the plant itself. The egg, larval and pupal periods occupy respectively 3–5, 15 and 6–8 days. The adult lays 30-253 eggs and they hatch in 3–6 days. The larval period lasts for 12-15 days and the larvae pupate in plant debris or soil. The pupal period is 6-17 days, the emerged adults are short lived for 1-3 days.

Whitefly – *Bemisia tabaci* (Aleyrodidae: Hemiptera)

Distribution

Whitefly is found in various locations of Andaman & Nicobar Islands. Its other hosts are tomato, chilli, okra, etc., the infestation on these crops is sporadically severe.

Nature of Damage

White, tiny, scale-like insects may be seen darting about near the plants or crowding in between the veins on ventral surface of leaves, sucking the sap from the infested parts. The pest is more active during the dry season and its activity decreases with the onset of rains. As a result of their feeding the affected parts become yellowish, the leaves wrinkle and curl downwards and are ultimately shed. Besides the feeding damage, these insects also exude honeydew which favours the development of sooty mould. In case of severe infestation, this black coating is so heavy that it interferes with the photosynthetic activity of the plant resulting in stunted growth. This whitefly also acts as a vector, transmitting the leaf curl virus.

Life History

Eggs are pear-shaped, light yellowish in colour, about 2 mm long and can be seen standing upright on leaves, being anchored by a tail-like appendage inserted into the stoma of leaves. On hatching, the nymphs crawl a little, settle down on a succulent spot on the same leaf and never move

again during that stage. Nymphs are oval, scale-like and greenish-white in colour. Adults are minute insects, about one mm long, covered completely with a white waxy bloom. Incubation period is 3–5 days in summer extending up to 33 days during winter. Nymphal development takes 9–14 and 17–81 days in summer and winter, respectively and pupal period lasts for 2–8 days being longer during winter than in summer. A life-cycle may be completed in as little as 14 days or it may even be prolonged up to 107 days. There are about 12 overlapping generations in a year.

Tomato

Tomato Fruit Borer *Helicoverpa armigera* (Noctuidae: Lepidoptera)

Distribution

Fruit borer is a major pest of tomato. This is widely prevalent and attacks many crops in Bay islands and extent of damage ranges 15-40 per cent . This is a polyphagous pest attacking chilli, cauliflower, okra, maize and pulses (French bean, cowpea etc.).

Nature of Damage

Eggs are laid singly, generally on leaves and flowers but sometimes on fruits as well. On hatching, the young larvae feed on tender foliage; advanced stage larvae attack the fruits. They bore circular holes and thrust only a part of their body inside the fruit and eat the inner contents. If the fruit is bigger in size, it is only partly damaged by the caterpillar but later it is invariably invaded by fungi and

bacteria and spoiled completely. The larvae move from one fruit to another and a single caterpillar may eat and destroy 2–8 fruits. It is capable of causing 10-60 per cent damage in tomato, chilliy and okra.

Life History

The eggs are yellowish-white, ribbed and dome-shaped, 0.4–0.5 mm in diameter. Freshly hatched larvae are yellowish-white in colour but gradually change and acquire greenish tinge. Full grown caterpillars are 40–48 mm long, apple-green in colour with whitish and dark-grey broken longitudinal stripes. Full grown caterpillars drop down from the plants and burrow in the soil where they pupate. Pupae are dark-brown in colour, 11–14 mm long and have a sharp spine at the anal end. Moths are medium sized, stout, ochreous with olive grey forewings in male and reddish brown forewings in female with dark brown circular spot in the centre and indistinct double waved antemedial lines. Hind wings are pale smoky-white with a broad blackish outer border. Fecundity of females is rather high ranging between 1200–1600 eggs. Incubation, larval and pupal stages last for 2–4, 15–24 and 10–14 days respectively and the entire life-cycle may be completed in 4–6 weeks.

Tobacco Cut Worm (*Spodoptera litura*), Noctuidae, Lepidoptera

Distribution

This is a polyphagous pest attacking chilli, cauliflower and okra also other than tomato.

Nature of Damage

It attacks seedlings, fruits and causes severe defoliation and damages up to 20-80 per cent .The caterpillars attack the crop from nursery stage itself causing defoliation.

Life History

The adult moth lays egg mass on the lower surface of leaf, each egg mass have 200-300 eggs. On hatching the tiny larvae scrape the green matter but older larvae defoliate the leaf. The caterpillars are initially greenish in colour but with maturity they grow dark coloured and become robust. The pest completes larval period in 6 instars. The pest pupates in the soil and crop debris. The adult moth is brown with a complex pattern of cream streaks criss-crossing the fore wings. The hind wings are silvery white and it has a wingspan of about 4 cm. The males have a blue-grey band from the apex to the inner margin of each fore wing.

Chillly

Fruit borer (*Helicoverpa armigera*) is the major pest found in Andaman and Nicobar Islands. The damage and management of this fruit borer is same as described for tomato fruit borer. Spiralling Whitefly, *Aleurodicus disperses* (Aleyrodiodae, Homoptera) and chilly mite, *Polyphagotarsonemus latus* (Tenuipalpidae, Acarina) are also found as occasional pest in Bay islands.

Cruciferous Vegetables (Cabbage, Cauliflower, Radish and Knolkhol)

Diamond Back Moth *Plutella xylostella* (L.) (Plutellidae: Lepidoptera)

Distribution

This pest is widely prevalent in Diglipur, North Andaman, where cole crops are cultivated over large areas and in South Andaman damage is very less.

Nature of Damage

The larvae bite holes in leaves and cause serious damage. The larvae cause blisters on the leaves which dry away in course of time. It attacks cabbage, cauliflower, radish and knolkhol.

Life History

The moth is grayish brown with narrow wings having pale white markings anteriorly which from diamond-like white patches dorsally when wings are folded over back at rest. It lays up to 57 eggs singly on the under surface of leaves along the veins. Egg period is 4 – 5 days. The larva is greenish with short thin hairs on the body. The larval period is 13 – 21 days. It pupates in a thin loose mesh of silken cocoon and the pupal period is 7 – 9 days. The period from egg to adult occupies 24–35 days.

Bhindi (Okra) (*Hibiscus esculentus* Moench)

Okra Shoot and Fruit Borer *Earias vittella* (Fabricius) (Arctidae: Lepidoptera)

Distribution

It is a major pest of okra in Bay islands. It is also found feeding on a large number of malvaceous plants, both wild as well as cultivated.

Nature of Damage

Eggs are usually laid singly on buds and flowers and occasionally on fruits as well, but in absence of these parts *i.e.*, at early stage of crop growth, the eggs are laid on shoot tips. When the crop is only a few weeks old, the freshly hatched larvae bore into tender shoots and tunnel downwards, these shoots wither, droop down and ultimately the growing points are killed, side shoots may arise

giving the plants a bushy appearance. With the formation of buds, flowers and fruits, the caterpilla bore inside these and feed on inner tissues. They move from bud to bud and fruit to fruit thus causi damage to a number of fruiting bodies. The damaged buds and flowers wither and fall down witho bearing any fruit whereas the affected fruits become deformed in shape and remain stunted in growt

Life History

Eggs are spherical in shape, about half mm in diameter, light bluish-green in colour and beautiful sculptured with 26–32 longitudinal ridges; the alternate ridges project upwards to form a crown. T moths emerge at dusk; mating takes place 2–3 days after emergence and oviposition commences aft 1–5 days of mating. A female lays on an average 400 eggs (65–695). Incubation, larval and pup periods last for 3–9, 9–20 (50–60 during winter) and 8–12 days respectively. A single life-cycle tak 22–25 days extending up to 74 days during winter and there may be 8–12 generations in a year.

Leaf Hopper *Amrasca biguttula biguttula* (Cicadellidae: Homoptera)

Distribution

Distributed throughout the locations of Andaman and Nicobar Islands. It is a polyphagous pest attacking okra, brinjal, beans, cucurbits and other malvaceous plants.

Nature of Damage

The nymphs and adults remain on the under surface of the leaves and suck the cell sap and while feeding inject their toxic saliva. As a result the plant become stunted, the leaves crinkle, turn yellowish and become cup shaped. Brownish or reddish colour may develop along the edges of the leaves. This is called the 'hopper burn'. This pest is very active from September to January.

Life History

The adult is a slender green insect. Elongate, yellowish eggs are laid singly inside the leaf vein on the under surface of the leaves. A female lays 15–30 eggs, leaves of 35 – 40 days old are preferred for egg laying. Egg hatch in 4 -10 days. Nymphs are green and wedge shaped. The first and second instar nymphs feed mostly near the base of the leaf vein. Then they distribute themselves throughout the leaf and feed from the under surface. The nymphs develop into adults in 7 – 21 days. They breed throughout the year.

Cucurbitaceous Vegetables (Cucumber, Ridge gourd, Bitter gourd, Snake gourd, Pumpkin and Bottle gourd)

Fruitflies, *Bactrocera cucurbitae* Coq. (Tephritidae : Diptera)

Distribution

It is commonly called melon fruit fly. It is widely distributed throughout the Andaman and Nicobar Islands. It is highly polyphagous having a wide range of host plants but its preferred hosts are bittergourd followed by cucumber, ridge gourd and snake gourd whereas pumpkin and bottle gourd and relatively less attacked.

Nature of Damage

B. cucurbitae is slightly bigger with brown marking on wings. The female flies puncture the soft and tender fruits with their stout and hard ovipositor and lay eggs below the epidermis @ 4–10 eggs per fruit each time. A single female can lay about 200 eggs in her life span of 8–10 weeks. A puncture made by one female is often used by others also for ovipositing and a single fruit may have more than one puncture made by one or more females. On hatching the maggots feed inside on the pulp of fruits and the infested fruits can be identified by the presence of brown resinous juice which oozes out of the punctures made by the flies for oviposition. These punctures also serve as an entry for various bacteria and fungi; as a result, the infested fruits start rotting, get distorted and malformed in shape and fall off from the plants pre-maturely.

Life History

In nature the population is generally low during dry weather and increases rapidly with adequate rainfall. Pre-oviposition, egg, maggot and pupal periods last for 9–21, 1–1½, 3–9 and 6–8 days respectively. During winter the larval and pupal stages are extended up to 3 and 4 weeks, respectively. A single life-cycle is completed in 10–18 days but it takes 12–13 weeks to complete a single life-cycle in winter. Adult longevity is 2–5 months; females live longer than males. Eggs are 1.0–1.5 mm long, whitish in colour, elongate, cylindrical in shape, slightly curved and tapering at both ends. Full grown maggots are 5–10 mm long, cylindrical in shape, tapering anteriorly, blunt at posterior end and

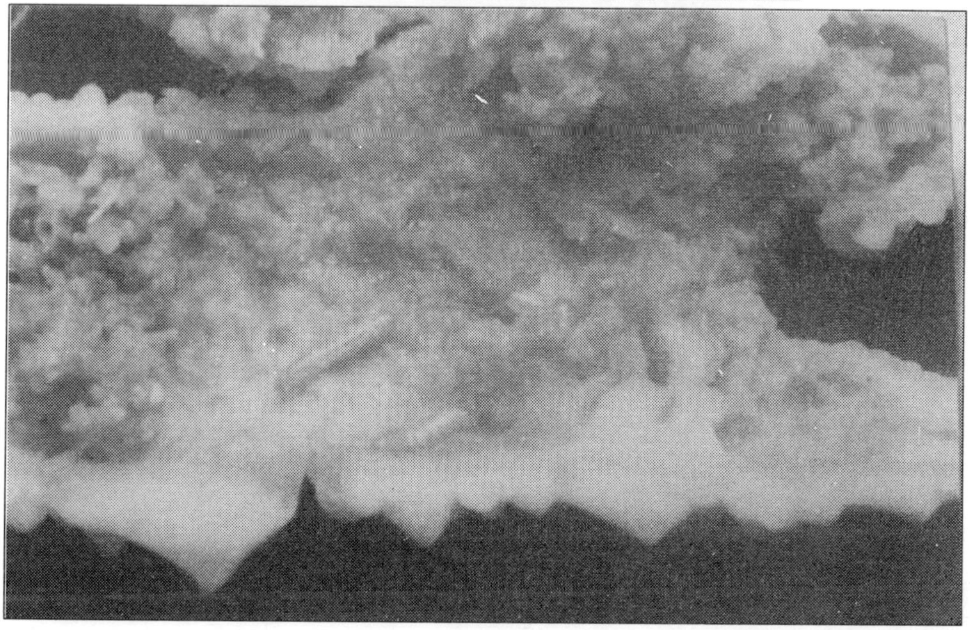

pale-white in colour. Pupae are 5–8 mm long, barrel shaped and brown to ochraceous in colour. The fully fed maggots come out of the fallen fruits and pupate 10–15 cm deep in the soil. Where the fruits do not fall, the maggots pupate inside the fruits (which are not common) or come out, drop down and pupate in the soil.

Pumpkin Beetle *Raphidopalpa foveicollis* (Lucas), (Galerucinae : Coleoptera)

Distribution

It is widely distributed throughout the Andaman and Nicobar Islands.

Nature of Damage

The beetles bite irregular holes on leaves and also feed on flowers. They prefer young seedlings and tender leaves and the damage at this stage may kill the seedlings. The roots as well as the stem and fruits that come in contact with the soil are damaged by the grubs. Pumpkin is preferred by *R. foveicollis*.

Life History

Adult beetles of *R. foevicollis* are 6 – 8 mm long having glistening yellowish red to yellowish brown elytra that are uniformly covered with fine punctures. They are active from March to October and the peak period of activity in April to June. The yellowish pink spherical eggs are laid in the soil which turns orange after two days and a beetle may lay 150–300 eggs. The egg period is 5–8 days. The grubs become full grown in 13–25 days and pupate in the soil. The prepupal period is 2–5 days. The pupal period ranges from 7–17 days. Total life-cycle occupies 32–65 days. In a year there may be 5–8 generations of the insect.

Future Strategies

The major pesticide consumption is in vegetables that necessitate switching over to IPM giving least importance to pesticides in vegetables. Though the usage of pesticides is still a dominating tool

to combat pests, it is conceded that the use of pesticides alone is not a satisfactory solution to the pest problem and some suitable alternative methods should be developed to minimize the pest incidence for sustainable production of vegetables in Bay Islands.

Plant origin insecticides also have a great advantage by being compatible with other low risk options which are acceptable for insect management, such as pheromones, oils, detergents, entomopathogenic fungi, predators and parasitoids, among others, which greatly increase probabilities of being integrated in IPM programs (Bhanukiran *et al.*, 2005). There is a need to ensure the supply of effective, biocontrol agents and plant products by private sectors. Egg parasitoids are widely recommended but they are not available to the farmers. There is also a need to promote the supply of quality bioagents and in sufficient quantity and use of light trap and trap crops helping in minimizing pest incidence should be promoted. Planning should be done before planting because many pest strategies require steps or inputs, such as beneficial organism habitat management, that must be considered well in advance (Jose *et al.*, 2008). Attempting to jump-start an IPM programme in the beginning or middle of a cropping season generally does not work.

There is a need to educate farmers about the importance of cultural and maintenance practices including summer ploughing, pruning, cleaning and maintenance of plants. With this, Islands can move one step forward towards organic farming. This will reduce the use of dangerous chemicals which may cause health hazards and also reduce cost of cultivation.

Why Move to Biorational IPM?

Biorational IPM incorporates ecological and economic factors into agricultural system design and decision making, and addresses public concerns about environmental quality and food safety. The benefits of implementing biorational IPM can include reduced chemical input costs, reduced on-farm and off-farm environmental impacts, and more effective and sustainable pest management. An ecology-based IPM has the potential of decreasing inputs of fuel, machinery, and synthetic chemicals—all of which are energy intensive and increasingly costly in terms of financial and environmental impact. Such reductions will benefit the grower and society.

Pest Identification

A crucial step in any IPM programme is to identify the pest. The effectiveness of both proactive and reactive pest management measures depend on correct identification. Misidentification of the pest may be worse than useless; it may actually be harmful and cost time and money.

Monitoring

Monitoring involves systematically checking crop fields for pests and beneficials, at regular intervals and at critical times, to gather information about the crop, pests, and natural enemies. Sweep nets, sticky traps, and *pheromone traps* can be used to collect insects for both identification and population density information.

Economic Injury and Action Levels

The economic injury level (EIL) is the pest population that inflicts crop damage greater than the cost of control measures. Because growers will generally want to act before a population reaches EIL, IPM programmes use the concept of an economic threshold level (ETL or ET), also known as an action threshold. The ETL is closely related to the EIL, and is the point at which suppression tactics should be applied in order to prevent pest populations from increasing to injurious levels.

Components of Biorational IPM

An important difference between conventional and biointensive IPM is that the emphasis of the latter is on proactive measures to redesign the agricultural ecosystem to the disadvantage of a pest and to the advantage of its parasite and predator complex. At the same time, biointensive IPM shares many of the same components as conventional IPM, including monitoring, use of economic thresholds, record keeping, and planning.

Good planning must precede implementation of any IPM programme, but is particularly important in a biointensive program. Planning should be done before planting because many pest strategies require steps or inputs, such as beneficial organism habitat management, that must be considered well in advance (Jose *et al.*, 2008). Attempting to jump-start an IPM programme in the beginning or middle of a cropping season generally does not work.

When planning a biointensive IPM program, some considerations include:

- ✰ Options for design changes in the agricultural system (beneficial organism habitat, crop rotations).
- ✰ Choice of pest-resistant cultivars.
- ✰ Technical information needs.
- ✰ Monitoring options, record keeping, equipment, etc.

Proactive Strategies (Cultural Controls)

- ✰ Healthy, biologically active soils (increasing belowground diversity)
- ✰ Habitat for beneficial organisms (increasing aboveground diversity)
- ✰ Appropriate plant cultivars

Cultural controls are manipulations of the agroecosystem that make the cropping system less friendly to the establishment and proliferation of pest populations. Although they are designed to have positive effects on farm ecology and pest management, negative impacts may also result, due to variations in weather or changes in crop management.

Other Cropping Structure Options

Multiple cropping is the sequential production of more than one crop on the same land in one year. Depending on the type of cropping sequence used, multiple cropping can be useful as a weed control measure, particularly when the second crop is interplanted into the first.

Intercropping is the practice of growing two or more crops in the same, alternate, or paired rows in the same area. This technique is particularly appropriate in vegetable production. The advantage of intercropping is that the increased diversity helps "disguise" crops from insect pests, and if done well, may allow for more efficient utilization of limited soil and water resources. Disadvantages may

relate to ease of managing two different crop species–with potentially different nutrient, water, and light needs, and differences in harvesting time and method–in close proximity to each other.

Other Cultural Management Options

- ☆ Use of disease-free seed and nursery stock is important in preventing the introduction of disease.
- ☆ *Resistant varieties* are continually being bred by researchers. Growers can also do their own plant breeding simply by collecting non-hybrid seed from healthy plants in the field.
- ☆ Sanitation involves removing and destroying the over wintering or breeding sites of the pest as well as preventing a new pest from establishing on the farm If, however, sanitation involves removal of crop residues from the soil surface, the soil is left exposed to erosion by wind and water.
- ☆ *Spacing of plants* heavily influences the development of plant diseases and weed problems. The distance between plants and rows, the shape of beds, and the height of plants influence air flow across the crop, which in turn determines how long the leaves remain damp from rain and morning dew. Generally speaking, better air flow will decrease the incidence of plant disease. However, increased air flow through wider spacing will also allow more sunlight to the ground, which may increase weed problems.

- *Optimum growing conditions* are always important. Plants that grow quickly and are healthy can compete with and resist pests better than slow-growing, weak plants. Too often, plants grown outside their natural ecosystem range must rely on pesticides to overcome conditions and pests to which they are not adapted.
- *Mulches*, living or non-living, are useful for suppression of weeds, insect pests, and some plant diseases. Hay and straw, for example, provide habitat for spiders. Research in Tennessee showed a 70 per cent reduction in damage to vegetables by insect pests when hay or straw was used as mulch. The difference was due to spiders, which find mulch more habitable than bare ground. Mulching helps to minimize the spread of soil-borne plant pathogens by preventing their transmission through soil splash. Mulch, if heavy enough, prevents the germination of many annual weed seeds. Winged aphids are repelled by silver- or aluminum-coloured mulches.

Biological Controls

Study and utilization of natural enemies of insect like predators, parasites and pathogens by man to manage pest population below economic injury level is called biological control of insect pest.

- *Predators*: Lady bird beetle: This insect feeds on aphids. *Chrysoperla carnea*: The larvae of this insect feeds on all soft bodied insects like aphids, jassids, white files, mealy bug etc.

- ☆ *Parasites*: Egg parasite: *Trichogramma chilonis* parasites egg of *Helicoverpa armigera.*
- ☆ *Larval parasite*: Bracon hibitor parasites larvae of *H. armigera*. Egg larval parasite: *Copidosoma kohleri* parasites egg of potato tuber moth and comes out at larval stage by killing the pest.
- ☆ *Pathogens Bacteria*: *Bacillus thuringiensis* develops disease in many lepidopterous pests (Sharma and Odak1996).
- ☆ *Virus*: Though there are many reports of entomopathogenic viruses Nuclear Polyhedrosis Virus (NPV) and Granulosis Virus (GV) are commonly used in insect pest control.
- ☆ *Fungus: Beauveria bassiana* is used for control of lepidopterous pests.

Chemical Controls

Included in this category are both synthetic pesticides and botanical pesticides.

Synthetic pesticides comprise a wide range of man-made chemicals used to control insects, mites, weeds, nematodes, plant diseases, and vertebrate and invertebrate pests. These powerful chemicals are fast acting and relatively inexpensive to purchase. Pesticides are the option of last resort in IPM programs because of their potential negative impacts on the environment, which result from the manufacturing process as well as from their application on the farm. Pesticides should be used only when other measures, such as biological or cultural controls, have failed to keep pest populations from approaching economically damaging levels. If chemical pesticides must be used, it is to the grower's advantage to choose the least-toxic pesticide that will control the pest but not harm non-target organisms such as birds, fish, and mammals. Pesticides that are short-lived or act on one or a few specific organisms are in this class. Examples include insecticidal soaps, horticultural oils, copper compounds (*e.g.*, bordeaux mixture), sulphur, boric acid, and sugar esters.

Biorational Pesticides

Although use of this term is relatively common, there is no legally accepted definition. Biorational pesticides are generally considered to be derived from naturally occurring compounds or are formulations of microorganisms. Biorationals have a narrow target range and are environmentally benign. Biorational pesticides are those synthetic, organic, or inorganic pesticides that are both low toxicity and exhibit a very low impact on the environment. They also have minimal impact on species for which they are not intended (non-target species). Biorational pesticides include oils, insecticidal soaps, microbials (such as *Bacillus thurengienesis* and entomopathogenic nematodes), botanical (plant-based) and insect growth regulators. Biorational pesticides should therefore be your first choice whenever a pesticide is needed.

Botanical pesticides are prepared in various ways. They can be as simple as pureed plant leaves, extracts of plant parts, or chemicals purified from plants. Pyrethrum, neem formulations, and rotenone are examples of botanicals (Bhanukiran *et al.*, 2005). Some botanicals are broad-spectrum pesticides. Others, like ryania, are very specific. Botanicals are generally less harmful in the environment than synthetic pesticides because they degrade quickly, but they can be just as deadly to beneficials as synthetic pesticides. However, they are less hazardous to transport and in some cases can be formulated on-farm. The manufacture of botanicals generally results in fewer toxic by-products.

A successful biointensive IPM programme takes time, money, patience, short- and long-term planning, flexibility, and commitment. Be aware that some IPM strategies, such as increasing beneficial insect habitat, may take more than a year to show results. A well-run biointensive IPM system may require a larger initial outlay in terms of time and money than a conventional IPM programme. Direct pesticide application costs are saved and equipment wear and tear may be reduced. In the long run, however, a good biointensive IPM programme should pay for itself.

References

Bhanukiran, V. and Panwar, V.P.S., 2005. Bioefficacy of neem products and insecticides against maize stalk borer, *Chilo partellus* (Swinhoe). *Indian J. Ent.*, 67: 24–28.

Deepthi, Jose, Shekharappa and Patil, R.K., 2008. Evaluation of biorational pesticides for the management of stem borer, *Chilo partellus* Swinhoe in sweet sorghum. *Karnataka J. Agric. Sci.*, 21(2): 293–294.

Sharma, M.L. and Odak, S.C., 1996. Efficacy of *Bacillus thuringiensis* separately and in combination with chemical insecticide against sorghum stem borer. *Indian J. Ent.*, 58: 354–358.

Shivlingaswamy, T.M. and Satpathy, S., 2007. Integrated pest management in vegetable crops. In: *Entomology: Novel Approaches*, (Eds.) P.C. Jain and M.C. Bhargava. New India Publishing Agency, New Delhi, pp. 353–373.

Srinivasan, K. and Krishna Moorthy, P.N., 1991. Indian mustard as a trap crop for management of major lepidopterous pests on cabbage. *Tropical Pest Management*, 37: 26–32.

Srinivasan, K. and Krishna Moorthy, P.N. 1992. Develpoment and adoption of Integrated Pest Management for major pests of cabbage using Indian mustard as trap crop. Diamondback moth and other crucifer pests. In: *Proceedings of the Second International Workshop*, Tainan, Taiwan, 10–14 Decmber, 1990. Asian Vegetable Research and Development Centre, AVRDC Publication Centre, AVRDC Publication No. 92–368. p. 511–521.

Plant Protection, Diversity and Conservation (2011) *Pages* **600–609**
Editors: **D. Prasad & Satya Kumar**
Published by: **BIOTECH BOOKS, NEW DELHI**

Chapter 45

Influence of Socio-Personal Traits of Vegetable Growers on their Knowledge Index regarding IPM technologies in Cabbage Production: An Empirical Study

Chowdhury Sarthak* and Ray Prabuddha**
Department of Agricultural Extension, Agricultural Economics and Agricultural Statistics, Palli Shiksha Bhavana (Institute of Agriculture), Sriniketan, Visva-Bharati, P.O.- Sriniketan, Dist. – Birbhum, Pin - 731 236, West Bengal

Integrated Pest Management (IPM) is a scientific paradigm (Perkins, 1982) which is now of global significance. Its basic concern is with designing and implementing pest management practices that meet the goals of farmers, consumers and governments in reducing pest loses while, at the same time, safeguarding against the longer term risk of environmental pollution, hazards to human health, and reduced agricultural sustainability. While the philosophy and ideas of IPM are now widely accepted in the political and scientific arena, the practical implementation of IPM has proved far more difficult to achieve. Over two decades, attempts to develop and disseminate IPM technologies in the developing countries have met with limited success (Kiss and Meerman, 1991; Yudelman *et al.*, 1998). A number of reasons for this limited success can be suggested including:

* E-mail: sarthak_chowdhury1@rediffmail.com; ** E-mail: prabuddha_ray1@rediffmail.com

Insufficient extension resource to serve the needs of the farmers who wish to employ IPM (Brader - 1979).

Commercial pressures on farmers to use pesticides, and the idea that pesticides companies disrupt IPM research and implementation activities could be important in specific cases, but a more important reason for a lack of IPM adoption is that the IPM 'product' is often not appropriate to farmers' need (Norton - 1976, Goodell - 1984). Farmers may perceive IPM as too complex, too expensive, too risky, and just not appropriate to their farming system. At the same time, the researchers must keep in mind that if the farmers are to manage pests in the best way, there is certain set of knowledge and information they need to be aware of, including conceptual and technical knowledge of IPM, as well as the 'Know-how' to carry out certain practices keeping the environmental factors in mind. Without this knowledge, the farmers are certain to avoid the adoption of IPM technologies which are new to them. In this regard, the broader question is that whether there is any insufficiency in the knowledge level of the farmers regarding the adoption of IPM.

But lack of knowledge is not the sole factor, though one of the major factors behind the non-adoption of IPM technologies. Here all the factors *viz.* the economic, social, political, environmental and philosophical aspects of our society play important roles. So, it is necessary to understand the relationship among various socio-economic factors that influence and determine the knowledge level of the farmers in relation to the IPM technologies for the selected vegetables. So, it is of utmost necessity for the scientific community to find out and analyze the various socio-economic factors that influence and determine the knowledge level of the farmers in relation to the IPM technologies especially in the vegetable cultivation in the third world countries where excessive and indiscriminate use of the pesticides were reported by various researchers like Rashid *et al.* (2003), IPM DANIDA Project (2004), Kim and Park (2005), Baral *et al.* (2006) etc.

Objectives of the Study

1. To find out the distribution pattern of the respondents on the basis of the various categories of Knowledge Index in relation to the IPM technologies for the selected vegetable, and
2. To examine the relationship between the socio-personal traits of the selected vegetable growers with their Knowledge Index in relation to the IPM technologies for the selected vegetable.

Two way direct communications is a "must" to achieve good rapport in any research work, which involves farmers as the unit of study. Keeping these in view, researchers' own state namely, West Bengal in India was selected for the present study.

The state of West Bengal has nineteen districts. Among these districts, the district of Bardhaman is one of the most agriculturally advanced and prosperous district of the state apart from being the home district of the present investigators. So, this district was purposively selected for the present study.

The district of Bardhaman has thirty one (31) Community Development (CD) Blocks. Among these CD Blocks, the Katwa - I Block was purposively selected for the present study, because the soil of Katwa - I Block is alluvial soil which attains an enormous thickness in the low level plains. This alluvial soil is formed of alluvium brought down by the rivers Ajay, Damodar, Bhagirathi and numerous other rivers. These soils are sandy, well drained and slightly acidic in nature. As a result this CD Block of the district of Bardhaman is famous for its various types of vegetable production. This was supported by the fact that nearly 600 hectares and 1000 hectares of land were under the vegetable cultivation in the summer and winter seasons respectively in the Katwa - I Block in the year 2005-06 (Source:-

Department of Agriculture, Govt. of West Bengal). This CD Block also consumed a large amount of pesticides for plant protection purposes. The Katwa - I Block consumed nearly 8000 litres of pesticides in the year of 2005-06. Apart from these facts, the Katwa - I Block was home block of one of the present investigators.

All the villages, eighteen (18) in numbers falling under the five (5) kilometers radius of Katwa Town were selected for the present study, as these villages have sizeable population who grew the above mentioned four vegetables in more than 0.33 acre of farm land were taken into consideration for the present study. At present 150 such farmers were there. So, all the 150 vegetable growers were selected as the sample population of the present study.

Among the summer and winter season, the production of Cabbage (*Brassica oleracea* L. var. *capitata*) was 7900 MT and this topped the list of the produced vegetables in the selected CD Block in 2004 – 2005 (Source: State Department of Agriculture, Govt. of West Bengal). So, Cabbage was selected for the present study.

The data was collected with the help of a structured pre-tested schedule developed for the study and through the personal interview method. The data was collected from June, 2005 to May, 2006 at the selected villages. After completion of data collection, thorough checking was made on the filled up schedules and then the schedules were numbered. The numbered schedules were tabulated according to their numbers. The collected data were analyzed with the help of the standard statistical tools.

It was thought worthwhile to measure the knowledge level of the respondent farmers regarding the IPM technologies in Cabbage cultivation through Knowledge Index (KI). Knowledge Index in the present context was conceptualized by the following general formula adopted by Shaikh *et al.* (1993) and subsequently used by Singh *et al.* (1999), Ananth (2001), Atchutaraju *et al.* (2004), Balmurugan and Vetriselvan (2004), Hanif and Ganesan (2004), Singh and Kaur (2004).

$$\text{Knowledge Index (KI)} = \frac{\text{Actual level of knowledge}}{\text{Recommended level of knowledge}} \times 100$$

Here, the Knowledge Index (KI) of the respondents regarding the IPM technologies for the Cabbage crops was calculated by the following specific formula:

$$\text{Knowledge Index (KI) regarding the the IPM technologies in Cabbage cultivation} = \frac{K_m}{P_m} \times 100$$

where,

P_m: Possible maximum number of the correct answers given by the respondents on the knowledge items regarding the IPM technologies to be applied for the Cabbage crops and

K_m: Actual number of the correct answers given by the respondents on the knowledge items regarding the IPM technologies to be applied for the Cabbage crops.

The knowledge test was administered to the respondents in the study area. Scores of one were given to correct responses and zero to incorrect responses or part thereof respectively as suggested by Shaikh *et al.* (1993), Singh *et al.* (1999), Arya *et al.* (2003), Atchutaraju *et al.* (2004), Hanif and Ganesan (2004).

There were thirty (30) numbers of items of knowledge, regarding the IPM technologies to be applied for the Cabbage crops as recommended by Department of Agriculture, Government of West

Bengal, upon which the knowledge level of the respondents regarding the IPM technologies to be applied for the Cabbage crops was calculated. These knowledge items were:

Cultural Methods

1. Knowledge about the effect of synchronized planting,
2. Knowledge about the summer ploughing,
3. Knowledge about clipping of Cabbage seedlings,
4. Knowledge about leaving the vegetable land fallow,
5. Knowledge about the ideal interval of rogue spacing,
6. Knowledge about significance of rogue spacing,
7. Knowledge about the significance of mixed cropping,
8. Knowledge about the synchronization of irrigation application and pesticide application,

Physical/Mechanical Methods

9. Knowledge about purpose of using light traps,
10. Knowledge about the pests controlled by light traps,
11. Knowledge about the methods to control the soil borne pests,

Biological Methods

12. Knowledge about the pests controlled by the bio-agents,
13. Knowledge about the beneficial insects (natural enemies of the pests) to the crops,
14. Knowledge about the correct process of using the bio-agents,

Chemical Methods

15. Knowledge about the Economic Threshold Limit (ETL) of the particular pest of selected vegetables (for chemical spray)
16. Knowledge about the proper type of the pesticides to be applied,
17. Knowledge about the proper dose of the pesticides to be applied,
18. Knowledge about the proper pesticide application procedure,
19. Knowledge about the precautions to be taken in the pesticide application,
20. Knowledge about the ideal gap in days between two consecutive pesticide applications,
21. Knowledge about the ideal gap in days between last pesticide application and harvesting of the vegetables,
22. Knowledge about pesticide rotation of the applied pesticides,
23. Knowledge about the biopesticides.

Miscellaneous

24. Knowledge about the ideal monitoring procedure of the crop regarding the attack of the pests,
25. Knowledge about the local indigenous knowledge (IK) system regarding the pest management practices,

26. Knowledge about crop stage (vegetative growth stage of the crop) of the pest attack,
27. Knowledge about the month of high incidence of the pests,
28. Knowledge about the specific symptoms of the particular pest attack,
29. Knowledge about the consciousness of the neighbouring farmers about the pesticide application on a nerby vegetable plot and
30. Knowledge about the negative effects of the pesticide application on the beneficial insects. So, here $P_m = 30$.

On the basis of the above mentioned thirty (30) numbers of knowledge items, Knowledge Index (KI) regarding the Integrated Pest Management (IPM) practices of the Cabbage crop of each individual respondent was calculated.

The individual Knowledge Index in relation to the IPM technologies of each of the selected 150 (one hundred fifty) respondents was measured by the formula mentioned before. The mean and Standard Deviation (SD) of the individual Knowledge Indexes (KI) regarding the IPM technologies of all the 150 respondents was calculated. Then the respondents were categorized into three categories of respondents whose respective Knowledge Indexes were High [More than (Mean + SD)], Medium [(Mean – SD) to (Mean + SD)] and Low [Less than (Mean – SD)]. This type of categorization was earlier performed by Singh *et al.* (1991), Singh *et al.* (2002), Arya *et al.* (2003).

Table 45.1: Description of Socio-personal Variables

Sl.No.	*Symbol of the variable*	*Description of socio-personal variables*	*Scoring techniques used*
1.	X_1	Age of the respondent	Schedule developed for the present study
2.	X_2	Caste of the respondent	SES Scale developed by Trivedi (1963)
3.	X_3	Educational Status of the respondent	Schedule developed for the present study
4.	X_4	Total monthly income of the family of respondent	Schedule developed for the present study
5.	X_5	Total monthly income of respondent	Schedule developed for the present study
6.	X_6	Type of family of respondent	SES Scale developed by Trivedi (1963)
7.	X_7	Size of family of respondent	Schedule developed for the present study
8.	X_8	Type of dwelling of respondent	SES Scale developed by Trivedi (1963)
9.	X_9	Material Possession of respondent	Modified Index of Kulshrestha (1980)
10.	X_{10}	Respondent's total cultivable land	Schedule developed for the present study
11.	X_{11}	Type of farm power used by the respondent in vegetable cultivation	Schedule developed for the present study
12.	X_{12}	Experience of respondent in vegetable cultivation	Schedule developed for the present study
13.	X_{13}	Social participation of respondent	SES Scale developed by Trivedi (1963)
14.	X_{14}	Respondent's family contact with extension agencies	Scaling developed by Selltiz *et. al* (1976)
15.	X_{15}	Extension participation of respondent	Scaling developed by Selltiz *et. al* (1976)
16.	X_{16}	Mass Media participation of respondent	Scaling developed by Selltiz *et. al* (1976)
17.	X_{17}	Exposure of the respondent to mass media sources	Scaling developed by Selltiz *et. al* (1976)

The Table 45.1 revealed the selected seventeen (17) socio-personal traits (variables) of the respondents (vegetable growers) mentioned along with the scoring techniques to measure the variables.

The relationship between the selected 17 (seventeen) socio-personal traits of the respondents and their Knowledge Index in relation to the IPM technologies were examined by using simple correlation of co-efficient and stepwise multiple regression tools.

The Distribution of the Respondents on the Basis of the Knowledge Index of the Respondents Regarding the IPM Technologies for the Cabbage Cultivation

Individual knowledge index regarding the IPM technologies for Cabbage cultivation for each of the respondents was calculated. The mean and Standard Deviation (SD) of the individual Knowledge Indexes (KI) regarding the IPM technologies in Cabbage cultivation of all the 150 respondents were calculated, which were 22.930 and 8.462 respectively. Then the respondents were categorized into three categories of respondents whose respective Knowledge Indexes were High [More than (Mean + SD)] (Knowledge Index of 31.393 and higher), Medium [(Mean – SD) to (Mean + SD)] (Knowledge Index of 13.468 to 31.392) and Low [Less than (Mean – SD)] (Knowledge Index up to 13.467).

The categorization of the respondents according to their knowledge index regarding the IPM technologies in the Cabbage cultivation was given in Table 45.2.

Table 45.2: Distribution of the Respondents According to their Respective Knowledge Index Regarding the IPM Technologies for Cabbage Cultivation

Sl.No.	Category of Knowledge Index Regarding the IPM Technologies for the Cabbage Cultivation	Frequency	Percentage
1.	High Knowledge Index	24	16.00
2.	Medium Knowledge Index	79	52.67
3.	Low Knowledge Index	47	31.33
	Total	150	100.00

From the Table 45.2, it was observed that a majority of the respondents had medium Knowledge Index regarding the IPM technologies for the Cabbage cultivation which was visible from the fact that 52.67 percent of the respondents had medium knowledge index. The Table also showed that nearly one-third of the respondents *i.e.* 31.33 percent of the respondents had low knowledge index regarding the IPM technologies. However, it was interesting to note that a small portion of the respondents (16.00 percent) had high knowledge index regarding the IPM technologies for the Cabbage cultivation. However, we must keep in mind that the mean of the individual Knowledge Indexes regarding the IPM technologies for the Cabbage cultivation was 22.930 which were very low itself. This fact indicated that the average Knowledge Index regarding the above mentioned aspect among the respondents was quite low and the knowledge level of the respondents was quite low, though the categorization of the respondents according to their individual Knowledge Index regarding the IPM technologies for the Cabbage cultivation gave a somewhat different picture due to the dependence of the categorization both on the mean and standard deviation (SD) of the individual Knowledge Index.

Relationship Between Socio-personal Traits of Respondents and their Knowledge Index Regarding the IPM Technologies for the Cabbage Cultivation

The correlation of coefficient between the seventeen (17) independent socio-personal- variables and the dependent variable (Y) *viz.* Knowledge Index regarding the IPM technologies for the Cabbage cultivation was given in the Table 45.3.

Table 45.3: Correlation of Co-efficient between Socio-Personal Traits (Independent variables) and Knowledge Index Regarding the IPM Technologies for the Cabbage Cultivation (Y) (Dependent variables)

Sl. No	*Variables*	*r-value*	*Sl. No*	*Variables*	*r-value*
1.	X_1	-0.062	10.	X_{10}	-0.053
2.	X_2	0.739**	11.	X_{11}	0.696**
3.	X_3	0.190*	12.	X_{12}	0.747**
4.	X_4	0.785**	13.	X_{13}	0.788**
5.	X_5	0.799**	14.	X_{14}	0.818**
6.	X_6	0.181*	15.	X_{15}	0.054
7.	X_7	0.007	16.	X_{16}	0.100
8.	X_8	0.223**	17.	X_{17}	0.789**
9.	X_9	0.715**			

* Correleation is significant at 0.05 level of significance.

** Correlation is significant at 0.01 level of significance.

The Table 45.3 showed that variables X_2 (caste of respondent), X_3 (educational status of the independent respondent), X_4 (total monthly income of the family of respondent), X_5 (total monthly income of respondent), X_6 (type of family of respondent), X_8 (type of dwelling of respondent), X (material possession of respondent), X_{11} (type of Farm power used by respondent in vegetable cultivation), X_{12} (experience of respondent in vegetable cultivation), X_{13} (social participation of respondent), X_{14} (respondent's family contact with extension agencies) and X_{17} (exposure of the respondent to mass media sources) were significantly and positively correlated with the dependent variable *i.e.* Knowledge Index of the vegetable growers in relation to the IPM technologies for the Cabbage cultivation.

The independent and the dependent variables were again judged with the help of the Stepwise Multiple Regression and the calculation was done up to the fourth step. The results of Stepwise Multiple Regression were given in Table 45.4.

Table 45.4: Result of the Stepwise Multiple Regressions Between the Socio-personal Traits of the Respondents (Independent variables) and (Y) (Knowledge Index Regarding the IPM Technologies for the Cabbage Cultivation) (Dependent variables)

Sl.No.	*Variables*	*R*	*R^2*
1.	X_{14}	0.818	0.753
2.	X_{14}, X_5	0.834	0.783
3.	X_{14}, X_5, X_{13}	0.837	0.787
4.	$X_{14}, X_5, X_{13}, X_{17}$	0.838	0.790

The Table 45.4 reveals that out of 17 independent variables, 4 variables *viz.* X_5 (total monthly income of respondent), X_{13} (social participation of the respondents), X_{14} (respondent's family contact with extension agencies) and X_{17} (exposure of the respondent to mass media sources) variables had explained an overwhelming 79.00 per cent of the variance in dependent variable (Knowledge Index regarding the IPM technologies in the Cabbage cultivation). R value *i.e.* multivariate equivalent indicated a very strong relationship between the combination of independent variables X_5 (total monthly income of respondent), X_{13} (social participation of the respondents), X_{14} (respondent's family contact with extension agencies) and X_{17} (exposure of the respondent to mass media sources) and dependent variable Y (Knowledge Index regarding the IPM technologies for the Cabbage cultivation). Here the most important point that is to be noted down that the X_{14} variable *i.e.* the respondent's family contact with extension agencies solely contributes 75.3 per cent variance in the dependent variable that is why X_{14} variable is the most important determinant for the respondents' Knowledge Index regarding the IPM technologies in the Cabbage cultivation

Conclusion

From the above-mentioned facts, it was observed that nearly half of the respondents (52.67 per cent) had medium level of knowledge regarding the IPM technologies for the Cabbage cultivation. It was quite interesting to note that nearly one-third of the respondents (31.33 per cent) had low level of knowledge, while only a minority of 16.00 per cent had higher level of knowledge regarding the IPM technologies for the Cabbage cultivation.

The study showed that the independent variables (socio-personal traits of the respondents) were significantly and positively correlated with the dependent variable Knowledge Index of the vegetable growers in relation to the IPM technologies for the Cabbage cultivation. Among these variables, the step down multiple regressions revealed that a very strong relationship between the combination of independent variables *viz.* X_5 (total monthly income of respondent), X_{13} (social participation of the respondents), X_{14} (respondent's family contact with extension agencies) and X_{17} (exposure of the respondent to mass media sources) and the dependent variable *i.e.*, the Knowledge Index regarding the IPM technologies for the Cabbage cultivation. Here the most important point that is to be noted down that the X_{14} variable *i.e.* the respondent's family contact with extension agencies solely contributes 75.3 per cent variance in the dependent variable that is why X_{14} variable is the most important determinant for the respondents' Knowledge Index regarding the IPM technologies in the Cabbage cultivation. To enhance the knowledge level of the farmers regarding the IPM technologies, the change agent should encourage more contacts between the farmers and the extension agencies and mass media sources through participatory technology development and extension programmes aiming to develop skill and knowledge of the vegetable growers regarding IPM.

References

Ananth, P.N., 2001. Role of farm research system, technology transfer system and clientele system in achieving farm production. *Unpublished Ph.D. Thesis*, Tamil Nadu Agricultural University (TNAU), Coimbatore.

Arya, S., Singh, G.P. and Sharma, P., 2003. Knowledge of farmers regarding improved sugarcane production technology. *Indian Journal of Extension Education*, 39(1 and 2): 111–114.

Atchutaraju, K., Babu, J. Dilip and Reddy, M.S., 2004. Knowledge on eco-friendly practices among the farmers of watershed area. *Journal of Extension Education*, 15(4): 3671–3677.

Balamurugan, V. and Vetriselvan, M., 2004. Knowledge level of trained and untrained paddy farmers on IPM. *Journal of Extension Education*, 15(4): 3688–3694.

Baral, K., Roy, B.C., Rahim, K.M.B., Chatterjee, H., Mondal, D., Ghosh, D. and Talekar, N.S., 2006. *Socio-economic Parameters of Pesticide Use and Assessment of Impact of an IPM Strategy for the Control of Eggplant Fruit and Shoot Borer in West Bengal, India – Technical Bulletin No. 37*, AVRDC – The World Vegetable Center, Shanhua, pp. 2–3, 12–13, 16.

Brader, L., 1979. Integrated pest control in the developing world. *Annual Review of Entomology*, 24: 225–254.

Goodell, G., 1984. Challenges to international pest management research and extension in the third world: Do we really want IPM to work? *Bulletin of the Entomological Society of America*, 30: 18–26.

Hanif, N.A.K.A. and Ganesan, R., 2004. Knowledge level of rice farmers on pest management practices. *Journal of Extension Education*, 15(2 and 3): 3628–3632.

IPM DANIDA Project, 2004. *Pesticides-Health Survey: Data of 124 Farmers in Chaiprakarn Chiang Mai, Thailand*, May, IPM DANIDA 64, November, Thailand, pp. 3–10.

Kim, J. and Park, H., 2005. *Slow Suicide: How Korean Farmers are Poisoning Themselves with Pesticides.* PAN Asia and the Pacific, Penang, pp. 19, 23–28.

Kiss, A. and Meerman, F., 1991. *Integrated Pest Management in African Agriculture.* World Bank Technical Paper 142, World Bank, Washington DC.

Kulshrestha, S.P., 1980. *Manual for Socio-economic Status Scale [Rural].* Department of Teaching Education, DAV College, Dehradun, National Psychological Corporation, Agra.

Norton, G.A. and Mumford, J.D., 1982. Information gaps in pest management. In: *Proceeding of the International Conference on Plant Protection in the Tropics*, Kuala Lumpur, 1st–4th March, pp. 589–597.

Norton, G.A., 1976. Analysis of decision making in crop protection. *Agro-Ecosystems*, 3: 27–44.

Perkins, J.H., 1982. *Insects, Experts and the Insecticide Crisis: The Quest for New Pest Management Strategies.* Plenum Press, New York.

Rashid, M.A., Alam, S.N., Rouf, F.M.A. and Talekar, N.S., 2003. *Socio-economic Parameters of Eggplant Pest Control in Jessore District of Bangladesh - Technical Bulletin No. 29*, AVRDC – The World Vegetable Centre, Shanhua, pp. iii, 9, 13, 14, 23.

Selltiz, C., Wrightsman, L.S. and Cook, S.W., 1976. *Research Methods in Social Relations.* Holt, Rinehart and Winston, New York.

Shaikh, A.A., Arya, H.P.S. and Nikam, T.R., 1993. Knowledge level of farmers on poultry production technology. *Indian Journal of Extension Education*, 29(3 and 4): 92–93.

Singh, A.K., Singh, G.P. and Singh, B., 2002. Correlates of farmers' knowledge of improved chickpea production technology. *Indian Journal of Extension Education*, 38(3 and 4): 162–167.

Singh, D. and Kaur, R., 2004. Knowledge level of the cotton growers regarding integrated pest management practices. *Journal of Extension Education*, 15(2 and 3): 3611–3618.

Singh, R.K., Kumari, K. and Singh R.P., 1999. 'Ratan', farmers' knowledge of late sown wheat production technology. *Indian Journal of Extension Education*, 25(3 and 4): 255–258.

Singh, S.P., Hudda, R.S. and Verma, H.K., 1991. Knowledge gap of citrus growers. *Indian Journal of Extension Education*, 27(1 and 2): 117–118.

Trivedi, G., 1963. Measurement and analysis of socio-economic status of rural families. *Ph.D. Thesis* (Unpublished), Indian Agricultural Research Institute (IARI), New Delhi.

Yudelman, M., Ratta, A. and Nygaard, D., 1998. Pest management and food production: Looking to the future, food. *Agriculture and the Environment Discussion Paper 25,* International Food Policy Research Institute (IFPRI), Washinton, DC.

Plant Protection, Diversity and Conservation (2011) *Pages* **610–626**
Editors: **D. Prasad & Satya Kumar**
Published by: **BIOTECH BOOKS, NEW DELHI**

Chapter 46

Ecofriendly Management of Nematode Pests of Vegetable Crops

Archana U. Singh* and Vijendra Singh
Division of Nematology, Indian Agricultural Research Institute,
New Delhi – 110 012

Plant-parasitic nematodes are microscopic invertebrate animals often called as threadworms, eelworms or roundworms. They are distributed all over the world in different kinds of habitats and found in nearly every biological niche that supports life. They damage the crops not only by feeding on plants but also by interacting with various other organisms. They cause severe losses to economically important crops like vegetables, cereals, pulses, oilseeds, fruit crops, etc. Many estimates have been made on crop losses due to nematodes by research workers. (Table 46.1) Annual loss worth as high as Rs. 4 crores in wheat and Rs. 3 crores in barely due to cereal cyst nematode (*Heterodera avenae*) in Rajasthan alone have been reported by Seshadri and Dasgupta, 1980. Sasser and Freckman (1987) have indicated an annual crop loss due to Phytonematodes on world wide basis to the tune of $100 billion. The yield of brinjal has 27.3 per cent loss respectively due to *Meloidogyne incognita* infestation under field conditions (Bhatti and Jain, 1977). Ogunfowora (1987) reported a yield loss of 20-59 per cent to cowpea crop by plant-parasitic nematodes.

A committee was constituted in 1971 on the estimation of crop losses in the U.S.A and estimated 6 per cent loss in field crops, 12 per cent loss in fruit and nut crops, 11 per cent loss in vegetable crops and 10 per cent loss in ornamental crops. Thus, recorded an annual monetary loss of 1590.7 million dollars due to nematodes. Losses estimated due to nematode attack to different cultivated crops all over the world by F.A.O. is around 400 million dollars (Anon, 1985).

Vegetables are the important component of human diet which contributes as a source of proteins, vitamins and minerals. The estimated area under vegetables is about 6.2 mha which is about 3 per cent of the total area under cultivation in the country with an annual production of 71 MT. India is the

* E-mail: arch_212@yahoo.com

econd largest vegetable producer in the world next to China which accounts for about 15 per cent of he world's production of vegetables.

Table 46.1: Estimated Loss in Yield of Various Agricultural Crops Due to Phytonematodes

Phytonematodes	*Crop*	*Loss (per cent)*	*Reference*
M. incognita	Okra	28.08	Parvatha Reddy and Singh, 1981
M. incognita	Okra	79.9	Bhatti and Jain, 1977
M. incognita	Cowpea	12.3	Patel *et al.*, 1993
M. incognita	Brinjal	27.3	Bhatti and Jain, 1977

Table 46.2: Major Vegetables Growing in India

Vegetables	*Area ('000ha)*	*Production ('000T)*
Potato	1640	30195
Onion	757	10655
Tomato	594	9878
Brinjal	564	9301
Cabbage	254	5724
Cauliflower	310	5524
Okra	398	4099
Peas	287	2370
Tapioca	255	8226
Sweet Potato	97	880

The destructive plant-parasitic nematodes are one of the major limiting factor in vegetable ›roduction throughout India. Low production and productivity of vegetables are mainly because of iotic and abiotic stresses, among the biotic stresses nematodes is also one of them. Large number of ›lant parasitic nematode is recorded from the rhizosphere of many vegetable crops like Root-knot Jematode (*Meloidogyne* spp.), Reniform nematode (*Rotylenchulus reniformis*), cyst nematode (*Heterodera* pp.), lesion (*Pratylenchus penetrans*) etc. Among all the plant parasitic nematodes, root-knot nematode *Meloidogyne* spp.) are the major phytonematodes causing damage to vegetable crops.

Different nematodes causing damage to crops are as follows:

Root-knot Nematodes (*Meloidogyne* spp.)

These are by far the most important pests of vegetable crops. The four most common species *viz.*, *A. incognita*, *M.javanica*, *M. hapla* and *M. arenaria* are by far most important and leads to the formation f conspicuous root galls. *M. incognita* and *M. javanica* are most wide spread in distribution and has a vide host range among vegetables, whereas *M. hapla* is encountered under temperate conditions and ttack potato and other vegetables while *M. arenaria* infects chillies and groundnut. The root-knot ematodes is the limiting factor in commercial production of vegetables and responsible for 15-60 per ent yield loss. Various researchers have reported a loss of 70-80 per cent in vegetable crops especially n okra, tomato, brinjal.

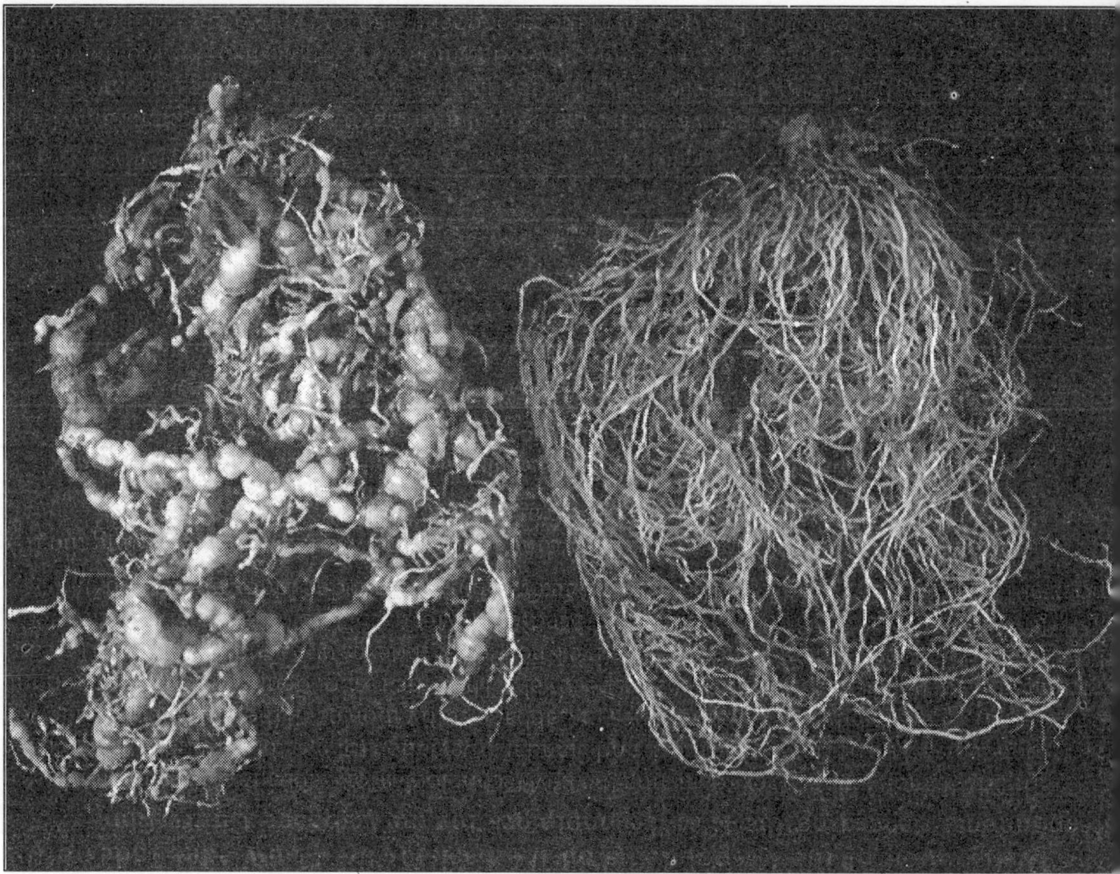

Figure 46.1: Root-knot Galls   Figure 46.2: Healthy Roots

Reniform Nematode (*Rotylenchulus reniformis*)

These nematodes infest tomato, brinjal, okra, cowpea, dolichos, French beans, parwal and othe vegetables. Two races of this nematode have been reported from India. It is polyphagous and is knowr to feed and multiply on a wide range of horticultural, vegetable and plantation crops. Some of th crops on which this nematode is recorded are cowpea, castor, tea, banana, sweet potato, soybean pigeonpea, pineapple, okra, tobacco, beans, eggplant, etc.

Root-lesion Nematodes (*Pratylenchus* spp.)

At present there are more than 68 species of this genus in world and 36 are reported from India At least five species *viz.*, *P.brachyurus*, *P. coffeae*, *P. penetrans*, *P. scribneri* and *P.vulnus* have been reportec to infect tomato. Root lesion infected plants show gradual decline or lack of plant vigour with stuntin chlorosis leading to rapid wilting. On the roots, there is formation of lesions and necrosis, whicl provided site for other micro-organisms to infect, grow and reproduce, thereby leading to other diseas complexes. The nematodes belonging to this genus are endoparasitic root feeders which migrate inte and intracellularly and feed in the cortical region leading to cell death and breakdown.

Spiral Nematode (*Helicotylenchus* spp.)

Okra, brinjal, tomato, chillies were found to be good host for this nematode.

Potato Cyst Nematode (*Globodera* spp.)

G. pallida and *G. rostochiensis* are the most important pests of potato. This nematode has been reported from Nilgiris and Kodai Hills of Tamil Nadu and Munar Hills of Kerala. Out of 9,000 ha under potato, 3,000 ha are infected by this nematode in Nilgiris. In Kodai Hills, about 200ha are infected. The life cycle processes are similar to the other cyst-forming species, with one life cycle being completed in 5-7 weeks. This is mainly a temperate area problem and thrives well under cool conditions. Under severe infestation conditions, the plants tend to wilt during the day hours but recover at night.

Stunt Nematodes (*Tylenchorhynchus* spp.)

Nematodes of this Genus are widely distributed and are root parasites. The most common species feeding on vegetables is *Tylenchorhynchus brassicae* causes poor germination and growth of cabbage and cauliflower. Besides cauliflower and cabbage, tomato, radish, sugarbeet and lettuce are also good hosts. Members of this group are ecto-parasites and rarely encountered in the cortical cells, one to two cell deep and feed on roots and root hairs. The optimum temperature for growth and reproduction is around 30 C with 25 -30 per cent moisture. Although, it is widely distributed both in temperate and tropical zones yet only a few species are known to be pathogenic to various vegetable crops. In southern USA, *T. claytoni* has been responsible for stunting of pea and *T. marioni* in sweet potato *T. brassicae* damages cabbage and cauliflower and *T. dubius* parasitizes cauliflower, pea, radish and turnip in Netherlands.

The degree of damage caused by nematodes depends upon the population density of nematodes, susceptibility of the crop, environmental condition such as soil fertility, moisture and also the presence of other pathogenic micro-organisms which may interact with nematodes as they are known to cause complex plant disease in association with fungi, bacteria and viruses. Crop losses due to nematode appear in different forms. It may be the loss in yield, deterioration in the quality of the product and increase in the cost of production, etc. Sometimes nematodes are involved in many disease complexes with fungi, bacteria and viruses, etc. It is now well established that the plant damage done by two or more pathogens is much more severe than what the pathogens cause independently. Continuous feeding of nematodes provides entry points to other pathogenic micro-organisms and breaks the physical as well as mechanical barriers of the host. In such a complex situation, nematodes may either act as wound causing agents, as resistance breaking agents or as modifiers of host tissues or as vectors. The losses in our agriculture production due to plant parasitic nematodes justify the need for adopting their different management practices to check their population and level of infestation.

Symptoms

The symptoms caused by the plant parasitic nematodes in vegetables are mainly divided into above ground and below ground symptoms.

Above Ground Symptoms

Above ground symptoms are non-specific and may be mistaken for the micronutrient deficiency in the soil. Typical above ground symptoms of nematode infections include stunting, yellowing, wilting and reduced yield. Patches of poor growth visible in the fields from a distance. Chlorosis, wilting, delay in flowering, fruit formation, reduction in plant size, yield etc. may be noted in above ground parts of infected plants.

Below Ground Symptoms

- ☆ Small galls/nodules seen on the roots which at a latter stage coalesce to form very large galls.
- ☆ Bushy appearance of secondary roots.
- ☆ Total reduction in root system.
- ☆ Other symptoms include Lesions and "Dirty Roots".

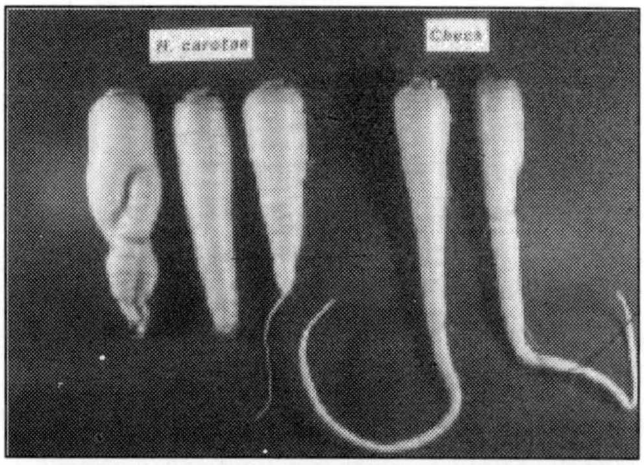

Figure 46.3: Carrot Damage by Carrot Cyst Nematode, *Heterodera carotae*

Figure 46.4: Northern Root-Knot Nematode *Meloidogyne hapla* Damage on Onion

Figure 46.5: Root-Knot Nematode *Meloidogyne incognita* Damage in Tomato

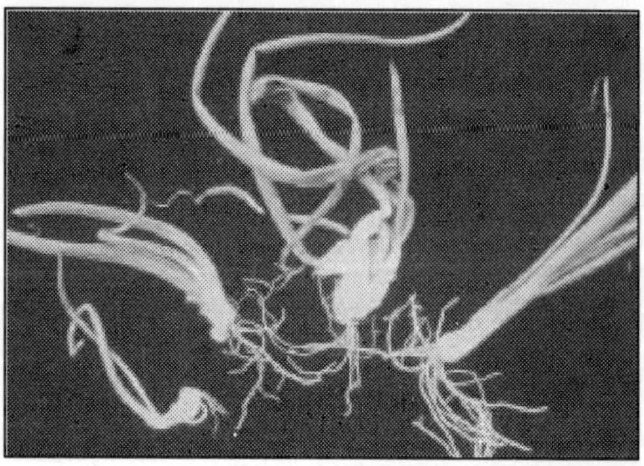

Figure 46.6: Stem and Bulb Nematode *Ditylenchus dipsaci* Damage on Onion

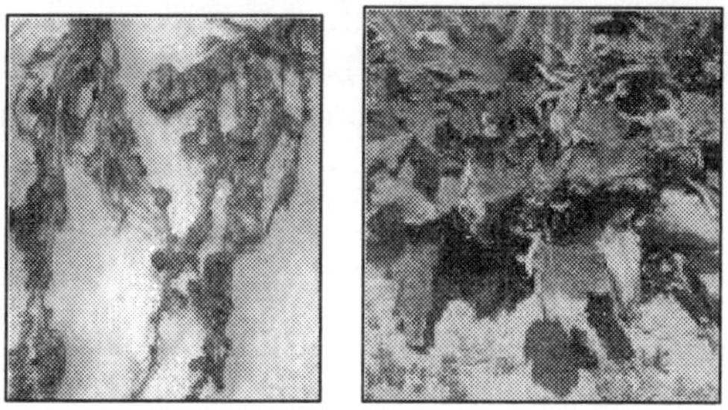

Figure 46.7: Root-Knot Nematode *Meloidogyne incognita* Damage in Brinjal

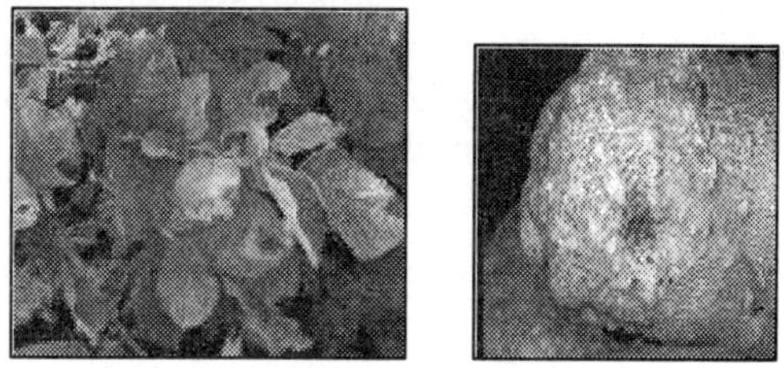

Figure 46.8: Potato Damage by Cyst Nematode

Table 46.3: Diseases Caused by Plant-Parasitic Nematodes in Different Agricultural Crops

Causal Organisms	*Disease*	*Host Plants*
Ditylenchus dipsaci (Stem nematode)	Onion bloat	Onion
Tylenchulus semipenetrans (Citrus nematode)	Citrus die back	Citrus
Globodera spp. (Potato cyst nematode)	Potato cyst	Potato
Rotylenchulus reniformis (Reniform nematode)	Reniform nematode	Pulse crops, Vegetables
Meloidogyne spp. (Root-knot nematode)	Root-knot nematode	Vegetables, Pulses, Ornamentals,
Ditylenchus destructor (Stem nematode)	Red rot	Potato

Because of the serious losses caused by the plant-parasitic nematodes principal management trategies i.e, Physical, Chemical, Cultural, Biological, Regulatory and IPM are undertaken to minimize he crop losses caused by nematodes.

Nematode Management

Nematode management is important for high yield and quality that are required by the high cost of modern crop production. The several disadvantages associated with chemical control like hazards to man, effects on non target organism, phytotoxicity, hazards of pesticide residues, upsetting the ecological balance and difficulty in application. Nematicides available in the world market (Sasser, 1987) may be classified as fumigants and non-fumigants, and now considerable research is focused on the development of novel compounds. To date, DBCP, EDP and DD have been withdrawn from United States markets through the action of regulatory agencies. In addition, 1,3-D has been removed from current use in California, and there is a risk that methyl bromide will be removed from the United States markets. The latter compound has been implicated as an agent that damages the ozone layer. Although a range of non-fumigant nematicides have acceptable levels of efficacy, most of these compounds have been detected in groundwater.

In spite of the problems associated with nematicides, they comprise an important tactic for many nematode IPM systems especially for high-value crops. As indicated by Sasser (1987) the procedure for selecting a nematicide should include procurement of information on efficacy and capacity to enhance yield of the primary crop or crops in a specific region. Sasser (1987) and Thomason (1987) provided details on nematicide safety and environmental issues, related economics and additional problems concerning factors in selecting and using these materials, their methods of application and dosage and use on annual or perennial crops.

Due to the ban of several nematicidal chemicals like DDT, Aldicarb etc. the Integrated Nematode Management have been originated. It is defined as an appropriate system of mixed technologies or pest control practices to reduce pest damage to tolerable level. The modern concept of pest management is based on ecological principles and involves the integration and synthesis of different components/control tactics into a pest management system. IPM is a component of the agro-ecosystem management technology for sustainable crop production (Dhaliwal *et al.*, 2004). A long/medium term strategies by combining low input components alone or together with other management practices to bring down the population level below threshold level should be the main aim for successful cultivation of crops. The management of phytonematodes can be achieved by the integration of different farming practices that fall into four broad strategies (Bridges, 1996)

1. Using direct, non-chemical, cultural and physical methods particularly crop rotation and soil cultivation,
2. Encouraging naturally occurring biological control agents by an understanding of cultivation methods and appropriate use of soil amendments,
3. Preventing the introduction and spread of nematode-free planting material, and
4. Enhancing the biodiversity inherent in traditional farming systems that use multiple cropping and multiple cultivars to increase the available resistance or tolerance to nematodes.

The different methods used in the management programme to bring down the nematode population significantly are as follows:

Cultural Methods

Fallowing

In many parts of India after harvest of *rabi* crops, the field is left fallow without host crops. This practice is widely used in Northern India which reduce the nematode population. If the fallow is coupled with summer ploughings, the decrease of endoparasite nematodes especially root-knot nematode was recorded to be as much as 40 per cent. The main principle behind this are starvation of nematodes due to non-availability of host crop and killed by solar heat and soil desiccation. Most of the phytonematodes do not survive in upper layers of soil. However, there are some problems with fallowing. It is a poor conservative practice and also cause monetary loss to the farmers.

Summer Ploughing

This is a good practice for controlling plant-parasitic nematodes. At least two summer ploughings during May and June, under the north Indian conditions, will go a long way in reducing the soil population of nematodes as well as other soil borne organisms which will help the farmer in getting a profitable crop. Haque and Prasad (1982) reported drastic reduction of plant parasitic nematode and significantly increased the yield with three deep ploughings.

Solarisation

The solar energy is preserved with the help of transparent polyethylene sheet to increase soil temperature enough to kill the nematodes present in soil profiles. This is only applicable in limited areas. Prasad and Yaduraju (1993) however indicated that 30 days solarisation is enough to achieve 95 per cent control of nematodes. Prasad *et al.* (1995) found about 90 per cent kill of root-knot and reniform nematodes when solarised for 30 day during May-June months. Solarisation coupled with triazophos application @ 1.5 l/ha completely erradicated the plant parasitic nematodes under field conditions. Kumar *et al.* (1993) reported significant decrease of weed seed emergence by over 90 per cent and reduced the population of plant parasitic nematodes by about 80 per cent. The growth of soybean was improved and seed yield increased by 78 per cent following solarisation. This method provided an economical alternative for multiple pest, disease and weed control in small areas like nursery beds, orchards, forestry plantation and high value crops in regions with hot summers. This practice is very effective in reducing plant-parasitic nematodes, fungus, weeds and soil insects population. Kumar *et al.* (1993) reported that nematodes population could be reduced to 90 per cent and weed population upto 75 per cent.

Rotation of Crops

Crop rotation can be one of the most effective means of controlling nematodes in low-input agricultural systems. Rotating crops in a sequential cropping system is the best known and most widely used cultural practice in traditional and susitainable agriculture. The basic principle of crop rotation for nematode management is to reduce initial populations of damaging nematode species to levels that allow the following crop(s) to become established and complete early growth before being heavily attacked. Mostly, rotation of plant crops (Table 46.4) is widely practised but rotations can also involve the sequential planning of plant crops with others like green manure, cover and trap crops, antagonistic plants, grass and other fallows.

Table 46.4: List of Selected Non-host Plants Useful in Crop Rotation

Non-hosts or Poor Hosts	*Nematodes spp.*
Lettuce, onion, radish	*Pratylenchus* spp.
Potato, tomato	*Tylenchorhynchus brassicae*
Cabbage, chilli, eggplant	*Hoplolaimus indicus*
Corn, greenbeans, red clover	*Globodera rostochiensis*
Castor bean	*M. hapla, M. thamesi*
Watermelon, tobacco	*Belonolaimus gracilis*
Corn, cowpea, potato, cotton	*Heterodera glycines*
Been, corn, onion, potato, alfalfa	*H. schachtii*

Antagonistic Plants and Trap Crops

The most widely reported antagonistic plants are *Tagetes* spp., mustard, *Asparagus*, sesame, sunhemp and neem. (Table 46.5) Antagonistic plants also often act as trap crops, *i.e.*, plants that reduce nematode population by allowing invasion but only partial development in the roots. Mixed cropping either with non-host or antagonistic crops such as *Tagetes*, The use of trap crops as green manure, not only reduce the nematode population but improves soil fertility in many ways and promotes antagonistic microbial activities. For example, sunhemp (*Crotolaria spectabilis*) or Dhaincha (*Sesbania*) which allow invasion by *Meloidogyne* spp. but does not support development and can serve as good trap crop (Mittal and Prasad, 2005).

Table 46.5: Nematicidal Principles Found in Botanicals and Effective Against Nematodes

Name of the Botanicals	*Family*	*Nematode Species*	*Active Principles*
Asparagus spp.	Liliaceae	*Trichodorus*	Glucoside
Christie	Asparagusic acid		
Azadirachta indica	Meliaceae	*M. incognita*	Nimbidine,
Carthamus tinctorius L.	Asteraceae	*Aphelenchoides*	3-cis, 11-transtrideca-1,3, triene, 5,7,9-triyne and 3 transis
Cymbopogaon	Poacease	*Meloidogyne* spp.	Citral, geraniol, Flexuosus wats
Nicotaiana tabacum L.	Solanaceae	*M. incognita*	Nicotine
Ocunium sanctum L.	Lamiaceae	*M. incognita*	Eugenol
Polygonum hydropiper L.	Polygonaceae	*M. incognita*	Formic acid, acetic acid, essential oil
Solanum	Solanaceae	*M. incognita*	Solamargine
Tagetes spp.	Asteraceae	*Meloidogyne* sp.	Polythienyes, X-terthienye
T. patula	Asteraceae	*Meloidogyne* sp.	Hydroxytrematone bithienye derivative

Resistant Varieties

It is cheap, easy to recommend and most practical method of phytonemtaode control. During th last few years, nematologists and breeders followed the conventional approaches of plant breeding t transfer nematode resistance genes into agronomically superior cultivated species. Crop cultivar resistant to phytonematodes can be most useful and cheapest means of nematode control for small scale farmers. This involves a cross between a resistant source usually a wild type related species wit a commercial type plant species. In subsequent steps, F_1 hybrids are back crossed with cultivate species to combine resistance factor however, it is now accepted that on long-term basis polygeni horizontal resistance to nematodes is more desirable than vertical resistance governed by mono gene Secondly, the presence of biotypes of *Meloidogyne* spp. which break the resistance of cultivated crops

It will provide an increasingly important contribution to the solution of many nematode problem (Roberts, 1992). With the enhanced availability of germplasm that has nematode resistance genes an rapidly advancing molecular-transfer techniques, resistant cultivars (Table 46.6) should become primary management tactic in nematode IPM. This approach will become more critical as chemica nematicides become more restricted. The striking returns from deployment of some nematode-resistar cultivars reflect the great potential of this management tactic.

Table 46.6: List of Nematode Resistant Varieties of Different Crops Against Nematodes

Nematode	*Crop*	*Resistant Varieties*
Meloidogyne spp.	Tomato	Patriot, VEN-8,NTR-1, Nematox, VFN Bush, Piersol, Radiant, Rpnit Anahu, Bresch, Helani, Campbell-25, Punuui, Arka Vardhan, Pelica Hawai7736, Hawai7747, Hisar Lalit, Y-220, Atkinson, Manalucie66N156 N-10, Lycopercicon peruvianum (LP3) NTDR-1, Karnataka Hybrid, Punja NR-7, VT-1, Healani, F-24-D-1
M.incognita	Brinjal	Giant of Banaras, Black Beauty,Gola, Vijaya Hybrid, Solanum sisymbrifoliu IIHR 29-5 PBR-91-2, Gachha Baigum
M. javanica	Chilli	CAP-63, Pusa Jwala, Jalandhari Malagachi Yellow, CA 1366, CA-205 B-70A, CA-2104

Contd.

Table 46.6–Contd...

Nematode	Crop	Resistant Varieties
Globodera rostochiensis	Potato	Kufri Dewa, Kufri Swarna, Accessions: CP3084, CP3091, CP3143, CP3162, CP3174 and CP3193
M. incognita	Pea	C-50, A-70, B-58
M. incognita	Cowpea	Barsati Mutant, Iron, New Selection, C-152, TVU 2439-P, 92-1-B, IC 9642-P, 82-1-B V-6, Sel A, Sel 1-2, Sel 4-1, 24-8-2, P-33-2-1 P-85-2(E-9) NP-2
M. incognita	French bean	Banat, Blue Lake String less, Bountiful Flat Brown Beauty, Cambridge Countess, Gallaroy Kenya-3, Pinto W5-114, Seafarer, Suttan's Master Piece,7-10,E.C.Nos.44779, 57257, 27258, 76266, 94470, 94473
M. incognita *M. javanica*	Muskmelon	Scarsol, S-445, France-D,Chemiari, HED 7-25-3, Goldstar, Perlita
M. incognita	Watermelon	Shehjanpuri
M. incognita	Ridgegourd	Panipati, Meerut Special
M. incognita	Ashgourd	Jaipuri, Agra
M. incognita	Pumpkin	Jaipuri, Dasna
M.incognita	China aster	Shashank-Resistant, Poornima-Mod-Resistant
M. incognita	Winged Bean	LBN C-3
M. incognita	Cluster bean	IL 1134, IL 1202 (germplasm lines)
M. incognita	Moth Bean	IL 1134, IL 1202 (germ plasm lines from+Indigenous collection)
M. incognita	Frenchbean	Springwater, Half runner, Coffee wonder, Blue lake, Brown Beauty, Cambridge, Kenya-3, Pinto WS-114, Seaferer,WS-114

Adjusting Planting Time/Escape Cropping

This is a practice of reducing nematode damage by growing crops at times that are not favourable for the nematodes, such as when temperatures is too low or too high for infection or development or when populations of the nematodes have declined to low thresholds. Sethi and Gaur (1986) found that adjustment of dates or seasons of planting can help to avoid nematode damage to crops for example crops such as tomato, egg plant and other susceptible vegetables are planted in the cooler winter (rabi) months to avoid root-knot nematode damage.

Organic Amendment

The use of organic amendment is a well established and age old practice. The three basic principles involved in management of phytonematodes with organic amendments are – habitat management, host modification and direct toxicity of the allelochemicals which are released through decomposition of organic materials of the different modes of action of organic amendments, the nematodes may be affected by a single-factor, but in nature the factors usually run concurrently. Farmers use plant and animal wastes in the form of compost and farmyard manure, crop residues, oilcakes, municipal refuse and industrial wastes have been used by farmers as they provide better media for plants to grow, result in better soil texture, increase water holding capacity, supply the nutrients to deficient soil and stimulate microbial population of actinomycetes, bacteria, fungi and elements of which might be antagonistic to nematodes. The decomposition of residues results in the accumulation in soil of specific compounds that may be nematicidal. But, the main drawback in using most soil amendments for nematode management is the large quantities required. The organic amendments have not only been found effective in normal cultivated crops but are also well suited in nurseries of vegetables fruits and ornamental plants (Table 46.7).

Table 46.7: Neem Pesticides Tested Against Plant-parasitic Nematodes

Plant-parasitic Nematodes	*Crops*	*Neem Pesticides*	*Methods*
Meloidogyne	Cowpea	Nimbicidine	Seed treatment
M. javanica	Tomato Brinjal, Okra, Cowpea	Neemark	Seed treatment Root dip, Nursery bed treatment
M. javanica *M. incognita*	Cowpea Okra, Brinjal, Tomato	Achook	Seed treatment, Root dip, Nursery bed treatment
M. javainca	Brinjal, Tomato	Neemgaurd	Seed treatment, Root dip, Nursery bed treatment
Rotylenchulus reniformis *Pratylenchus delattrei*	Tomato	Margocide	Seed treatment, Root dip, Nursery bed treatment
M. incognita	Okra	Nimbasol	Seed treatment

Physical Control

In this method of control, plant-parasitic nematodes are basically eradicated by steam sterilization, hot water treatment, electrical soil heating, radiation treatments, seed cleaning, heat treatment, etc. But, soil solarization is the most recent practice for control of phytonematodes. The nematodes are effectively controlled but the method is restricted to summer seasons and that too only in tropical and sub-tropical regions of the world.

Biological Control

Biological control or biocontrol is broadly defined as the "use of natural or modified organisms, genes or gene products" to reduce the effects of pests and diseases. The different approaches to biological control can be categorized simply into three strategies 1) Regulation of the pest population 2) exclusionary systems of protection and 3) systems of self-defense. This method of control is more inconsistent, less effective and slower acting than control normally achieved with chemicals. (Stirling, 1991) This method includes the use of predaceous and parasitic organisms such as fungi, bacteria, viruses, nematodes, tardigrades, mites, etc. (Goswami and Mittal, 2004)

Bacteria: Pasteuria penetrans, Pseudomonas fluorescens, Bacillus subtilis.

Nematophagous Fungi.

Trapping Fungi: *Dactylaria lobota* (Sticky knobs).

Dactylaria ellipsospors (Sticky branch).

Arthrobotrys oligospora (Sticky network)

Dactylaria candida (non constricting ring)

Dactylaria brachophaga (constructing ring)

Endozoic Fungi: *Harposporium anguillulae* (simple spore)

Catenaria anguillulae (Flagellated spore)

Egg-parasitic fungi: *Paecilomyces lilacinus, Sepedonium maheswarium*

Toxic fungi: *Aspergillus terreus, Aspergillus niger, Aspergillus fumigatus*

The biology, ecology and potential of biological control agents for nematodes have been extensively reviewed in recent years by Jatala (1983) and Kerry (1980). Nematologists have identified natural enemies with a range of modes of action similar to those currently studied by plant pathologists for the control of soil-borne diseases. Soil conditions, nematode species, rate of development density and host plant all have a considerable effect on the establishment and activity of biological control agents for nematodes. It is essential to understand these interactions for effective control. Work done by several researchers have proved that some microorganisms especially fungi produced metabolites that are toxic to root-knot nematodes. (Mittal and Goswami, 2004a; Mittal and Goswami, 2004b; Mittal and Goswami, 2004c) Thus, Mankau in 1968 reported nematicidal activity of culture filtrates of *Aspergillus niger*. Later, Singh *et al.* (1983) reported that culture filtrates of *Aspergillus niger, Curvularia lunata, Trichoderma viride* and *T. lignorum* obtained from rhizosphere of tomato plants proved to be nematoxic and inhibited hatching of *M. incognita* larvae. Tabreiz *et. al.* (1884) studied the effect of culture filtrates of 8 species of *Aspergillus* on hatching and mortality of *M. incognita* out of which *A. niger, A. terreus* and *A. fumigatus* were more toxic than the other species. Sharma and Saxena (1992) observed that culture filtrates of *Rhizoctonia solani* and *Trichoderma viride* adversely influenced hatching of *M. incognita* larvae. Both fungus showed relatively more toxicity.

In recent years with the development of a number of fungal bioagents isolated from egg masses of *M.incognita* infecting vegetables and categorization of two types i.e, a) toxic and b) egg parasitic ones.(Goswami *et al.*, 2005; Mittal and Goswami, 2001; 2005)) Attempts have been made to test the performance of two types alone as well as together. The combined effect of a toxic bioagent *Aspergillus terreus* with an egg parasitic one, *Paecilomyces lilacinus* showed better performance in reducing nematode population with improved plant health of tomato than when either of the above fungi was used alone. (Goswami and Sharma, 1999). Similar finding is observed by Goswami and Mittal (2000) when *A. fumigatus* was applied to the soil along with *Trichoderma viride* on tomato infested with *M. incognita*. The combined effect of both fungal bioagents *T. viride* and *P. lilacinus* for the control of root-knot nematodes is mainly based on the toxic effect of former followed by the egg parasitic ability of the latter. (Mittal, *et al.*, 2005; Mittal and Prasad, 2005) Thus, a reasonable number of the infective second stage juveniles occurring in the soil around the root zone of affected plants would be reduced. The remaining juveniles expected to be less in number as also possessing lower rate of infectivity due to the toxic effect of *T. viride* would invade the roots. Out of these reduced number of juveniles which in turn would invade the fresh eggmasses, very few would develop into adult which are attacked by *P.lilacinus* resulting in the production of mostly unviable eggs. Many of the eggs due to this fungus are found empty.

Regulatory Control

Regulatory methods are used to prevent the spread of nematodes from one place to other place and from one area to another area through infested material. This strategy involves the strict imposition of plant quarantine laws. This method is useful for checking the introduction of pests and pathogen but it can not check the activity of nematode species already well established in an area of country. It is also ineffective when nematodes are spread by wind, water, insects, etc. *Globodera rostochiensis* was introduced into Europe from South America through infested tubers, soon it found its way to Asia, North America.

Chemical Control

Combating of nematodes can be achieved by many ways but the use of nematicides are quite effective and cannot be replaced by any other method of nematode control. Chemicals are essentially used under such conditions where other methods offer far less certainity of killing the nematodes thus, the nematicides are not likely to be dispensed with altogether. No doubt, nematicides have brought tremendous benefits to mankind by increasing food and fibre production. The continuous use of such chemicals pose residual toxicities in plant/soil and cause environmental pollution. Therefore, efficient strategies are required for the management of nematodes so that there is maximisation of advantages and minimisation of disadvantages. Nematicides can be used before the infection takes place (prophylactic measures) or the infection is suspected to cause damage (curative measures). Prophylactic measures include seed pelleting, seed soaking and foliar sprays.

Integrated Nematode Management

Integrated nematode management deals with the combining of two or more strategies to manage one or more nematode species. The main aim of the integrated approach is:

1. To maintain the nematode population below the threshold level.
2. Maximising natural environmental resistance against plant-parasitic nematodes.
3. Providing maximum profits to growers with location and resource specific recommendation.
4. Utilizing several compatible and useful control techniques in combination which might have economical and social level.
5. Applying specific and drastic control measures only as and when necessary.

Integrated management strategies can be applied sequentially or simultaneously. The first approach includes the season to season or year to year integration of strategies and is particularly relevant to annual cropping cycles. The second major approach to INM involves simultaneous application of two or more strategies. It is one of the most feasible and reliable approach for the control of plant-parasitic nematodes. In the current agricultural scenario, INM have been emphasized due to the social, economical and environmental status in the agricultural society. Economic and social welfare of the INM depends upon a healthy environment and agricultural economy. (Kogan, 1998) INM programs developed so far have already contributed to a decline in chemical inputs on many crops. This decline has the dual effect of increased profitability and environmental protection. Increased public concern about nematicide and pollution of streams and groundwater, effects on farm worker health, impacts on non-target species and general food safety has generated interest in programs with the potential to reduce these factors while maintaining farm productivity and profitability. Integrated management for nematode requires: 1) determining whether pathogenic nematodes are present within the field 2) determining whether nematode population densities are high enough to cause economic loss and 3) selecting a profitable management option. Attempts to manage nematodes may be unprofitable unless all of the above Integrated management procedures are considered and carefully followed. Similarly, some management methods pose risk to people and the environment. (Jain and Bhatti, 1989).

Studies carried out to determine the effect of some fungal bioagents on root-knot nematode (*Meloidogyne incognita*) infecting Brinjal under field condition by Goswami and Mittal (2002). *Aspergillus niger* and *Paecilomyces lilacinus* along with mustard cake gave maximum reduction in root-knot nematode population in soil and root. (Table 46.8).

Table 46.8: Effect of *A. niger, P. lilacinus* Both Alone and in Combination Along with Mustard Cake on Biomass and Multiplication of Root-knot Nematode Infecting Brinjal

Treatments	*Biomass*	*No. of Galls/ Plant*	*No. of Egg Masses*	*Eggs/Egg Mass*	*Soil Population/ 500g Soil*
A. niger + Nematode(N)	10.2	30	16	262	440
P. lilacinus +N	10.6	40	25	160	570
An + *Pl*+ N	11.2	20	15	254	370
An+*Pl*+Mustard cake + N	13.2	16	12	235	290
N alone	7.0	78	57	328	2760
Control	15.2	00	00	000	000

Sharma *et al.*, 2007 studied the combined effect of inorganic supplements with *Trichoderma viride* and Kalisena against root-knot nematode infecting okra. They found that plant growth was observed to be best with NPK with *T. viride while* zinc showed better plant growth in presence of Kalisena.

Senthilkumar and Ramakrishnan (2004) studied compatibility of *Pseudomonas fluorescens, Trichoderma viride* and Carbofuran (3G) and their influences on *Meloidogyne incognita* in okra and reported that *P. fluorescens* and Carbofuran treated plant recorded the highest fruit weight/plant (165g) and lesser gall index (1.25) (Table 46.9).

Table 46.9: Effect of *Psuedomonas fluorescens, Trichoderma viride* and Carbofuran Singly/Combination on Yield and *M. incognita* in Okra

Treatment	*Dose (kg/ha)*	*Nematode Population*	*Gall index*	*No. of Fruits/ Plant*	*Fruit Weight/ Plant (g)*
P. fluorescens	2.5	128.75	1.75	10.0	150.0
T. viride	2.5	177.05	3.25	7.5	75.0
P. fluorescens + *T. viride*	1.25+1.25	135.75	2.25	9.0	117.0
P. fluorescens + Carbofuran 3G	1.25+0.5	113.5	1.25	11.0	165.0
T.viride +Carbofuran 3G	1.25+0.5	155.5	2.50	8.2	91.0
Carbofuran 3G	1.0	124.0	1.5	8.7	105.0
Untreated control	_	286.25	4.25	7.2	58.0

Future Line of Work

1. Intensive and extensive surveys on the presence of parasitic nematodes infesting agricultural crops should be conducted.
2. Use of cultural practices should be emphasized depending on area/locality.
3. Resistant lines/sources should be identified.

4. Emphasis should be given on the integration of nematode management practices by adoption of cultural practices, resistant varieties, nematicides, organic amendments, biological control agents, etc.
5. Role of secondary organisms in causing diseases along with nematodes should be studied and nematode management practices be evolved.
6. INM will provide solution to pest and disease problems in an eco-friendly manner.
7. There is great need to develop varieties, which are resistant or tolerant to nematodes.
8. Research on the use of cultural methods and judicious use of nematicides along with bioagents should be evaluated in the INM programme.
9. Development of integrated nematode management strategies for vegetable crops should be taken up.
10. Research on the use of cultural methods such as deep summer ploughings, use of plastics, intercropping and crop rotations should be intensified.

References

Anonymous., 1985. *Annual Report, AICRP on Tobacco*. ICAR, New Delhi, p. 90–91.

Bhatti, D.S. and Jain, R.K., 1977. Estimation of loss in okra, tomato and brinjal yield due to *Meloidogyne incognita*. *Indian J. Nematol.*, 7(1): 37–47.

Bridge, J., 1996. Nematode management in sustainable and subsistence agriculture. *Ann. Rev. Phytopathol.*, 34: 201–205.

Dhaliwal, G.S., Arora, R. and Dhawan, A.K., 2004. Crop losses due to insect pests in Indian agriculture: An update. *Indian J. Ecol.*, 31: 1–7.

Goswami, B.K. and Mittal, A., 2002. Effect of some fungal bioagents on root-knot nematode, *Meloidogyne incognita* infecting brinjal. *Pakistan Journal of Nematology*, 20: 1, 5–59.

Goswami, B.K. and Mittal, A., 2004. Management of root-knot nematode infecting tomato by *Trichoderma viride* and *Paecilomyces lilacinus*. *Indian Phytopath.*, 57(2): 235–236.

Goswami, B.K., Mittal, A. and Singh, S., 2005. Management of disease-complex caused by root-knot nematode, *Meloidogyne incognita* and root rot fungus, *Rhizoctonia solani* on Balsam, *Impatiens balsamina* by fungal bio-agents. *Pesticide Res. J.*, 17(1): 19–22.

Goswami, B.K. and Sharma, S.B., 1999. Management of root-knot disease of tomato with integration of fungal bioagents (*Aspergillus terreus* and *Paecilomyces lilacinus*) Abstract. In: *Nat. Symp. on Pl. Dis. Scenario under Changing Agroecosystems*, Oct. 29–30, HPKV, Palampur, H.P., p. 42.

Haque, M.M and Prasad, D., 1982. Effect of agronomical practices and DD application to nematodes and crop yield. *Indian J. Nematol.*, 13: 226–229.

Jain, R.K. and Bhatti, D.S., 1989. An integrated approach for the management of root-knot nematode (*Meloidogyne javanica*) in tomato. *Ind. J. Nematol.*, 19: 159–161.•Jatala, P., 1983. Biological cont rol

of plant parasitic nematodes *Ann. Rev. Phytopathol.*, 24: 453–489.•Kerry, B.R., 1980. Biocontrol: Fungal arasites of female cyst nematodes. *J. Nematol.*, 12: 253–259.•Kogan, M., 1

98. Integrated pest management: Historical perspectives and contemporary development. *A. Rev. Ent.*, 43: 243–270.

Kumar, B., Yaduraju, N.T., Ahuja, K.N. and Prasad, D., 1993. Effect of soil solarization on weeds and nematodes under tropical Indian conditions. *Weed Res.*, 33: 423–429.

Mankau, R., 1968. Reduction of root-knot disease with organic amendments under semifield conditions. *Plant Dis. Reptr.*, 52: 313–319.

Mittal, A. and Prasad, D., 2005. *Paecilomyces lilacinus*: Fungal Bioagent for controlling phytonematodes. In: *Crop Protection Management Strategies*. Daya Publishing House, New Delhi, p. 549–560.

Mittal, A. and Goswami, B.K., 2001. Role of undi oil seed cake for the management of disease complex caused by *Fusarium solani* and *Meloidogyne incognita* on brinjal. *Pakistan J. Nematol.*, 19(1/2): 87–90.

Mittal, A. and Goswami, B.K., 2004a. Effect of culture filtrates of *Aspergillus* species on mortality and hatching in *Meloidogyne incognita. J. Mycol. Pl. Pathol.*, 34(2): 281–282.

Mittal, A. and Goswami, B.K., 2004b. *Myrothecium roridum*: An opportunistic fungal parasite on *Meloidogyne incognita. J. Mycol. Pl. Pathol.*, 34(2): 284.

Mittal, A. and Goswami, B.K., 2004c. Hatching inhibition in black egg-masses of *Meloiodgyne incognita* caused by Hyphomycetous fungi. *J. Mycol. Pl. Pathol.*, 34(2): 283.

Mittal, A. and Goswami, B.K., 2005. Interaction of *Meloidogyne incognita*, root-knot nematode of tomato and soil fungi. *Pakistan J. Nematol.*, 23(2): 259–264.

Mittal, A., Goswami, B.K. and Chowdhry, P.N., 2005. Hyphomycetous fungi from rhizopshere and rhizoplane of root-knot affected cucurbitaceous fungi from Trans-Yamuna river belt of Delhi. *Indian Phytopath.*, 58(4): 480–482.

Ogunfowora, A.O., 1976. Research on *Meloidogyne* at the institute of agricultural research and training. Univ. Ife. Moor Plantation Ibadan. In: *Proc. 1st IMP Res. Plant Conf. Root-knot Nematode, Meloidogyne spp.* Region IV. 7–11 June. Int. Institute. Trop. Agr., p. 9–14.

Prasad, D., Mittal, A. and Yaduraju, N.T., 1995. *(Abstr.) Nat. Symp. Perspectives in Eco-friendly Approaches to Plant Protection*. Soc. of Pl. Protect Sci., 7–8, Sept. IARI, New Delhi, p. 75.

Roberts, P.A., 1992. Current status of the availability, development and use of host plant resistance to nematodes. *J. Nematol.*, 24: 213–227.

Seshadri, A.R. and Dasgupta, D.R., 1980. *Handbook of Agriculture*, New Delhi, ICAR, p. 551–557.

Sethi, C.L. and Gaur, H.S., 1986. Nematode management: An overview. In: *Plant Parasitic Nematodes of India: Problems and Progress*, p. 424–445.

Sharma, M. and Saxena, S.K., 1992. Effect of culture filtrates of *Rhizoctonia solani* and *Trichoderma viride* on hatching of larvae of root-knot nematode, *Meloidogyne incognita. Curr. Nematol.*, 3: 61–64.

Sharma, H.K., Pankaj, Jagan Lal and Singh, S., 2007. *Indian Journal of Nematology*, 37: 192–195.

Sasser, J.N. and Freckman, D.W., 1987. A world perspective on nematology: the role of the society. In: *Vistas on Nematology*, (Eds.) J.A.Veech and D.W. Dickson. Hyattsville, Maryland, USA, Society of Nematologists, p. 7–14.

Stirling, G.R., 1991. *Biological Control of Plant Parasitic Nematodes.* Wallingford, UK, CAB International, p. 282.

Senthilkumar, T. and Ramakrishanan, S., 2004. *Ann. Pl. Protec. Sci.*, 12: 140–142.

Tabreiz, A., Khan, T.A., Hussain, S.I. and Azam, M.F., 1884. Effect of culture filtrates of eight species of *Aspergillus* on the hatching and mortality of *Meloidogyne incognita. Indian J. Nematol.*, 14: 50–74.

Thomason, I.J., 1987. Challenges facing nematology: Environmental risks with nematicides and the need for new approaches. In: *Vistas on Nematology*, (Eds.) J.A. Veech and D.W. Dickson. Hyattsville, Maryland, USA, Society of Nematologists, p. 469–476.

Plant Protection, Diversity and Conservation (2011) *Pages* **627–631**
Editors: **D. Prasad & Satya Kumar**
Published by: **BIOTECH BOOKS, NEW DELHI**

Chapter 47

Efficacy of Certain Rhizosphere Fungi against Root-knot Nematode (*Meloidogyne incognita*) and *Rhizoctonia solani* in Tomato Crop

***Rekha Arya*[1] *and S.K. Saxena*[2]**
[1]*Head, Botany Department, Tika Ram Kanya Mahavidyalaya, Aligarh – 202001, U.P.*
[2]*Retired Professor, Botany Department, A.M.U., Aligarh – 202002, U.P.*

Isolation of rhizosphere and rhizoplane mycoflora of tomato cv. *Pusa Ruby,* Technique used were Warcup's soil plate method and serial rot washing technique of Harley and Waid. Pure culture of fungi were maintained on potato dextrose broth. Sterilised seeds of tomato cv *Pusa Ruby* were sown in 15ml pts containing 100gm of sterilised soil. Soil was inoculated with *M.incognita, R.solani* and seven rhizosphere fungi separately and in different combinations and different dosages.

In a previous study the per cent germination of above mentioned tomato seedlings was observed after 5, 10, 20, 25 days. Out of these seven fungi four were selected for present studies in which their effect was observed on the growth of plant and nematode multiplication both at seedling emergence and transplantation stages. These fungi were *Epicoccum purpurascens, Penicillium vermiculatum, Trichothecium roseum* and *Trichoderma viridae.*

Transplantation was made in 6 inches pot containing 1 kg of sterilised soil. The soil was inoculated with 5gm of fungal suspension and 1000J_2 nematode larvae in mono, bi and tri pathogenic conditions. Various growth parameters and rate of nematode multiplication was observed after 45 days of transplantation.

Results presented in Tables 47.1 and 47.2 indicate that the growth was adversely affected to a varying degree in almost all the cases. Out of the 4 fungi screened *Trichothecium rosum* comes out most effective in mitigating the adverse effect of *R. solani* and *M.incognita*. The plants treated with *T.roseum* show a slight improvement in the growth in over the check both in bi and tri pathogenic conditions *P. vermiculatum* was effective against *R. solani* but not *M. incognita*. All the four fungi were effective in reducing the nematode population where *T. rosum* most effective followed by *T.viridae*. The effect of *T.roseum* was more pronounced when the plants were inoculated both at seedling emergence and transplantation stages. It is likely that this fungus is poor competitor in soil and infestation only at the seedline emergence state is not enough to have the desired effect. Subsequent infestation of soil with the fungus has much more desired effect.

Table 47.1: Change in Growth of Tomato cv '*Pusa Ruby*' as a Result of Infestation of Soil with *Epicoccum purpurascens, Penicillium vermiculatum, Trichothecimn roseum* and *Trichoderma viridae* with Root-knot Nematode *Meloidogyne incognita* (with or without *Rhizoctonia solani*)

Treatments	*Length (cm)*		*Fresh Weight (g)*		*Dry Weight (g)*	
	Inoculation at Seed Germination	*Inoculation at Seed Germination and Transplantation*	*Inoculation at Seed Germination*	*Inoculation at Seed Germination and Transplantation*	*Inoculation at Seed Germination*	*Inoculation at Seed Germination and Transplantation*
Nematode	–10.83	–21.03	–17.71	–30.72	–21.80	–32.07
R. Solani	–14.11	–19–68	–24.00	–33.10	–31.12	–36.17
E. Pupwascence	–11.48	–15.77	–16–12	–39.61	–30.17	–35.70
P. Vermiculatum	–08.84	–19.00	–09.33	–20.72	–02.21	–24.01
T. Roieum	–14.50	–00.55	–11.51	–08.69	–20.53	+09.32
T. vindae	–09.53	–05.44	–10.58	–08.67	–21.16	–19.11
R.s + Nematode	–16.15	–25.93	–26.47	–43.27	–35.07	–43.60
E.p + Nematode	–10.78	–14.72	–19.59	–29.03	–31.43	–32.38
P.v + Nematode	–02.20	–03.91	–05.95	–18.41	–14.06	–23.85
T.r + Nematode	+03.54	+03.54	–06.91	+16.67	–00,63	+13.27
T.v + Nematode	–12.26	–18.60	–14.57	–25.90	–21.32	–24.80
R.s + E.p	–15.10	–11.20	–28.00	–05.66	–29.38	–19.27
R.s + P.v.	–06.22	–16.37	–20.79	–14.79	–09.95	+/– 00.00
R.s + T.r.	–15.35	–14.32	–15.80	+01.06	–18.16	+01.89
R.s + T.v	–09.17	–17.98	06.40	–21.68	–17.06	–26.38
R.sl + E.p. + Nematode	–09.11	–15.09	+06.46	–12.67	–03.79	–13.74
R.s. + P.v. + Nematode	–03.59	–13.75	–12.98	–27.20	–17.21	–24.01
R.s + T.r. + Nematode	–12.03	–13.17	–08.50	–10.65	–18.16	–24.64
R.s + Tv. + Nematode	–13.86	–19.34	–17.95	–26.26	–21.01	–27.48

Decrease (–) / Increase (+) in growth plants.

Each reading is a mean of 10 replicates in a treatment with 10 pots in each treatment.

Table 4 .2: Effect of Infestation of *Epicoccum purpurascens*, *Penicillium vermiculatum*, *Trichotecium roseum* and *Trichoderma viridae* in Soil Together with Root-knot Nematode (with or without *Rhizoctonia solani*) on Nematode Multiplication on Tomato on Pusa Ruby

Treatment	No. of System Galls/Root		Root-knot Index		No. of Larvae/ Root System		No.of Females Root System		No. of Larvae/ kg Soils		Total Nematode No.	
	Plants Inoculated only at Germination	Plants Inoculated at germination and Transplantation	Plants Inoculated only at Germination	Plants Inoculated at germination and Transplantation	Plants Inoculated only at Germination	Plants Inoculated at germination and Transplantation	Plants Inoculated only at Germination	Plants Inoculated at germination and Transplantation	Plants Inoculated only at Germination	Plants Inoculated at germination and Transplantation	Plants Inoculated only at Germination	Plants Inoculated at germination and Transplantation
Nematode	25.27	107.50	3	5	356.50	2274.87	18.12	101.93	-	1241.87	554.62	3618.67
R.s. + Nematode	20.15	075.46	3	4	466.25	1588.81	16.31	069.06	-	0798.87	482.56	2456.74
E.p. + Nematode	21.50	098.25	3	4	535.00	2272.50	14.75	093.50	-	1277.50	549.75	3543.50
P.v. + Nematode	21.00	076.50	3	4	305.00	1280.00	19.25	106.00	-	0837.50	324.25	2223.50
T.r. + Nematode	01.25	034.50	2	4	002.75	0741.75	00.50	029.00	-	0485.00	003.25	1255.75
T.v. + Nematode	25.00	037.00	3	4	500.00	1515.00	14.25	040.50	-	0450.00	514.25	2005.50
R.s. + *E.p.* + Nematode	19.50	076.75	3	4	520.00	1762.50	14.50	074.50	-	0780.00	534.50	2617.00
R.s. + *P.v.* + Nematode	28.25	090.50	3	4	392.50	2190.00	26.50	134.25	-	1170.00	419.00	3494.25
R.s. + *T.r.* + Nematode	06.75	039.00	2	4	014.25	0858.25	03.50	036.25	-	0593.75	017.75	1488.25
R.s. + *T.v.* + Nematode	22.00	033.00	3	4	466.75	0720.00	23.75	29.25	-	0390.00	490.50	1139.25
C.D. at 5 per cent	02.054	004.823			031.623	0033.206	01.258	05.053	-	0023.153	023.237	0052.672

R.s.: *Rhizoctonia solani*; *E.p.*: *Epicoccum purpurascens*; *P.v.*: *Penicillium vermiculatum*; *T.r.*: *Trichothecium roseum*; *T.v.*: *Trichoderma viridae*.

Each reading is mean of ten replicates in each treatment with ten pots in each replicates.

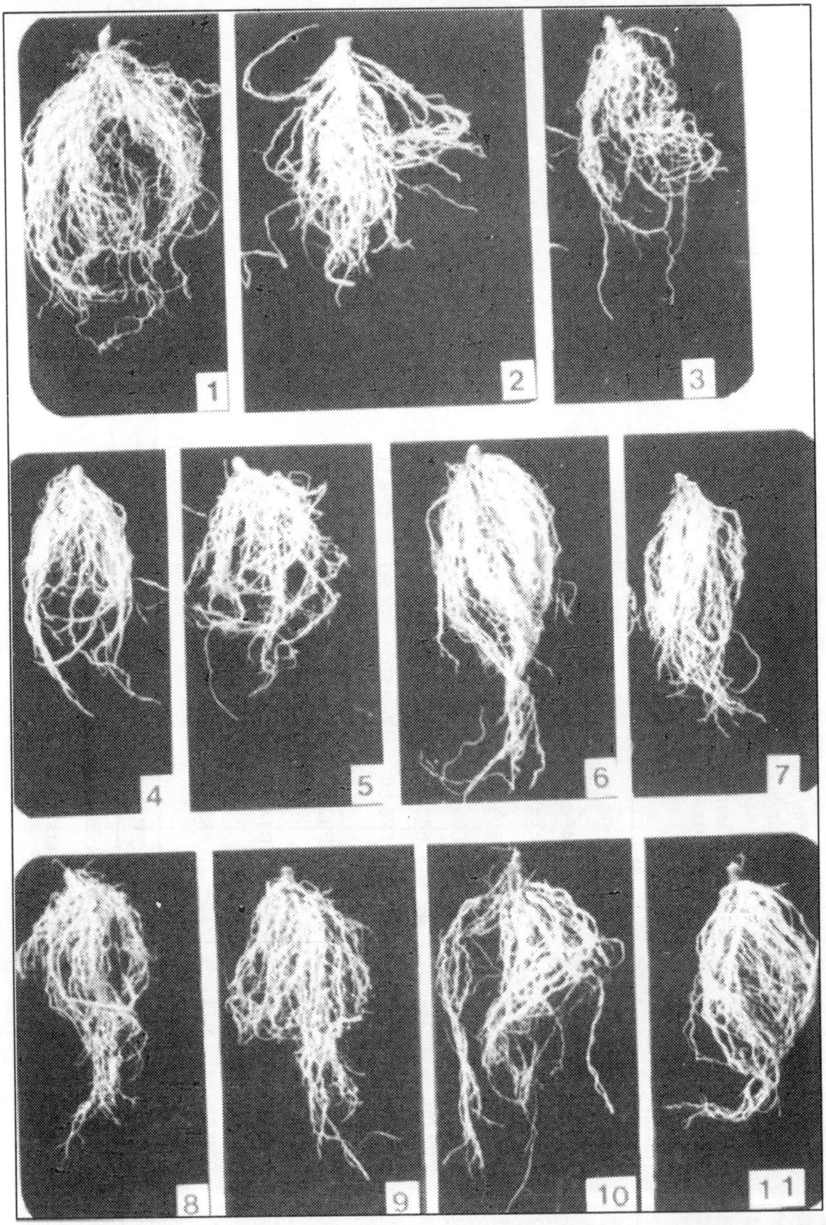

Figure 47.1: Roots of Tomato cv 'Pusa Ruby' Plants Growth in (1) Uninoculated soil; (2) Soil inoculated with nematode (1000 J_2/kg soil); (3) *Rhizoctonia solani* (5g/kg soil) + nematode (1000 J_2/kg soil); (4) *Epicoccum purpurascens* (5g/kg soil) + nematode (1000 J_2/kg soil); (5) *Penicillium vermiculatum* (5g/kg soil) + nematode (1000 J_2/kg soil); (6) *Trichothecium roseum* (5g/kg soil) + nematode (100J_2/kg soil); (7) *Trichoderma viride* (5g/kg soil) + nematode (100J_2/kg soil); (8) *R. solani* (5g/kg soil) + *E. purpurascens* (5g/kg soil) + nematode (100J_2/kg soil); (9) *R. solani* (5g/kg soil) + *P. vermiculatum* (5g/kg soil) + nematode (100J_2/kg soil); (10) *R. solani* (5g/kg soil) + *T. roseum* (5g/kg soil) + nematode (100J_2/kg soil); (11) *R. solani* (5g/kg soil) + *T. viride* (5g/kg soil) + nematode (100J_2/kg soil).

Plants were inoculated both at germination and transplantation stages.

Exact mechanism cannot be suggested for *T.roseum* in view of paucity of information. Fungus metabolites like trichothecenes was found effective against nematode at the concentration of 50m gm/ml (Bascikova, D., Betina, V., Nemec, P.1965, P.1965 as quoted by Chitwood 2002)

More over the fungus has been reported as the predator of the nematode (Mekhtieva, 1962; Cooke and Godfrey, 1964). *T.viridae* and *P.vermiculatum* are reported parasitising *Rhizoctonia* spp. (Boosalis, 1956; Bedlan, 1988)

The possibility of enzymes in the metabolitie of fungus affecting the body of nematode however cannot be ruled out. As in case of *E.purpurascens* in which the compounds like chrysin and p-hydroxy benzoic acid were found to be antagonistic against nematode (Singh, 1985).

Acknowledgements

Financial Assistance rendered by CSIR throught the studies is gratefully acknowledged.

References

Boosalis, M.G., 1956. Effect of soil temperature and green manure amendment of unsterilised soil on parasitism of *Rhizoctonia solani* by *Penicillium vermiculatum* and *Trichoderma spp. Phytopathology,* 46: 473–478.

Bascikova, D., Betina, V. and Nemec, P., 1965. Anti-helminthic activity of antibiotics. *Nature,* 206: 1371–1372.

Chitwood, D.J., 2002. Phytochemical based strategies for nematode control. *Ann. Rev. Phytopath.,* 40: 221–19.

Cooke, R.C. and Godfrey, B.E.S., 1994. A key to the nematode-trapping fungi. *Trans. Brit. Mycol. Soc.,* 47: 61–74.

Golden, J.K. and van Gundy, S.D., 1975. A disease complex of okra and tomato involving the nematode *Meloidogyne incognita* and soil inhabiting fungus *Rhizoctonia solani. Phytopathology,* 65: 265–273.

Mekhtieva, M. and Me, N.A., 1962. On the predacity of fungus *Trichothecium roseum. Bot. Zh. S.S.S.R.,* 47: 1362–1363.

Sen, B., 2000. Biological control: A success story. *Indian Phytopath.,* 53(3): 243–249.

Singh, R.S., 1985. Use of *Epicoccum purpurascens* as an antagonist against *Macrophomina phaseolina* and *Colletotrichum capsici. Indian Phytopath.,* 38: 258–262.

Plant Protection, Diversity and Conservation (2011) *Pages* **632–644**
Editors: **D. Prasad & Satya Kumar**
Published by: **BIOTECH BOOKS, NEW DELHI**

Chapter 48

Environmentally Safe Management Options of Plant-Parasitic Nematodes Associated with Oilseed Crops

Archana U. Singh* and D. Prasad
Division of Nematology, I.A.R.I, New Delhi – 110 012

Oilseeds are the backbone of our agriculture based economy from time immemorial. Oilseed area in the world is about 410 million tones, highest is of soybean with 232.9 million tones. The oilseeds production during 2007-08 is estimated to be 288.25 lakh tonnes which is about 45 lakh tonnes more than the oilseeds production during 2006-07. India having 10 per cent share in the world oilseed market is largest producer of Groundnut, Sesamum, Linseed and Castor. Oilseed crops are gaining great importance in the country owing to the increasing demands. India is the third largest oil economy in the world next to United States of America and China, harvesting 22 million tonnes of various oilseeds against the world production of 250 million tonnes. India accounts for 9.7 per cent of world's oilseed production, 8.7 per cent of world's production of major protein meal and 9.2 per cent of world's oil meal exports. In India, major oilseed crops namely groundnut, rapeseed, mustard, castor, soybean, sesamum and linseed cultivated to a lesser or greater extent in almost in all the states. Among all, groundnut occupies an important place with an area of about 12.0 million hectare and total production of 19.5 million tonnes. Thus, India has the largest area under the crop in the world but low productivity of the crop (847 kg/ha). However, one of the prime constraints in oilseed production is the insect pests, nematodes and diseases. But on a world wide basis, the loss in major oilseed crops comes to 32 per cent including 10 per cent by insect pests, 8 per cent by diseases, 8 per cent by weeds and 5 per cent by plant-parasitic nematodes. (Jayraj, 1993). Despite of so much production, per capita oilseed consumption is very low *i.e.*, 10.6 kg per annum which is much less than 20.8 kg per annum of Japan and 48 kg per annum of USA.

* E-mail: arch_212@yahoo.com

Distribution of area under various oilseed crops is as follows-

Crop	*Area*	*Production*	*Distribution in India*
Groundnut	7.5 mha	6 mt	Gujarat, TN, AP, KR, MH, Orissa
Rapeseed and Mustard	4 mha	2 mt	UP, PB, BH, HR, MP, RJ, WB and Orissa
Sesamum	2.5 mha	52 t t	UP, MP, RJ, MH, AP, TN and Orissa
Castor	4.48 lakh ha	1.40 lakh t	AP, KR, Gujarat and Orissa
Safflower	5.90 lakh ha	1.30 lakh t	MH, KR and AP
Niger	4.8 lakh ha	1.2 lakh t	MP, BR, TN, MH and Orissa
Sunflower	2.16 mha	1 mt	KR, AP, MH

A faunistic survey of the nematodes associated with different oilseed crops showed the presence of root-knot nematodes around the rhizopshere of sunflower and safflower in a varying degree (Prasad and Khan, 1990). Prasad (1989) reported 23-27 per cent loss in groundnut by *Meloidogyne arenaria* and ectoparasites cause 11-15 per cent and 40-45 per cent loss by Kalahesti Malady Disease and 30-40 per cent due to *Rotylenchulus reniformis* and *Meloidogyne incognita* together. Prasad and Ahmad (2002) studied the assessment of avoidable yield loss of sunflower due to *Rotylenchulus reniformis.* They reported that yield of sunflower seeds was found to be higher with the application of nematicide-carbofuran @ 2 kg a.i/ha.

Nematode Diseases on Oilseed Crops

Rotylenchulus reniformis on Groundnut and Sunflower

The reniform nematode was recorded from groundnut crop in Senegal and India. The groundnut cultivar 28-206 was found to be resistant to both the strains of this nematode. In a study in fields near Bareilly, Uttar Pradesh sunflower crop was found to be infested with *R. reniformis.*

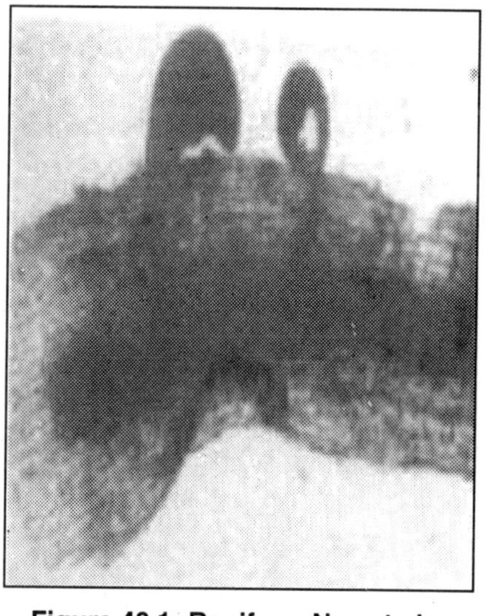

Figure 48.1: Reniform Nematode, *Rotylenculus reniformis*

M. arenaria on Groundnut

Plants infested with nematodes commonly develop enlarged roots and pegs which in turn develop to galls of numerous sizes. Pods also become infested and develop knots and protuberances. Galls on roots, pegs and pods sometimes begin to deteriorate on crop maturity. A typical symptom of stunting of groundnut plant with yellowing of top leaves and much reduced leaf size was observed under field conditions. Infested field have poor coverage of the ground surface at mid season and having restricted root system.

M. incognita on Soybean

In general plants show stunting, yellowing and sparse foliage under field conditions. The root system development is much reduced. Similar symptoms were observed under field infested with *R. reniformis.* Heavy infestation by reniform nematode drastically reduced the root system of soybean crop and plants can easily be pulled out from the soil. Prasad

(1999) reported that carbofuran 3G @ 3 kg a.i/ha controlled both root-knot and reniform nematode population and resulted in increased yield of soybean crop. The per cent increase over untreated control was calculated to be 18.7.

Tylenchorhynchus brevilineatus on Groundnut

Reddy *et al.* (1984) reported a severe disease caused by *T. brevilineatus* on groundnut and first time reported from Triputi region of A. P. Further, Prasad and Rangappa (1994) reported no foliage reduction or yellowing but the nematode affected the nut production with blackening of nutshell. They revealed the presence of *Tylenchorhynchus brevilineatus*, *T. nudus*, *Meloidogyne arenaria* and *Helicotylenchus dihystera* in large numbers with this disease of groundnut. The parasitic nematode attack the roots, pegs and pods of groundnut causing mechanical injury and pave way for fungus, *Macrophomina phaseolina* which may cause blackening on the pods.

Figure 48.2: Kalahasti Malady of Groundnut

Meloidogyne incognita on Sunflower

Survey work of IARI field, A.P, U.P. and Bihar states revealed the presence of *M. incognita* in high number. Stunting and yellowing were the common symptoms caused by root-knot nematode coupled with reduction in size of flower and number of seeds per flower. The per cent reduction in weight of flower head was found to be 25.3, 26.6 and 100 per cent in 100, 1000 and 10, 000 juveniles of *M. incognita*.

Heterodera cajani on Sesamum

The attacked Sesamum plants showed relatively stunted growth, sparse root-system with bunchy appearance and reduced pod production.

During the last three decades we are using chemicals as major plant protection method due to various advantages associated with its use are:

- ✰ The only practical control measure
- ✰ The means to get quick results
- ✰ Wide range of properties, uses and methods
- ✰ Cheap and offer greater result
- ✰ Benefit-cost ratio favourable

But along with its advantages some disadvantages are also there due to its excessive and indiscriminate use that are:

- ✰ Hazards to man
- ✰ Effects on non target organism
- ✰ Phytotoxicity
- ✰ Hazards of pesticide residues
- ✰ Upsetting the ecological balance
- ✰ Difficulty in application

So now we are shifting from chemical control methods to Eco-friendly approach of nematode management. These consists of following:

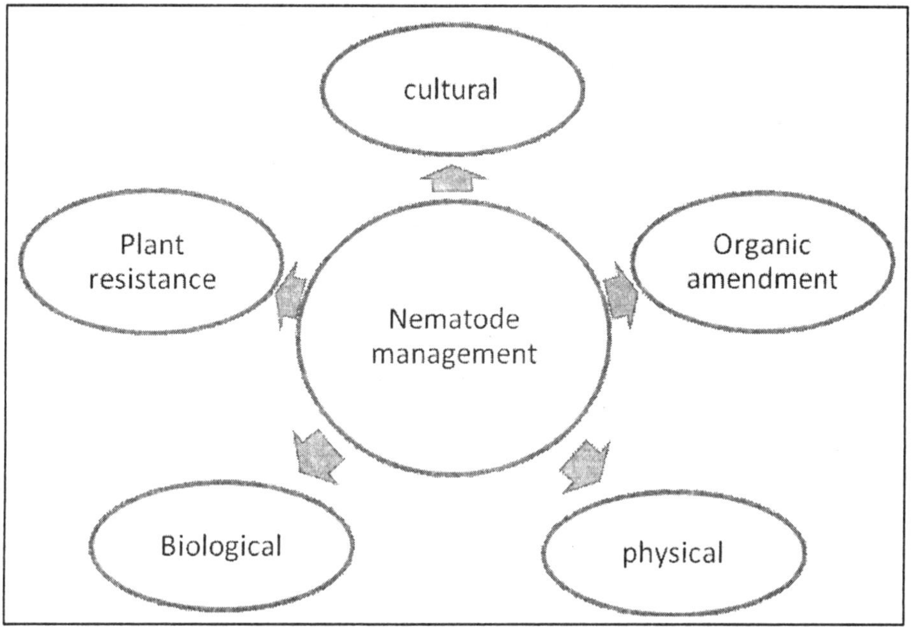

Figure 48.3

Cultural Methods

Fallowing

This is keeping the land free of any vegetation including weeds. It kills the nematodes due to starvation, desiccation and solar heat. In many parts of India after harvest of rabi crops, the field is left fallow without host crops. This practice is widely used in Northern India which reduces the nematode population. If the fallow is coupled with summer ploughing, the decrease of endoparasitic nematodes especially root-knot nematode was recorded to be as much as 40 per cent. The main principle behind this is starvation of nematodes due to non-availability of host crop and desiccation due to heat during summer.

Flooding

Nematodes are died due to starvation and asphyxiation.

Summer Ploughing

Exposure of nematodes to solar heat and desiccation. It accentuates the effect of fallowing and kills nematodes, other pests and weeds. Three or more ploughings during summer months exposes the nematodes to solar energy which are killed due to desiccation. The kill was found upto 60 per cent.

Time of Planting

Early or delay planting escapes nematode damage. It does not require any additional costs and no loss of cropping period.

Crop Rotation

Growing of poor or non host crops in alternation with susceptible crops. The advantages with crop rotation are no losses of cropping period as well as it improve soil fertility. Onion, garlic, asparagus are the non host crops and good for management of root-knot nematode.

Management of Kalahasti Malady (*T. brevilineatus*) of Groundnut through Crop Rotation

Naidu *et al.* (2000) done work to control *T. brevilineatus* in groundnut through crop rotation and they found that Rabi Groundnut preceded by summer rice and *kharif* rice recorded maximum decrease in nematode population (59.7 per cent), disease severity (42.6 per cent) and increase in pod yield (83.3 per cent).

This crop rotation system was closely followed by the two cropping sequences:

1. Rabi groundnut preceded by summer Rice and *kharif* sunhemp
2. Rabi Groundnut preceded by kharif Rice

They also found that Blackgram, Greengram and Maize proved to be susceptible host for the nematode (*T. brevilineatus*).

Intercropping

Mixed cropping either with non-host or antagonistic crops such as *Tagetes*, Sesame, mustard etc. have also been reported to reduce root-knot and reniform nematodes. The use of trap crops as green manure not only reduce the nematode population but improves soil fertility in many ways and promotes antagonistic microbial activity. For example, sunhemp or Dhaincha which allow invasion by *Meloidogyne* spp. but does not support development and can serve as good trap crop. However, the incorporation of harvested crop residues of marigold (*Tagetes erecta*), mustard (*Brassica campestris*) and or sunflower (*Helianthus annuus*) into soil proved effective in reducing the population build up of some plant-parasitic nematodes. Marigold residues were most effective followed by mustard and sunflower. However, chopped green leaves of *Ruta graveolens* were less effective than fenamiphos for the control of *M. javanica* on sunflower. The effect of *R. graveolens* was more evident on nematode population density than on plant growth parameters.

Eight plant species namely *Solanum surattense, Theviatia peruviana, Calotropis procera, Thuja sinesis, Parthenium husterophorus, Croton sparsiflorus, Colocasia antiguorum* and *Datura stramonium* were evaluated for the nematicidal properties against J2 of *Meloidogyne incognita* and pre-adults of *Rotylenchulus reniformis* in vitro by Prasad *et al.* (2002). Hundred per cent mortality of both the nematode species were found in S/16 dilutions except *S. surattense* fruits (unripened) whereas in S/32, the range was 50-100 per cent in case of *M. incognita* and 51-86 per cent in case of *R. reniformis.* They also observed that when plant extracts (S/16 and S/32 dilutions) were put around mung crop in pots, there was an increase in shoot length, maximum being in *Colocasia antiguroum* leaves extract. Similarly, a significant increase of shoot and root weights were also observed. However. maximum reduction of galls was noticed in *Calotropis procera* followed by *Datura stramonium* and *Solanum surattense* leaves extracts.

Naidu *et al.* (2000) studied the management of Kalahasti malady (*Tylenchorhynchus brevilineatus*) of groundnut through crop rotation in Andhra Pradesh, India, in 1991-94. *Rabi* groundnut preceded by summer rice and kharif rice recorded the greatest reduction in the nematode population (59.7 per cent) and disease severity (42.6 per cent) and the greatest increase in pod yield (83.3 per cent) compared to continous cropping with groundnut.

Antagonistic Crop

The nematodes are killed by action of root exudates and not allowed to mature. Antagonistic crops like African marigold, *Asparagus* and *Crotalaria* are used for the control of reniform nematode. Popular term in the litreature "mustard effect" is attributed to glucosinolate compounds contained in *Brassica* residues. Toxicity is attributed to enzymatically induced breakdown products of glucosinolates, a large class of compounds known as isothiocyanates and nitriels that suppress nematodes by interfering with their reproductive cycle. These glucosinolate breakdown products are similar to the chemical fumigant VAPAM® (metham sodium), which degrades in soil to methyl isothiocyanate. Allelochemicals are plant-produced compounds (other than food compounds) that affect the behaviour of other organisms in plant's environment. For example, sudangrass (and sorghum) contain a chemical, "Dhurrin" that degrades into hydrogen cyanide, which is a powerful nematicide. Jack Brown, a plant breeder specializing in *Brassica* at the University of Idaho, has released two biofumigant varieties, "Humus" of rapeseed and "IdaGold" of mustard, each containing elevated levels of glucosinolates. Other crops like *Tagetes* which contain α-terthienyl and *Asparagus* which contain "Asparagusic acid" are detrimental to plant parasitic nematodes.

Use of Organic Amendments for Nematode Management

It includes:

1. Oilseed cakes
2. Extracts of plant parts
3. Compost and FYM
4. Mature crop residue
5. Neem products
6. Green manures

Ononuju and Kpadobi (2008) in their attempt to control *Meloidogyne incognita* in Soybean with the help of plant leaf ash found that by using *Ecornia lotufolia* plant height and nodules of nitrogen fixing bacteria are maximum. With *Cymbopogon citrates* number of nematode galls are less and with *Azadirachta indica* no. of pods/plant are maximum.

Treatment (Ash)	Plant Height (cm)	No. of Galls	No. of Nodules	No. of Pods/Plant
Azadirachta indica	43.98	97.50	96.75	67.00
Ecornia lotufolia	59.98	72.75	134.00	59.00
Uvaria kemiyae	38.68	79.00	111.00	58.50
Cymbopogon citratus	46.70	28.75	88.50	58.25
Tetrepleura tetraptera	43.48	65.75	94.25	68.50
Nematicide (Counter)	39.00	21.75	102.25	67.50
Control	37.05	180.50	84.75	20.75

Prasad and Mittal (2004) studied the effect of different neem formulations on *Rotylenchulus reniformis* and growth of Sunflower. They found that Linanool @ 20 l/ha is most effective in controlling *R. reniformis* and increasing plant yield.

Treatment	Shoot Length (cm)	Root Length (cm)	Flower Weight (g)	R. reniformis pop. (100 g)
Azadirachtin @ 20 l/ha	15.6	3.6	1.1	14.0
Neem gold @ 20 l/ha	15.6	4.0	0.7	10.8
Bioneem @ 20 l/ha	24.0	4.7	1.0	12.6
Rakshak @ 20 l/ha	22.6	5.0	0.7	5.8
Linanool @ 20 l/ha	28.0	5.6	3.4	3.6
Nimbidine @ 20 l/ha	14.6	2.0	1.1	17.0
Sukrina @ 20 l/ha	21.6	5.6	1.8	15.5
Trizophos @ 2l/ha	22.6	4.0	2.6	6.3
Inoculated check	23.6	5.3	2.4	42.3

Joshi and Patel (1995) did experiment using different organic amendment in management of *Meloidogyne javanica* on groundnut. They found that fresh *Azolla* gave best result among all the treatments.

Treatment	Shoot Length (cm)	Root Length (cm)	Root-knot Index (0 – 5)
Azolla pinnata	13.8	10.7	1.60
Pink periwinkle	12.8	9.5	2.00
White periwinkle	12.5	9.1	2.06
Clerodendron inerme	12.7	10.3	2.00
Castor cake	11.35	8.70	2.00
Karanj cake	11.20	8.50	2.30
Mustard cake	15.25	8.60	2.00
Dry azolla	15.38	12.43	1.30
Fresh azolla	15.95	12.90	1.00
FYM	10.90	8.60	2.10
Poultry manure	11.27	8.40	2.80

Singh *et al.* (2001) studied the effect of aqueous extracts of plant products on hatching and penetration of *M. incognita* infecting sunflower and found that cotton cake is most effective in inhibiting hatching of RKN in sunflower crop.

Treatment	Dilutions of Stock Solution (N)				
	N	N:10	N:20	N:40	N:80
Neem leaves	0	88.8	157.7	162.8	251.8
Melia leaves	0	64.2	125	245	325.4
Calotropis leaves	0	30	135.4	131.2	221
Mustard cake	0	8.8	124.6	227.4	333.6
Cotton cake	0	4.2	58	101.8	155.4

Anver *et al.* (2006) studied the combined effect of oilseed cakes and ploughing on the control of plant parasitic nematodes associated with linseed and found that in both the cases *i.e.*, normal and deep ploughing, neem cake is effective. Neem cake with deep ploughing gave best result in terms of plant growth and nematode control.

Verma (1999) found that combination of saw dust and NPK @ 25 q/ha is most effective in increasing yield of groundnut and reducing Kalahasti malady disease caused by *T. brevilineatus*

Treatment	*No. of Pods/Plant*	*100 pod wt. (g)*	*Yield (q/ha)*	*% Disease Index*
Check	8.60	72	22.60	68.50
Neem cake @ 10 q/ha	6.90	80	25.67	48.00
Castor cake @10q/ha	9.25	77	24.00	52.00
Groundnut cake @10 q/ha	9.50	81	25.33	49.00
Poultry manure @ 50 q/ha	10.15	84	25.34	49.00
Farm yard manure@100 q/ha	9.70	82	26.33	48.00
Saw dust + NPK @25 q/ha	9.85	83	26.90	47.00
Press mud @ 25 q/ha	9.60	82	24.66	54.00
Carbofuran @ 4 kg ai/ha	10.45	87	28.00	47.00

Plant Resistance

Plant resistance is an important eco-friendly nematode management method. It prevents nematode from infection so no increase in the population, but there is always possibility of break-down or evolution of resistance breaking biotypes. Resistant varieties like 'Forrest' – cultivar of soybean in U.S.A. against *R. reniformis*, 'Tirupathi-2' and 'TPT-3' varieties of Groundnut for kalahasti malady disease control. The high-yielding soybean cultivars with potential to suppress reniform nematode are 'HY574', 'Padre', 'DP7375RR' and 'NK83-30'.

Dahiya and Gupta (1992) screened 24 cultivars/lines of sunflower and found no variety to be resistant. Three cultivars of *Helianthus annuus* i.e, Giza 1, Makhatt, Miak were found to be susceptible to *M. javanica*, *M. incognita* and *R. reniformis*. The lowest nematode multiplication occurring with *R. reniformis* on cv. Giza and Miak (Montasser *et al.*, 1985). Raut and Sethi (1980) reported that soybean cultivars like Forest, JS-2, Selection DS74-20-1 were found resistant against *M. incognita*. Among oilseed crops, Prasad (1989) screened about 500 varieties/lines of groundnut crop against *M. arenaria* and exhibited two lines as resistant namely C-41(NRCG-31) and NCAC-2196(NRCG-1060).

Biological Control of Nematodes

Use of parasites, predators and pathogens for controlling nematodes. Fungi (*Trichoderma harzianum*, *T. viridae*), bacteria (*Pasturia penetrans* and *Pseudomonas fluorescence*) and some VAM fungus are also used to manage the nematode population in the field. The biological control of *M. arenaria* on groundnut under greenhouse conditions by using *Paecilomyces lilacinus* and *Verticillium chlamydosporium* has been reported by Godoy *et al.* (1983). Presence of *M. arenaria* in the pods of groundnut led to the increased invasion of shells and kernels by *Aspergillus flavus*. However, the damage by root-knot nematodes did affect *A. flavus* infection. Wilting period of groundnut plants was reduced by 13 to 16 days on inoculation with *Fusarium solani* and *M. arenaria*. Also *Aspergillus flavus* along with *M. arenaria* showed synergistic affect on the growth of groundnut (Patel *et al.*, 1985).

Chemical Control

Combating of nematodes can be achieved by many ways but the use of nematicides are quite effective and cannot be replaced by any other method of nematode control. Chemicals are essentially used under such conditions where other methods offer far less certainity of killing the nematodes. No doubt, nematicides have brought tremendous benefits to mankind by increasing food and fibre production. The continuous use of such chemicals pose residual toxicities in plant and cause environmental pollution. Therefore, efficient strategies are required for the management of nematodes so that there is maximisation of advantages and minimisation of disadvantages. Nematicides can be used before the infection takes place (Prophylactic measures) or the infection is suspected to cause damage (curative measures). Prophylactic measures include seed pelleting, seed soaking and folia sprays.

Prasad (1990) studied the control by seed pelleted with carbofuran (40 flowable) and aldicarl sulfone (75WP) at the @ 3 and 6 per cent w/w. Results obtained revealed no deleterious effect or germination or on the growth of seedlings. The penetration of *M. arenaria* larvae was significantly low with the duration of time of observation and increased dosages. Carbofuran seed treatment togethe with soil treatment of carbofuran, phorate or ethoprophos were evaluated for the control of *M. arenari* on groundnut in field trials in Uttar Pradesh. Carbofuran (25 ST w/w) as seed treatment combine with soil treatment of phorate (10G) at @ 2kg/ha gave best control of *M. javanica* and increase in yielc

Sharma *et al.* (2001) worked on the management of *Meloidogyne incognita* on groundnut throug nematicides like carbofuran, phorate and sebufos at 2.0 kg/ha in pots. Results indicated that al treatments were significantly superior to the untreated control with respect to plant growth. A simila trend was recorded with regard to nematode reproduction.

Prasad *et al.* (1992) studied the management of root-knot nematode, *Meloidogyne arenaria* infecte groundnut by seed treatment with oxime carbamates namely Marshal 25 STD and Oncol 50 SP @ 1, 2 3 and 4 percent on seed weight basis under pot culture conditions. Plant growth characters recorde after 45 days of inoculation indicated better growth in lower doses (1 and 2 per cent) than higher one (3 and 4 per cent) of both the nematicidal seed treatment. As against this, gall index was low in highe dose. With respect to the plant growth and nematode control, 2 per cent is quite effective dose i groundnut.Pre-sowing application of benfuracarb and basudin on soybean @ 1, 2, 3, 4 and 5 kg a.i/h resulted better growth in lower doses of both the chemicals. Out of these two chemicals, basudin wa found to be quite effective at 2 kg which was at par with 3 kg of benfuracarb application (Prasad an Chawla, 1992).

Prasad (1998) undertaken a study to evaluate benfuracarb and triazophos at 100, 200 and 40 ppm as seed soak on the penetration by juveniles of *M. incognita* on soybean. In general both th nematicides reduced the penetration of root-knot nematode; the effect increased with time. Out of tw nematicides used as seed soak, triazophos was slightly better than benfuracarb in reducing th penetration of *M. incognita* juveniles into soybean roots.

Mittal and Prasad (1999) studied the effect of chemicals used as seed soaking on the developme of *Meloidogyne incognita* and growth of soybean cv. Pusa 16. Three chemicals *viz.*, Triazophos (40EC carbosulfan (25 EC) and monocrotophos (36 per cent SL) @ 400, 200 and 100ppm were tested as see soaking against *M. incognita* infecting soybean. In general, higher concentrations of all the chemica (*i.e.*, 400ppm) was found most effective in reducing nematode population both in root and soil.

The effect of soil drench with carbosulfan 25 STD and triazophos on penetration and developme of *Meloidogyne incognita*. Race –1 on sunflower (cv. EC 68414) was studied in pots by Prasad ar

Narayana (2000). Development of root-knot nematode at lower concentration of nematicides was found to be quicker than at higher concentrations, 2nd stage changed to 3rd and 4th stages. On 7th day of observation both the chemicals were found to be quite effective in reducing the total penetration and subsequent development. This, in turn also affected the fecundity of the nematodes.

Sharma and Prasad (1995) under taken a study to evaluate the efficacy of deoiled cakes of cotton, sunflower and soybean (1, 5 and 10 per cent w/w) and leaves of datura and eucalyptus (5 and 10 per cent w/w) for the management of *Meloidogyne incognita* on soybean cv. MACS–13 in plots. The results clearly showed reduction in penetration of 2nd stage juveniles and further development of galls and egg-mass with the increase in dose of botanicals. However, penetration and galling was least in cotton cake (10 per cent w/w). Penetration, galling and egg-mass were found to be reduced by 94, 84 and 90 per cent respectively with cotton cake (10 per cent). The cotton cake application was most effective followed by eucalyptus, sunflower and datura.

Kalaiarasan *et al.* (2008) used organic crab shell chitin for suppression of root-knot nematode *Meloidogyne incognita* in groundnut and found that chitin @ 1.2 per cent is most effective in suppressing nematode population while chitin @ 1.6 per cent is best in terms of yield of groundnut.

Physical Method of Nematode Control

It includes:

- ☆ *Sanitation* of field and destruction of infected crop residue – It is important as residues of crop plants and contain nematodes and also weeds serve as alternate hosts to nematodes.
- ☆ *Solar heat* - Expose the soil to direct heat or sun by summer ploughing and fallowing. Mulching the moist soil with thin polythene sheet for at least six weeks in hot summer month. Most effective in loamy to clayey soil in arid and semi arid condition. It is a good alternative for multiple pests, disease and weed control but expensive to use over large field. The solar energy is preserved with the help of transparent polyethylene sheet and increases soil temperature, enough to kill the nematodes present in a particular field. This is only applicable in limited areas. Prasad *et al.* (1995) found about 90 per cent kill of root-knot and reniform nematodes when solarised for 30 days during May-June months. Solarisation coupled with triazophos application @ 1.5 l/ha completely erradicated the plant parasitic nematodes under field conditions. This method provides an economical alternative for multiple pest, disease and weed control in small areas like nursery beds, orchards and forestry plantation and high value crops in regions with hot summers.
- ☆ *Hot water treatment* – Easier to use, suitable for disinfecting the seeds and other planting materials.

Integrated Nematode Management

Investigations were also carried out on INM for control of nematodes associated with groundnut crop. Prasad and Khan (1995) used cotton seed cake, neem cake @ 0.5 per cent applied one week before sowing with triazophos 1000 μg/ml sprayed on the foliage of groundnut. Results thus obtained indicate an increase of plant growth parameters and maximum was recorded in treatment where all three combinations were applied together with lowest gall index. Prasad and Khan (1990) found that integration of aldicarb, neem cake and soobabul leaves under field conditions resulted in better growth and higher yield of soybean. All three combination resulted in maximum yield of soybean besides maintaining low nematode population and maximum reduction being at 15 days of sampling.

Prasad (1999) conducted a field trial at State Govt. farm, Bilwa, Bareilly (U. P) during season o 1994 for management of plant-parasitic nematodes in groundnut. Various treatments significantly reduced population of *M. arenaria, T. vulgaris* and *Hoploliamus indicus* sampled just after 20 days o sowing. There was a drastic reduction of *M. arenaria* from 113 J2 to 7 in 20 days and 6 in 60 days o sampling in plots receiving summer ploughings + mustard cake 10g/ha + triazophos @ 2 litres a.i/ ha. Treatments resulted in significant increase in yield, however, maximum yield was observed ir mustard cake which account about 78.2 per cent increase over control.

Prasad *et al.* (1997) studied an integrated approach for the management of plant parasitic nematodes in groundnut. He found that all treatments were effective in reducing root-knot and reniform nematodes significantly and increased the yield of groundnut. However, maximum increase was found in chopped madar leaves applied 15 days before sowing over untreated control.

Prasad and Mittal (2004) gave recommendation for integrated control of *R. reniformis* in soybean cv. Pusa-24. They used various treatment and their combination which are as follows- *Calotropis procera* @ 600 g per 4m², *Trichoderma harzianum* @ 5 kg/ha, Carbofuran 3G@ 1.5 kg/ha. They found tha combination of all the three treatments were effective in managing the nematode population. Maximum yield and nematode control was achieved in plots receiving all three treatments (*Calotropis* + *Trichoderma* + Carbofuran) followed by *Calotropis* + Carbofuran.

Prasad (2007) studied the management of *Rotylenculus reniformis* on soybean through integrated approach and found that the combination of all three components reduced the population and increased the yield of soybean crop.

Treatment	*R. Reniformis Population (250 cc soil)*		*% Decrease in population*
	Initial	*Final*	
T1 – chopped *Calotropis*@0.4 kg/6 m²	1033	803	22.2
T2 – Carbosulfan seed treatment @ 1.5 per cent	1566	933	40.4
T3 – Triazophos soil app. @ 1.01 ai/ha	1300	730	43.8
T4 = T1 + T2	1533	703	54.1
T5 = T1 + T3	1666	666	60.0
T6 = T2 + T3	1500	733	51.1
T7 = T1 + T2 + T3	1200	583	51.4
T8 - control	1500	3303	-

Conclusion

- ☆ Eco-friendly nematode management methods are safe to human being, economical, easily applicable and environment friendly.
- ☆ These methods are not solely sufficient to control plant parasitic nematodes but avoid crop losses
- ☆ Integration of these eco-friendly methods with judicious use of chemicals are best options to control plant parasitic nematodes in any crops
- ☆ Intensive and extensive surveys on the presence of plant-parasitic nematodes infesting oilseed crops should be conducted.

- ☆ Use of cultural practices should be emphasized depending on area / locality.
- ☆ Resistant lines / sources should be identified.
- ☆ Emphasis should be given on the integration of nematode management practices by adoption of cultural practices, resistant varieties, nematicides, organic amendments, biological control agents, etc.
- ☆ Role of secondary organisms in causing diseases along with nematodes should be studied and nematode management practices be evolved.

References

Anver, S., 2006. *Indian J. Nematol.*, 36: 192–195.

Dahyia, R.S. and Gupta, D.C., 1992. *Indian J. Nematol.*, 22(1): 9–10.

Godoy, G., Rodriguez-Kabana, R. and Morgen-Jones, G., 1983. *Nematropica*, 13: 301–213.

Jayraj, S., 1993. *National Seminar on Oilseed Research and Development in India: Status and Strategies.* ISOR, Hyderabad, p. 49.

Joshi, P. and Patel, H., 1995. *Indian J. Nematol*, 25: 76–78.

Kalairasan, P., Lakshmanan, P. and Samiyappan, R., 2008. *Indian J. Nematol.*, 38: 196–202.

Mittal, A. and Prasad, D., 1999. *J. Oilseed Res.*, 16(1): 157–158.

Montasser, S.A., Al-Sayed, A.A. and Kher, A.M., 1985. *Pak. J. Nematol.*, 3(2): 91–94.

Naidu, P., Moses, G. and Sitaramairh, K., 2000. *Indian J. Nematol.*, 30: 58–61.

Ononuju, C. and Kpadobi, U., 2008. *Indian J. Nematol.*, 38: 1–4.

Patel, H.R., Vaishnav, M.U. and Dhruj, I.U., 1985. *Indian J. Nematol.*, 15(1): 98–99.

Prasad, D., 1990. *Pak. J. Nematol.*, 7: 97–101.

Prasad, D., 1990. *Curr. Nematol.*, 1: 43–46.

Prasad, D. and Khan, E., 1990. *(Abst.) Nat. Sym. and Prospects of Bot. Pesticides in IPM*, 21–22 Jan, CTCRI, Rajamundry, A.P., p. 44–45.

Prasad, D. and Chawla, M.L., 1992. *Curr. Nematol.*, 3(2): 201–202.

Prasad, D., Mittal, A. and Yaduraju, N.T., 1995. *(Abst.) Nat. Sym. on Perspectives in Eco-friendly Approaches to Plant Protection.* Soc. of Pl. Protec. Sci., 708 Sept. IARI, New Delhi, p. 75.

Prasad, D. and Khan, E., 1995. *(Abst.) Global Conference on Advance in Research, Plant Diseases and their Management*, 12–17, Feb, at RAU, Rajasthan, Udaipur.

Prasad, D. and Khan, E., 1990. *(Abst.) Nat. Symp. Problem and Prospects of Bot. Pesticides in IPM*, 21–22. Jan. CTCRI, Rajamundry, A.P., p. 44–45.

Prasad, D., 1998. *Ann. Pl. Prot. Sci.*, 6: 195–197.

Prasad, D., Dubey, K.N. and Mittal, A., 1997. *Ann. Pl. Protec. Sci.*, 5(2): 217–219.

Prasad, D., Chawla, M.L. and Khan, E., 1992. *Curr. Nematol.*, 3(2): 139–142.

Prasad, D. and Narayana, R., 2000. *Ann. Pl. Protec. Sci.*, 8(1): 58–61.

Prasad, D., Ram, D. and Ahmad, I., 2002. *Ann. Pl. Protec. Sci.*, 10(2): 360–364.

Prasad, D. and Ahmad, I., 2002. *Ann. Pl. Protec. Sci.*, 10(1): 172–173.

Prasad, D and Chawla, M.L., 1992. *Curr. Nematol.*, 3(2): 127–132.

Prasad, D., 1999. *Nature and Biosphere*, 4(1&2): 16–19.

Prasad, D. and Rangappa, T., 1994. *Ann. Pl. Protec. Sci.*, 2(2): 37–42.

Prasad, D. and Mittal, A., 2004. *Ann. Pl. Protec. Sci.*, 12: 465–467.

Prasad, D. and Mittal A., 2004. *Ann. Pl. Protec. Sci.*, 12: 234–236.

Raut, S.P. and Sethi, C.L., 1980. *Indian J. Nematol.*, 10: 166–174.

Reddy, D.D.R., Subramniyam., S., Reddy, G.H.R.R.C. and Siva Rao, D.V., 1984. *Pl. Disease*, 68: 526–529.

Sharma, S., Siddiqui, A.U. and Parihar, A., 2001. *Indian J. Nematol.*, 31(1): 79–80.

Sharma, H.K. and Prasad, D., 1995. *Ann. Agric. Res.*, 16(4): 473–476.

Singh, R., Chhabra, H. and Kaul, V., 2001. *Indian J. Nematol.*, 31: 34–37

Verma, K., Sitaramaiah, K., Naidu, P. and Reddy, K., 1999. *Indian J. Nematol.*, 29: 59–68.

Plant Protection, Diversity and Conservation (2011) *Pages* **645–652**
Editors: **D. Prasad & Satya Kumar**
Published by: **BIOTECH BOOKS, NEW DELHI**

Chapter 49

Herbicidal Potential of Allelopathic Plants against *Parthenium hysterophorus* L.

***Moolchand Singh*[1], *T.V. Prasad*[1], *Arun Kumar*[2] *and P.S. Chandurkar*[3]**
[1]*National Bureau of Plant Genetic Resources, Pusa, New Delhi – 110 012*
[2]*Directorate of Maize Research, Pusa campus, New Delhi – 110 012*
[3]*Directorate of Plant Protection, Quarantine and Storage, N.H. IV, Faridabad – 121 001, Haryana*

In allelopathy, one plant exerts detrimental effect on another plant through the production of germination and growth inhibiting substances in crops, grasses, weeds and trees. It is an important ecological process in vegetational composition and agricultural sciences and governs the plant productivity (Rice, 1984). It plays important role in plant diversity, adaptation of species, organization of communities and in species evolution (Jeffery *et al.*, 1989). It is a tool of self-defense in plants, helping to defend against invading insects and other neighbouring plants (Lovett, 1991). Recent advances in plant biology have explained the plant invasion and plant-plant communication in rhizosphere. Allelochemicals are secondary metabolites produced via the acetate and shikimic acid pathways and produces phenolic acids, coumarins, flavonoides, terpenoides, alkaloids and sulphides (Narwal, 1994). They affect all basic plant processes *viz.* photosynthesis, chlorophyll production, respiration, hormonal balance, protein synthesis and plant water relations. They enter the soil through rain leachates from foliage, root exudation and decomposition of plant residues or as microbial by-products of residue decomposition and selectively inhibit the growth of other plants and soil microorganisms.

Parthenium (*Parthenium hysterophorus* L.), an annual invasive weed of family Asteraceae, a native to tropical America and has invaded North America, South America, the Caribbean, and many parts of Africa, Asia and Australia (Navie *et al.*, 1996). It is a very prolific seed producer and produces up to 25,000 seeds per plant, leading to large seed bank of about 200,000 seeds m^{-1} in fields (Joshi, 1991). Furthermore, seeds germinate throughout the year given suitable moisture levels and remain viable for long time and thrive under very harsh environmental conditions (Williams and Groves, 1980). The weed also has a very high regenerative potential. The weed is highly adaptive to adverse environmental conditions, and is a successful invader of any open land. The plants commence flowering, when 4 to 8 weeks old and may flower for several months. Under drought stress the weed can germinate, grow, mature and set seeds in 4 weeks. It was accidently introduced to India in the mid-1960s through imported food grains and has since spread throughout the India (Chandra and Vartak, 1970). Now it has become naturalized and grows luxuriantly in open spaces, grasslands, wastelands, around the fields and sometimes also in cultivated fields. Its invasive nature is evident from its ability to form huge monocultural stands with no other plant in the vicinity. It causes numerous environmental and agricultural problems like loss of crop productivity, fodder scarcity, biodiversity depletion and health problems for human beings and livestock (Evans, 1997). In native Australian grasslands, open woodlands, riverbanks and floodplains, parthenium invasion has totally changed the habitat (McFadyen, 1995). Similar invasion of national wildlife parks has occurred in south India(Evans, 1997). In artificial feeding tests, buffalo bull calves accepted the weed, alone or in mixture with green fodder, with severe consequences. The majority developed severe dermatitis and toxic symptoms and died within 8-30 days. Lesions were found subsequently in the gastrointestinal tract, liver and kidneys (Narsimhan *et al.*, 1977). Changes in blood chemistry and inhibition in liver dehydrogenases, as well as degenerative changes in both the liver and kidneys have been reported in buffalo and sheep (Ahmad *et al.*, 1988). The milk of livestock may also be tainted by parthenin. Its various parts contain water soluble phenolics including caffeic, ferulic, vanicillic, anisic and fumaric acids, and sesquiterpene lactones including parthenin and coronopilin and acetylated pseudoguaianolides (Das *et al.*, 2007).

Chemical herbicides like 2,4-D sodium salt, MCPA, MSMA, atrazine and glyphosate are effective against parthenium (Javaid, 2007).However, in recent years the use of chemicals has increased consumer's concern and their use is becoming more restrictive due to various problems *viz.* carcinogenic effects, residual toxicity problems, environmental pollution, and occurrence of microbial resistance. Increasing public concern has led the scientist to develop alternative pest management system which are less pesticide dependant or based on natural compounds. Use of biological control agents could be environmentally safe strategy to control this weed but only few potential biological control agents have been reported. Thus weed biologists have screened various allelopathic plant species for their herbicidal effects against this weed, so that effective ingredients from them can be isolated to develop natural compound based herbicidal formulations to control it. This review presents the herbicidal avtivities of various plants against the parthenium weed.

Plant Species

Numerous plants are allelopathic and these should be used in agriculture for weed management. Herbicides based on natural products are attractive because they exhibit bioactivity at low concentrations. Besides, their mode of actions are different than commercial herbicides, making them ideal lead compounds for new herbicide discovery. Crude extracts, plant residues and purified compounds of allelopathic plants have shown their herbicidal activity against germination and growth of parthenium (Table 49.1).

Table 49.1: Phytotoxic Effects of Allelopathic Plants on Parthenium Weed

Name of Crop/Weed	*Plant Part*	*Allelochemicals Present*	*Suppression/Inhibitory Effect*	*References*
Crops				
Oryza sativa	Root and shoot	Momilactones A and B, phenolic acids, 3-isopropyl-5-acetoxycyclohexane-2-1 and 5,7,4-trihydroxy-3, 5-dimethoxyflavone	Reduced germination and shoot/root growth	Javaid *et al.* (2006), Kato-Noguchi *et al.* (2008)
Sorghum bicolor	Root and shoot	Benzoic acid, p-hydroxybenzoic acid, vanillic acid, gallic acid, caffeic acid, furulic acid and chlorogenic acid	Reduced germination and shoot/root growth	Javaid *et al.* (2006)
Helianthus annuus	Root and leaves	Phenols and terpenoids	Reduced germination and root growth	Javaid *et al.* (2006), Macias *et al.* (2002)
Grasses				
Imperata cylindrica	All parts	Caffeic, ferulic, p-hydroxybenzoic, p-coumaric, vanillic, chlorogenic and syringic acids	Reduced germination and shoot/root growth	Anjum *et al.* (2005), Hussain and Abidi (1991)
Broad-leaf weeds				
Breynia retusa			Chlorosis, necrosis and defoliation	Rani *et al.* (2006)
Cassia tora	Leaves		Reduced vegetative and reproductive growth	Thapar and Singh (2006)
Cannabis sativa	Leaves		Reduced germination, biomass, protein and pigment content	Singh and Thapar (2003)
Withania somnifera	Leaves and roots	Withaferin A	Reduced germination and plant growth	Javaid *et al.* (2009)
Trees				
Alstonia scholaris	Leaves	Indole alkaloids	Reduced germination and plant growth	Javaid *et al.* (2006), Kam *et al.* (1997)
Eucalyptus citriodora	Leaves	Phenolic acids, tannins, flavonids and eucalyptus oils	Reduced germination	Javaid and Shah (2007), Shiva and Bandopadhyay (1985)
Eucalyptus globus	Leaves	Monoterpenes (cineole, citronellol and linalool	Reduced germination and chlorophyll content	Kohli *et al.* (1998), Singh *et al.* (2002)
Azadirachta indica	Leaves	Gallic, benzoic, p-coumaric, p-hydroxybenzoic,vanillic and transcinamic acid	Reduced germination and dry biomass	Safique *et al.*(2005), Xuan *et al.* (2004)

Crops

Rice

Rice (*Oryza sativa* L.) is a well known allelopathic crop and its aqueous root and shoot extracts reduces the germination and root growth of parthenium weed (Javaid *et al.*, 2006). Different rice varieties exhibit variable phytotoxic effects against germination and growth of parthenium. Similar to that of rice extracts, incorporation of rice residue also reduced the germination and growth of parthenium(Javaid *et al.*, 2008). The living and dead rice plants release different allelochemicals. Allelochemicals from living rice plants are: momilactones A and B, 3-isopropyl-5-acetoxycyclohexane-2-1 and 5,7,4-trihydroxy-3, 5-dimethoxyflavone. Allelochemicals from rice residues are momilactone *B* and phenolic acid (Kato-Noguchi *et al.*, 2008).

Sorghum

Aqueous shoot and root extracts of sorghum (*Sorghum bicolor* L.) are highly toxic to germination and seedling growth of parthenium. Shoot extract was more toxic than root extract. Sorghum contains nine allelochemicals *viz.*, benzoic acid, p-hydroxybenzoic acid, vanillic acid, gallic acid, caffeic acid, furulic acid and chlorogenic acid, which are toxic to weeds. Sorgholeone exuded from the sorghum roots suppresses the growth of weeds (Netzly and Butler, 1986).

Sunflower

The shoot and root extracts (5-25 per cent) of sunflower (*Helianthus annuus* L.) significantly reduced the germination and root growth of parthenium(Javaid *et al.*, 2006). Several phenols and terpenoides have been reported as potential allelochemicals in various cultivars of sunflower. The heliannuols are a promising group of phenolis allelochemicals isolated from sunflower which may be responsible for reduction of germination and growth of parthenium (Macias *et al.*, 2002).

Grasses

Certain grass species *Dicanthium aristatum*, *Digitaria milanjiana*, *Cenchrus ciliaris* and *Bothriochloa insculpta* displaced the parthenium weed (O'Donnel and Adkins, 2005). The frequency and density of parthenium was significantly lower in *Imperata cylindrica* and *Desmostachya bipinnata* dominating localities than nearby areas without infestation of these allelopathic grasses (Anjum *et al.*, 2005). Further studies revealed that aqueous shoot and root extracts of *I. cylindrica* and *D. bipinnata* and other allelopathic grass *viz.*, *Dicanthium annulatum*, *Cenchrus pennisetiformis* and *Sorghum halepense* reduce the germination and seedling growth of parthenium (Javaid and Anjum, 2006). The inhibitory effects were concentration dependant and shoot extracts were generally more inhibitory than root extracts. Thus, water- soluble chemicals of allelopathic grasses are herbicidal to the parthenium weed. The allelopathic grasses contain various phenolics which may suppress the germination and growth of parthenium.

Broad-leaf Weeds

Cassia tora is a legume and grows wild in tropics and is considered a weed in many places. Its powdered leaves adversely affected the growth of parthenium upto 60 days after sowing. It reduced the plant height, leaf size and number of branches, capitula and seeds per plant. The total sugars and lipids were decreased both in leaf and stem of parthenium plants supplied with *Cassia tora* residues (Thapar and Singh, 2006).

The leaf residues of *Croton bonplandianum* decreased the leaves size and seedling height of parthenium. The number of leaves, branches, capitula and seeds per plant were decreased. The inhibition of growth was due to decrease in chlorophyll, sugars, proteins and lipids content, while organic and amino acids were increased (Thapar and Singh, 2006).

Singh and Thapar (2003) evaluated the phytotoxic effects of leaf leachates of female plants of *Cannabis sativa* on morphological and biochemical paremeters of parthenium. Dry leaf leachates of *C. Sativa* caused maximum reduction in biological activities in parthenium. The leachates reduced the seed germination, fresh weight and pigment content of the test plant. The protein content and nitrate reductase activity followed a decreasing trend with an increase in concentration of leaf leachates.

Foliar application of water extract of *Breynia retusa* caused chlorosis and necrosis in the parthenium plants. The crude leaf extracts acted as contact herbicide. The extract initially killed only the parts of the plant that were sprayed/exposed to extract but later the entire plant was defoliated (Rani *et al.*, 2006).

Aqueous and methanol extracts of roots and leaves of *Withania somnifera*-a medicinal plant markedly suppressed the germination, root and shoot growth of parthenium in laboratory bioassay. Foliar spray of these extracts (10 per cent w/v) significantly reduced shoot length and biomass. Incorporation of its dry leaves in soil at 1, 2 and 5 per cent can reduce parthenium seed germination by 43-89 per cent. The major chemical constituents reported from *W. Somnifera* are withanolides and the most important one is withaferin A, a highly oxiginated steroidal lactone. *W. Somnifera* also possesses other secondary metabolites *viz.*, flavonol glycosides, glycowithanolides, sterols and phenolics. One or more of these compounds may be responsible for herbicidal activity of *W. Somnifera* against parthenium weed (Kannan and Kulandaivelu, 2007).

Trees

Aqueous leaf extracts of *Alstonia scholaris* inhibited the germination and growth of parthenium both in laboratory bioassays and in pot experiment (Javaid *et al.*, 2006). Its leaves contains many indole alkaloids which may be responsible for the herbicidal activity of its aqueous extracts. Its leaves collected from India, Indonesia, Malaysia, Pakistan, Philippines and Thailand showed diverse alkaloidal patterns. The trees from India, Pakistan and Thailand contains picrinine type alkaloids, while those from Indonesia and Philippines contain alkaloids based on the angustilobine skeleton (Kam *et al.*, 1997).

Eucalyptus citriodora is a well known allelopathic tree. Aqueous leaf extracts of *E. citriodora* and *E. camaldulensis* reduced the germination, root and shoot growth of parthenium due to the allelochemicals present in their leaves (Shiva and Bandopadhyay, 1985). Eucalypt oils from *E. globules* and *E. citriodora* exerted deleterious effects on parthenium weed (Kohli *et al.*, 1998). The weed germination was inhibited and the chlorophyll content and cellular respiration of the mature plants exposed to eucalypt oils were reduced drastically, followed by complete wilting of plants after 15 days. The E. *citriodora* oil was more effective than *E. globules* oil (Kohli *et al.*, 1998). The monoterpenes, the constituents of eucalypt oil, decreased the photosynthesis by reducing chlorophyll content (Singh *et al.*, 2002). Interference of monoterpenes with respiratory ability decreases the the germination and growth of plants (Abrahim *et al.*, 2000). The insolubility of monoterpenes, in water limits their toxicity to other plants. However, monoterpenes are generally active at concentrations far below their maximum solubility and can be brought into solution form with some surfactants to manage parthenium and other weeds.

Safique *et al.*(2005) assessed the herbicidal activity of aqueous extracts (2, 4, 6, 8, 10 per cent) of dry leaves of *Ficus bengalensis, Azadirachta indica, Melia azadarach, Mangifera indica* and *Syzygium cumini* against parthenium seeds germination in lab bioassays. The 8 and 10 per cent extracts of these tree species significantly suppressed the germination of parthenium seeds. Extracts of *F. bengalensis* and *M. indica* were found most inhibitory and those of *M. azadarach* were least inhibitory to radical and plumule growth of parthenium.

Conclusion

Parthenium (*Parthenium hysterophorus* L.) is world's worst weeds. It is native of tropical America and has rapidly spread in Asia, Africa, Australia and Europe. It can be controlled by herbicides, but their use adversely affects the environment and human health. Hence, alternate eco-friendly, cost effective and bioefficaceous method of weed control are needed. Efforts are being made to use natural plant products for weed management. It has been shown that extracts, residues and essential oils of many allelopathic plant species effectively reduce the germination and growth of parthenium. There is a need to isolate and identify the active herbicidal ingredients from plants. These chemical constituents may provide the structural lead to prepare natural product based environment friendly herbicides to manage this weed.

References

Abrahim, D., Braguini, W.L., Kelmer Bracht, A.M. and Ishi-Iwamoto, E.L., 2000. Effect of four Monoterpenes on germination, primary root growth and mitochondrial respiration of maize. *Journal of Chemical Ecology*, 26: 611–623.

Anjum, T., Bajwa, R. and Javaid, A., 2005. Biological control of parthenium. I. Effect of *Imperata cylindrica* on distribution, germination and seedling growth of *Parthenium hysterophorus* L. *International Journal of Agriculture and Biology*, 7: 448–450.

Chandra, G.S. and Vartak, V.D., 1970. Problems caused by *Parthenium hysterophorus* L. in Maharashtra Region, India. *PANS*, 16: 212–214.

Das, B., Reddy, B.S., Krishnaiah, M., Sharma, A.V.S., Kumar, K.R., Rao, J.V. and Sridhar, V., 2007. Acetylated pseudoguaianolides from *Parthenium hysterophorus* L. and their cytotoxic activity. *Phytochemistry*, 68: 2029–2034.

Evans, H.C., 1997. *Parthenium hysterophorus*: A review of its weed status and the possibilities for biological control. *Biocontrol News and Information*, 18: 89–98.

Hussain, F. and Abidi, N., 1991. Allelopathy exhibited by *Imperata cylindrica*. *Pakistan Journal of Botany* 23: 15–25.

Javaid, A., 2007. Efficacy of some chemical herbicides against *Parthenium hysterophorus*. *Pakistan Journal of Weed Science Research*, 13: 93–98.

Javaid, A. and Anjum, T., 2006. Control of *Parthenium hysterophorus* by aqueous extracts of allelopathic grasses. *Pakistan Journal of Botany*, 38: 139–145.

Javaid, A., Shafique, S., and Bajwa, R., 2006. Effect of aqueous extracts of allelopathic crops on germination and growth of *Parthenium hysterophorus* L. *South African Journal of Botany*, 72: 609–612.

Javaid, A., Bajwa, R., Shafique, S. and Shafique, S., 2006. Herbicidal potential of aqueous extracts of *Alstonia scholaris*. In: *Abstracts 1st International Chemistry Conference, Recent Challenges in Chemistry*, November 1–3, Govt. College, University of Faisalabad, Pakistan, p. 22.

Javaid, A. and Shah, M.B.M., 2007. Phytotoxic effects of aqueous leaf extracts of two *Eucalyptus* spp. against *Parthenium hysterophorus* L. *Science International*, 19: 303–306.

Javaid, A., Shafique, S., Shafique, S. and Riaz, T., 2008. Effect of rice extracts and residue incorporation on *Parthenium hysterophorus* management. *Allelopathy Journal*, 22: 353–362.

Javaid, A., Shafique, S. and Shafique, S., 2009. Management of alien invasive parthenium weed by *Withania somnifera*. In: *Abstracts 9th Weed Science Conference*, June 28–30, NWFP Agricultural University Peshawar, Pakistan, p. 7.

Jeffery, D., Weidenhamer, D., Harnett, C. and Romeo, J.T., 1989. Density dependant phytotoxicity: Distinguishing resource competition and allelopathic interference in plants. *Journal of Applied Ecology*, 26: 613–624.

Kam, T.S., Nyeoh, K.T., Sim, K.M. and Yoganathan, K., 1997. Alkaloids from *Alstonia scholaris. Phytochemistry*, 45: 1303–1305.

Kannan, N.D. and Kulandaivelu, G., 2007. Novel method to isolate withaferin *A* from *Withania somnifera roots and its bioactivity. Allelopathy Journal*, 20: 213–220.

Kato-Noguchi, H., Ota, K. and Ino, T., 2008. Release of momilactone A and B from rice plants into rhizosphere and its bioactivity. *Allelopathy Journal*, 22: 321–328.

Kohli, R.K., Batish, D.R. and Singh, H.P., 1998. Eucalyptus oils for the control of *Parthenium hysterophorus* L. *Crop Protection*, 17: 119–122.

Lovett, J.V., 1991. Changing perceptions of allelopathy and biological control. *Biological Agriculture and Horticulture*, 8: 89–100.

Macias, F.A., Ascension, T., Galindo, J.C.G., Varela, R.M., Alvarez, A.J. and Molinillo, J.M.G., 2002. Bioactive terpinoides from sunflower leaves cv. Peredovick. *Phytochemistry*, 61: 687–692.

McFadyen, R.E., 1995. Parthenium weed and human health in Queensland. *Australian Family Physician* 24: 1455–1459.

Narsimhan, T.R., Ananth, M., Mangala, A. and Rao, P.V.S., 1977. Toxicity of *Parthenium hysterophorus* L. *Current Science*, 46: 15–16.

Narwal, S.S., 1994. *Allelopathy in Crop Production*. Scientific Publishers, Jodhpur, India, pp. 288.

O'Donnell, C. and Adkins, S.W., 2005. Management of parthenium weed through competitive displacement with beneficial plants. *Weed Biology and Management*, 5: 77–79.

Rani, P.U., Sudheer, S.D. and Devanand, P., 2006. Herbicidal potential of *Breynia retusa* leaf extracts on *Calotropis gigantea, Parthenium hysterophorus, Datura metal* and *Tridax procumbence. Allelopathy Journal*, 17: 65–79.

Rice, E.L., 1984. *Allelopathy, 2nd Edn*. Academic Press, Orlando, FL, p. 422.

Shafique, S., Javaid, A., Bajwa, R. and Shafique, S., 2005. Biological control of parthenium IV: Suppress ability of aqueous leaf extracts of some allelopathic trees against germination and early seedling growth of *Parthenium hysterophorus* L. *Pakistan Journal of Weed Science Research*, 11: 75–79.

Shiva, V. and Bandopadhyay, J., 1985. *Eucalyptus* in rainfed farm forestry. Prescription for desertification. *Economic Policy Weekly*, 40: 1667–1688.

Singh, H.P., Batish, D.R., Kaur, S., Ramejani, H. and Kohli, R.K., 2002. Comparative phytotoxity of four monoterpenes against *Cassia occidentalis*. *Annals of Applied Biology*, 141: 111–116.

Singh, N.B. and Thapar, R., 2003. Allelopathic influence of *Cannabis sativa* on growth and metabolism of *Parthenium hysterophorus*. *Allelopathy Journal*, 12: 61–70.

Thapar, R. and Singh, N.B., 2006. Phytotoxic effects of *Cassia tora* on growth and metabolism of *Parthenium hysterophorus*. *Allelopathy Journal*, 18: 255–266.

Williams, J.D. and Groves, R.H., 1980. The influence of temperature and photoperiod on growth and development of *Parthenium hysterophorus*. *Weed Research*, 20: 47–52.

Xuan, T.D., Eiji, T., Mitsuhiro, M., Khanh, T.D. and Chung, I.M., 2004. Evaluation on phytotoxicity of neem (*Azadirachta indica*) to crops and weeds. *Crop Protection*, 23: 335–345.

Plant Protection, Diversity and Conservation (2011) *Pages* **653–672**
Editors: **D. Prasad & Satya Kumar**
Published by: **BIOTECH BOOKS, NEW DELHI**

Chapter 50
Herbicide-Resistance and Weed-Resistance Management

Ozair Chaudhry

Biology/Environmental Science,
Albert Campbell Collegiate Institute (Con. Ed.), Toronto, Ontario, Canada

In the past two decades, evolution of newer herbicides provided wider user choice. Selection of most promising prudent products intensified in use followed by genetically induced herbicide resistant crops. This is how the broad-spectrum herbicides created a great deal of impact on the stakeholders. It has been an established fact that weeds reduce farm yields and farm income drastically. Among all other weed control practices, herbicides alone is easy prompt, most effective and economically acceptable mean; therefore, herbicides are overwhelmingly used by the farmers. Agriculture world over should have experienced a drastic shortfall in overall crop yields should herbicide availability be limited. Now, when the herbicide use gained a momentum and has been popularized as a formal input tool particularly in mechanized large holdings and commercial farming, the loss of herbicide effectiveness due to selection of herbicide-resistant weed populations has a negative impact on farmers. Herbicide resistance is the inability of a herbicide to effectively control a weed species that was previously controlled by the same herbicide. Herbicide resistance is detected when a biotype within a weed species possessing a resistant trait increases in abundance while susceptible biotypes are controlled by use of the same herbicide. The resistant trait is inheritable and therefore, is passed from one generation to the next. Once a herbicide-resistant population has been selected for, the likelihood of the weed population reverting back to a population dominated by the susceptible biotype is low. Resistant weed population becomes a serious constraint because it develops far faster (in 3 to 5 years) than the time and money investment on research, testing and registration for another newer chemical that meet modern environmental and health regulatory standards. As a result, herbicide with a new mode of action will not likely serve as a solution for herbicide-resistant weed populations. Therefore, it is absolute important that the herbicide options presently available be maintained through sound product stewardship. Stewardship implies that whoever produces, sells, or uses herbicide, exercises all precautions for minimizing any undesirable effects of the herbicide, including selection of resistant biotypes.

Agrochemical Industry has founded an international body, Herbicide Resistance Action Committee (HRAC) primarily aiming to collect information and prepare a database on resistant biotypes. It is a sort of cooperative project in collaboration with Weed Science Society of America (WSSA) lead by Dr. Ian Heap (WSSA.com). There is however, scant recognition by the largely land holding growers, the herbicide producers, the industry counterparts and the policy making partner *i.e.* the state and federal governments regulatory agencies. This reflection can be assessed from the fact that there is no regulatory binding for posting remedy or precaution on the label that demonstrate any instruction on resistance management. Recognition of the resistance concern is absolutely non-existent in today's marketing plans of the herbicide-producing sector. In US Indiana, growers perspective study on herbicide resistance revealed that 65 per cent respondents expressed moderate or low level of concern about weeds developing resistance to glyphosate. Only the growers who own 800 ha or more holding were the only willing to adopt resistance management strategies (Johnson and Gibson, 2006). In this scenario, one can imagine what will be response of developing and under developed countries where agriculture represent a large economic sector and is heavily input driven. Tons of herbicides are used beyond North America and Europe where absolutely meager or no awareness or resistance management actions are in place as a combat strategy in the years ahead.

In this chapter, emphasis is drawn on the overall review of the weed types and crops where mechanism of resistance imposed by either or in combination of genetic modification, environmental interaction and cultivation practices have been identified exerting pressure on plants for herbicide resistance. The contents of this chapter are segmented into different profiles to the extent of covering basic interest of stakeholders; growers, policy makers, professional scientists, farm technologists, and graduate students. Herbicide resistance is a growing concern that warrants efforts in the developing and implementing new input based cultivation strategies.

For years farmers have been fighting against their farm pests: rodents, insects, micro-pathogen and weeds that tolled heavy crop yield losses dwelling a segment of world population from malnutrition to starvation. According to a recent Cornell University report, the overall effects of weeds and invasive plants on the US agriculture, water quality, wildlife and recreation have been estimated to cost U.S. $34.7 billion annually (WSSA.net). As research based knowledge accumulated, concepts of modern farm technologies shaped up and new choices replaced the older cost inefficient farm practices. Attempts were made to kill the pests by spraying pesticides and herbicides consequently, hundreds of pest species have become resistant to chemical pesticides. Although resistance of insects to chemicals was recognized as old as 78 years ago, the problem peaked up and was reported in several parts since 1940 after the use of synthetic organic pesticides was increased. Herbicide resistance was first reported against 2,4-D (Phenoxy group) in 1957 from Hawaii (Hilton, 1957). In 1968, first confirmed report of herbicide resistance in common groundsel *Senecio vulgaris* against triazine herbicide was documented (Ryan, 1970). Consequently, several other reports confirmed resistance developed against dozens of other herbicides in four decades (Table 50.1)

Glyphosate became a prominent herbicide in agriculture about 12 years ago when it was discovered that glyphosate resistance genes could be inserted into crops using biotechnology. Now, glyphosate resistant corn, cotton, soybean, canola and sugar beets are common. Glyphosate being broad spectrum can kill most or all unwanted weeds while crops remain unharmed. Glyphosate thus became the dominant weed control method on many farms in North America and abroad, and quickly replaced other weed control practices. Glyphosate's effectiveness as a broad spectrum herbicide left many growers relying on it frequently and even exclusively in their battle to control weeds. Unfortunately, once a naturally resistant weed appears in a field, it can escape and multiply into a serious problem in the next few years. In 2000, marestail (Horseweed) *Cornyza canadensis* surfaced resistance to glyphosate in soybean fields of Delware where glyphosate was consecutively used over 3 years. (Figure 50.1). Within 3 years the resistance spread in 100,000 of states 560,000 acres of cropland. Now nine other states including Indiana have either confirmed or suspect the presence of glyphosate resistance marestail. This rapid widespread has been due to the fast airborn property of its seeds *i.e.*, the seeds can travel 1/4th mile per mild wind speed of 10 MPH (Barnes, J *et al.*, 2003).

Over the past several years, we have seen the list of glyphosate resistant weeds grow to almost one dozen species (Table 50.1), which are scattered across at least 20 states (WSSA. net). A consultant's report from argentine endorsed that johnsongrass *Sorghum helepense* has shown resistance against glyphosate (Valverde and Gressel, 2006). Farmers are being challenged to control glyphosate resistant weeds like Palmer amaranth, *Amaranthus palmeri* and giant ragweed *Ambrosia trifida* in certain crops. We urgently need to slow the development of resistance before glyphosate's value to farmers is diminished. Monsanto commercialized Glyphosate as RR (Roundup-Ready) with their respective genetically modified (GM) transgenic crop seed. This RR- seed package in soybean, corn, canola and cotton where offered a weed free high farm yield simultaneously, did evolve herbicide resistant biotypes,

Figure 50.1: Glyphosate Resistant Marestail (Source: Barnes *et al.*, 2003)

a challenge that has now become a key issue for all stakeholders. GM transgenic herbicide resistant crops are becoming volunteer weeds are also associated with segregation and introgression of herbicide resistant traits in weed population that has ecological impact on plant communities (Owen and Zelaya, 2004).

Table 50.1: Worldwide Herbicide Resistance Against Widely Used Types of Herbicide Group

Herbicide	*Year of Resistance found*	*Year of Reporting*
2,4-D	1945	1963
Dalapon	1953	1962
Atrazine	1958	1988
Picloram	1963	1973
Trifluralin	1963	1982
Diclofop	1977	1982
Trialate	1962	1987
Chlorsufuron	1982	1987
Glyphosate	2003	2006

Source: LeBaron, 1991; Valverde and Gressel, 2006.

Most recent world wide updated *WSSA* inventory provides comprehensive information on herbicide resistant weeds by country and by mode of action is compiled by WSSA (Appendix-I).

Resistance is a phenomenon that develops rapidly. When a pest population starts showing resistance, it responds favorably to a change in tactics for only a small period of time after detection. Resistance progress within season(s) until it leads to its climax where change in control strategy becomes imperative. It has been generally observed that if a pest population is resistant to one or more pesticides, it will develop resistance to other compounds especially when the compounds have similarity in mode of action. Most pests can retain inherited resistance to pesticides for longer time therefore, it is logically attractive that different and newer compounds varying in their mode of action are used provided it comply with cost efficiency and regulatory approval. While developing new compounds, the application philosophy and principles of integrated pest management (IPM) can be a strength. As such delaying the duration of spread of resistance will help gain more time to monitor the resistance episode.

Pesticide resistance is solely a technical problem that can be readily overcome with the right type of new pesticide or by devising appropriate adjustments within the conventional use of pesticide. It is imperative because experience has revealed that resistance episode will enlarge unexpectedly.

As research organizations and industrial groups concentrate on monitoring resistance issue, it is necessary that deeper understanding of resistance is taken collectively between applied and academic biology. The key idea to this concerted effort is to step forward and identify practical solution to pesticide resistance problem. It is however, understood that resistance is potentially a powerful pervasive natural phenomenon that can be minimized, eliminated or managed by human actions. Inadvertent neglect or intended defer in dealing with resistance can lead to explosions in weed pest population resulting crop in failure consequently driving nations in to grave food crisis that have already been escalated by global trade issues since 2007 (Ozair, 2007 and 2008).

Resistance verses Tolerance

Herbicide resistance is an induced inherent ability of some plant species to survive and reproduce after receiving a lethal dose of herbicide (Prather *et al.*, 2000). In contrast herbicide tolerance can be defined as the inherent ability of plant to survive and reproduce with a herbicide treatment at a normal use rate (Vargas and Wright, 2004). In other words tolerance is the ability to compensate the damaging effects of the herbicide with no physiological mechanisms involved (Menalled and Dyer, 2006)

In plants, herbicide resistance is developed either by random mutation or it is self-induced by genetic engineering. In nature, it is infrequent and there is no evidence of herbicide induced mutation at any point. Where herbicide cause selection pressure *i.e.*, susceptible plants are killed whereas the resistant plants survive to reproduce without confronting any competition from the susceptible plants.

Types of Resistance

Herbicides target attack at one or more location in a weed plant. These locations can either be enzyme proteins, other non-enzyme proteins, cell division path etc. are called site of action.

One such example is of ALS [acetolactosynthase also called AHAS acetohydroxy acid synthase] leads making branched chain amino acids in plants. Herbicides like; [sulfonylurea, imidazolinone and pyrimidinyloxybenzoate] bind to this enzyme. As such, when this enzyme is complexed with herbicide, it is no more available for the normal synthesis of certain amino acids and consequently, protein deficiency and ultimate death of plant result. The chemical structure of above set of herbicides is different but their target site is same. The plants that resist to ALS/or AHAS herbicides have altered the ALS enzyme in such a way that it does not bind with the herbicide,

Now, the resistant weed biotype that has been evolved by selection pressure from one ALS attacking herbicide will be resistant to all herbicides that act on this particular site. Such resistance of plant where one class of herbicide within one group or to several herbicide classes within one group is called Cross-Resistance. A population of yellow starthistle, *Centaurea solstitialis* evolved resistance to synthetic auxin, picolinic acid that promoted Picloram herbicide in Washington State. When this population was tested with another picolinic acid herbicide, Clopyralid, the plant showed same resistance (Prather *et al.*, 2000).

Another example of cross-resistance reported from Australia where a biotype of wild oats *Avena fatua* became resistant to fenoxiprop (an ACCase *i.e.* Acetyl Co-enzyme-A Carboxylase) also became resistant to several other ACCase inhibiting herbicides. (Powles and Holtum, 2008).

On the contrary, a Multiple Resistance is said to have occurred when resistance to several groups of herbicides with different biochemical target such as triazines acting on PS system as photosynthetic-e- inhibitor and sufonylure inhibit ALS *i.e.* Acetolactosynthase enzyme (Menalled and Dyer, 2006). In southern Australia 3 herbicides, diclofopmethyl (postemergent), sulfonylurea such as Chlorsulfuron and Triasulfuron, had been used against annual ryegrass *Lolium rigidum* for almost 10 years 1978-89. The Diclofop-methyl resistant biotype revealed multiple cross-resistance to other groups as listed below (Powles and Holtum 1990):

Aryloxyphenoxypropionates

Diclofop-methyl, Fluazifop-butyl, Fenoxyprop-ethyl, Haloxyprop-methyl, Quizalofop-ethyl.

Cyclohexanediones

Alloxidim, Sethoxidm, Tralkoxydim

Sulfonylureas

Chlorsulfuron, Metasulfuron-methyl, Triasulfuron

Dinitroanilines

Trifluralin

Triazines

Atrazine, simazine

The mechanism of multiple resistance caused in rye grass has been investigated not only due to change at the site of action of herbicide alone but the detoxification of herbicides by strong oxidising enzyme, cytochrome-P450 in plants. This enzyme is also called mixed-function oxidase (MFO). Similar enzyme develops resistance to insecticides in insects (Prather *et al.*, 2000).

Oxidation rate of similar monooxygenase enzymes-P450 in mammal has been reported to show a broad specificity towards foreign lipophilic foreign chemicals including herbicides. A human P450 CYP1A1 metabolizes various herbicides with different structures and mode of action, was introduced in to rice plant. The transgenic rice showed broad cross-resistance towards various herbicides such as Atrazine, Chlortoluron, Norflurazon etc. (Kawahigashi *et al.*, 2006)

Gravity of Resistance

The herbicide resistance issue became a more serious issue when noxious weed plant species were identified emerging with resistance against selective herbicides. In recent years, the appearance of herbicide resistance in plants is increasing exponentially as compared with the case of other pests such as insects and fungi. Besides 61 species of triazine resistance, there are more than 200 species surfaced resistant to 16 other classes of herbicides. At least one weed species has emerged resistant to herbicide in each country of the world where herbicide is used as farm input. Country wise declared resistant weed species against site of action based classified herbicides are appended in Appendices-I. Resistance however, evolves rapidly where monoculture cropping is prevalent. Weeds belonged to one dozen genera have shown developed resistance at almost a dozen or more locations worldwide (Table 50.2)

Table 50.2: Most Common Genera of Weeds Developing Resistance to Herbicides Worldwide

Genus	*Common Name*	*Number of Documented Occurrence of Herbicide Resistance*
Amaranthus	Pigweed	42
Chenopodium	Lambsquarters	25
Conyza	Fleabane or horseweed	22
Lolium	Ryegrass	21
Setaria	Foxtail	17
Avena	Wild oat	15
Echinochloa	Barnyardgrass or watergrass	15
Alopecurus	Blackgrass	13
Senecio	Groundsel	12
Polygonum	Knotweed or smartweed	12
Solanum	Nightshade	11

Source: Vargas and Wright, 2004.

Most recently, in 2003-2006, Monsanto's very potent broad spectrum herbicide glyphosate has been reported resisted against some weeds such as grassy weed Johnson grass, *Sorghum helepense* in Argentina (Valverde and Gressel, 2006) and Rigid rye grass bio-type *Lolium rigidum* have exhibited resistance in almond orchard of northern California.(Vargas, 2001). Both hairy fleabane *Conyza bonariensis* and buckhorn plantain *Plantago lanceolata* have been reported glyphosate resistant in South Africa. Hairy fleabane has been difficult to control with glyphosate in California production system indicating possible resistance. Similarly, reports of poor or ineffective control of *Chenopodium Sp.* lambsquarters in Roundup Ready cotton system have surfaced in the last two years. Recently, Roundup resistance horseweed (marestail) *Conyza canadensis* has been confirmed in the eastern US. (Vargas and Wright, 2004). Until 2006, WSSA updated worldwide Roundup (glyphosate) resistant species enlarged to 11. (Figure 50.2).

Similarly, little seeded canary grass *Phalaris minor* and Jungle rice, *Echinochloa colona* have reportedly developed resistance against Isoproturon and propanil, respectively, within 3 years 1990-93. As a result, crop yield losses are reported in the tune of 40-60 per cent in the affected areas of India (Malik and Sing, 1995). *P. minor* is steadily developing resistance against alternate used herbicides such as Clodinofop and Sulfosulfuron (Mahajan and Brar, 2001).

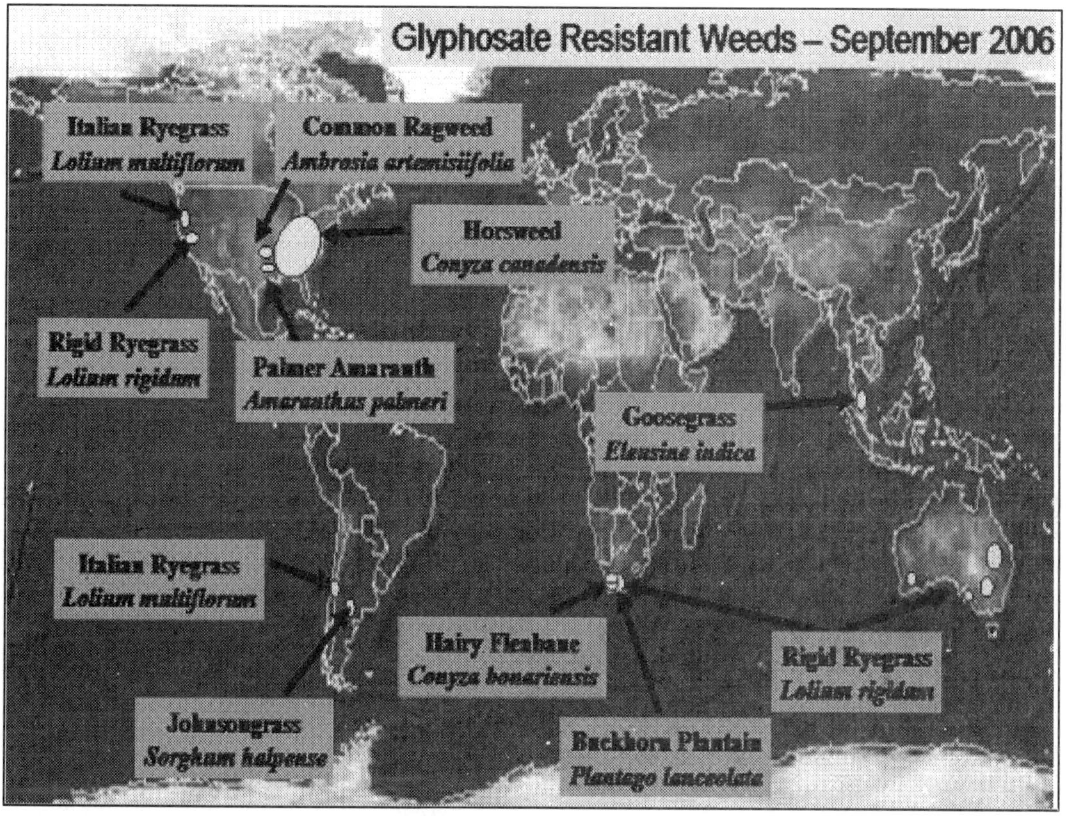

Figure 50.2: Worldwide Reported Glyphosate Resistant Dicot. and Monocot. Weeds (*Source*: WSSA.com)

A consortium of expert committee: Herbicide Resistance Action Committee (HRAC), the North American Herbicide Resistance Action Committee (NAHRAC) and the Weed Science Society of America (WSSA) founded by Agro-chemical Industry, has jointly focused to monitor the evolution of herbicide-resistant weeds and assess their impact throughout the world (Heap, 2007). Global collaboration between weed scientists made the survey that claims dramatic number of weed species has developed resistance against variety of herbicides since 1980 (Figure 50.3). As to date, 319 resistant biotypes that belonged to 185 species (111 dicots. + 74 monocots.) are spread over internationally over 290, 000 fields (Appendix-I).

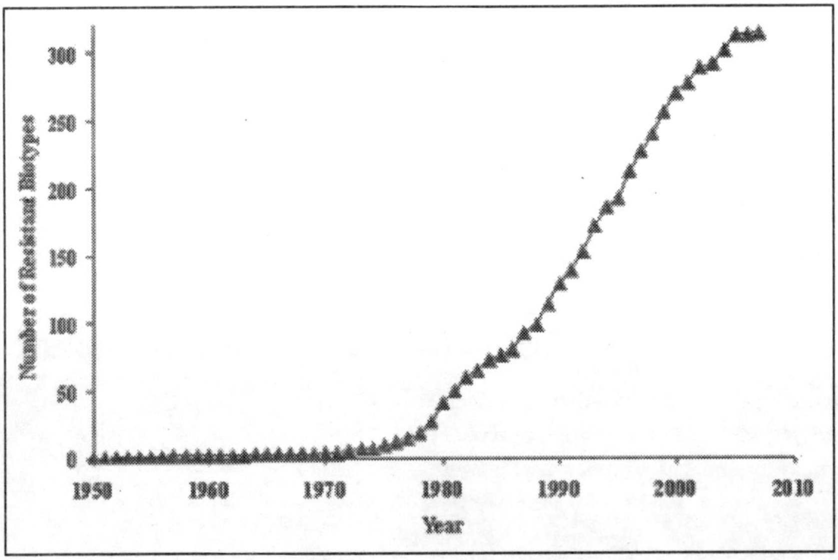

Figure 50.3: World Wide Chronological Increase in the Number of Herbicide Resistant Weeds (*Source*: Heap, 2007)

In two decades (1955-1975), the average annual rate in term of number of introducing herbicide in to the farm was six to seven and this number has now declined to as low as one to two herbicides. In the wake of introduction of high value transgenic crop seeds the package of crop selective herbicides gained commercial hegemony in the last 12 years. This high cost farm technology has further constrained the number of newer molecules. On the other hand, the requirement of re-registration of some herbicides in many crops reduced the number of choices for alternate use or rotation of chemicals. Repeated use of already dwindling types of molecules is itself becoming potential for promoting resistance in weeds. Also, it has been noticed that there are meager herbicides for minor crops that offer relatively less or no choice for different site of action herbicide use in rotation. Consequently, probability of herbicide resistance increases particularly, in countries where registration of herbicides for minor crops is further narrow. Therefore, it is not only the need to research newer herbicides, it is also vital to device a mechanism that ensure reduced repetition of the already available herbicides and switch over reliance on alternate weed control strategies so that selection pressure on resistance is minimized. Herbicides from 8 different group that are most selectively target-acting in plants and have evolved resistance in weeds, have been gradually increasing resistance trend, most pertinent from 1990 onward (Figure 50.4).

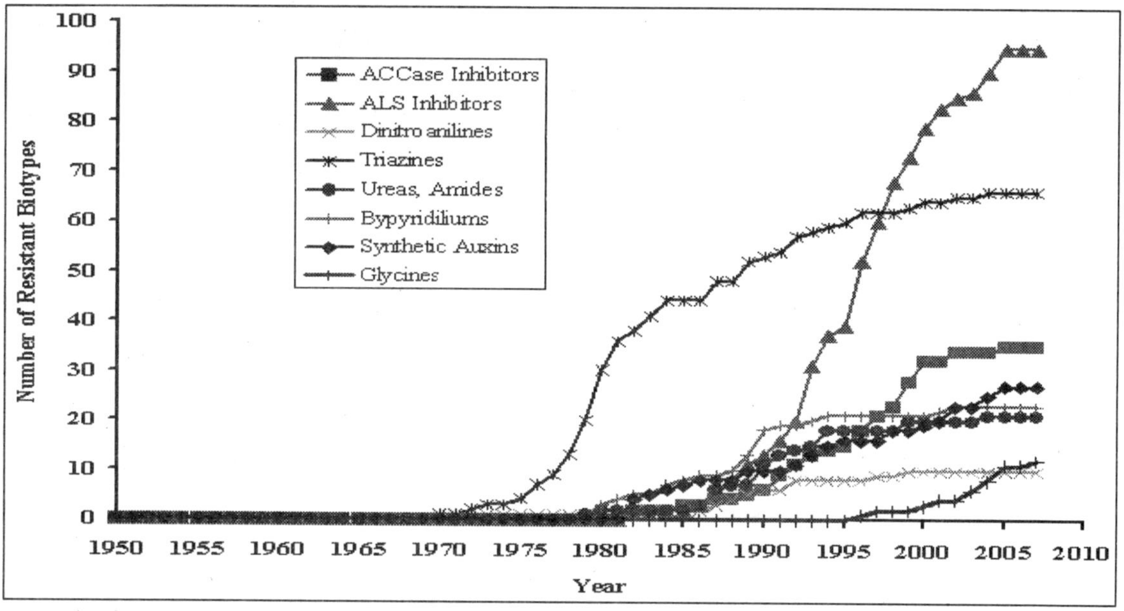

Figure 50.4: The Chronological Increase in the Number of Herbicide Resistant Weeds for Several Herbicide Classes (*Source*: Heap, 2007)

Sources of Inducing Resistant Weeds

As evident from the definition of resistance, it is not due to the mutation caused by the herbicide as chemical, rather resistance appears from the selection of natural mutation that exist as small fraction of population of resistant plants. Herbicide-resistant plant bio-types are believed to be emerging from only one or a few plants that are already present in a population. It may be a single plant in a population of several millions. Although they look morphologically identical, minor invisible genetic differences do exist among them that confer inherent resistance against herbicides. Such a minute number of resistant plants continue grow and expand by generation over time in seasons. Generally, it is hard to predict resistance merely from visual inspection until at least 25 per cent or small patches of such resistant plants are observed. When we apply a herbicide continuously for consecutive seasons, the susceptible plants of a weed type decrease drastically and those resistant bio-types increase gradually to the extent that we find that the herbicide appears to be ineffective at one point. At this stage we say that the weed has developed resistance against a herbicide or in other words called selection pressure of herbicides reached to maximum (Duke *et al.*, 1991). This whole process is presented in Figure 50.5.

Factors Influencing Herbicide Resistance

Biodiversity is a product of evolution and natural selection. Plants being directly exposed to external environment are vulnerable to variety of stresses, therefore many plants, particularly weeds, contain enormous genetic potential to survive such variations. Both weed as well as the herbicide characteristics influence in the development of herbicide resistance in plants. Most weed species contain adequate genetic variations that allow them to survive under variety of environmental stresses. As discussed earlier, the selection pressure imposes resistance in plants. This selection pressure can be generated either by repeated use of one herbicide, or use of long residual soil applied pre-emergence herbicide or due to repeated application of same post emergence herbicide. Factors that stimulate the development of herbicide resistance are many folds; however, the key factors include weed characteristics, chemical properties of the herbicide and cultural practices that are discussed below briefly (Vargas and Wright, 2004).

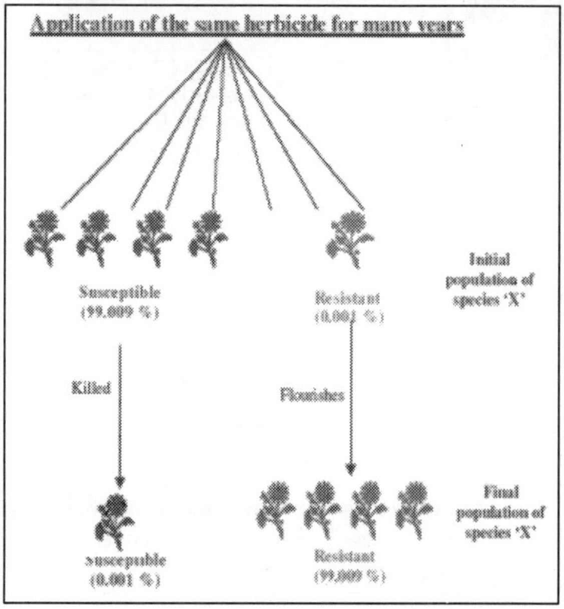

Figure 50.5: The Evolution of Herbicide Resistance (per cent values are arbitrary) (*Source*: Tharayil Santhakumar, 2003)

Weed Characteristics

The most likely weed characteristics that favor increase resistance against a particular herbicides can be as following:

- ☆ Initial frequency of resistant biotype. If in a particular population some species has relatively more individuals that have specific inherent resistance, then the chances for the resistance to surface are more.
- ☆ Annual growth habit. Annual plants complete their life cycle in a relatively shorter time and produce tremendous amounts of tiny seeds that have more rapid dissemination over perennials mode of growth. Perennials being proliferated vegetatively provide less chance of mutation and genetic variability.
- ☆ High seed production. Relatively rapid turnover of seed bank to high percentage of seed germination each year or in other words a little or low seed dormancy will lead to death of most susceptible plants by herbicide. Consequently the remainder resistant though fewer, will have brighter chance to grow with vigor and reproduce with increased resistant trait.
- ☆ Several reproductive generations in each growing season
- ☆ Extreme susceptibility to particular herbicide. It is also called hypersensitivity of weeds to a particular herbicide. Due to hypersensitivity, a single application of herbicide can eradicate most (90-95 per cent) population consequently, high selection pressure will allow the resistant bio-types to prevail and thrive best to stand fit in the field.
- ☆ Some weeds have high frequency of resistant gene that develop higher and rapid rate of resistance *e.g. Lolium rigidum*

Herbicide Characteristics

The following properties of herbicide molecule build the resistance in weeds to label them as different bio-types:

- ☆ A herbicide that has single site of action favors resistance far more faster than the herbicide with several mode of actions. The resistant biotype can easily endorse resistant against the herbicide that has only one site of action rather than the herbicide has multiple sites of actions.

- Herbicides that are subjected to enhanced metabolism in weeds have least chance to endure resistance in plants than weeds expressing resistance due to change at site of action. However, metabolism-based resistance to ACC-inhibiting herbicides is much less known although this type of resistance seems to be wide-spread (Delye, 2005).
- Broad spectrum control
- Long residual activity in soil will keep susceptible biotypes eliminated or suppressed for longer time giving competition free growth autonomy to the resistant to flourish and reproduce.

Cultural Characteristics

Farm cultural practices can also shift selection pressure on weeds. Because persistently using similar cultural practices such as below increase chances to develop herbicide resistant weed biotype are:

- A move from old multi-cropping pattern to monocrop culture where single crop is grown every season or where a fixed horticultural orchard is in place receiving similar input every season.
- Using reduced soil cultivation, or zero tillage system.
- Using single herbicide or combination (tank mix) that have same mode of action in every season persistently. This example refers to situations where transgenic herbicide resistant seeds are planted with a package of specific tolerant herbicide.
- Dose violation *i.e.*, use of low or very high doses in relation to the optimum rates prescribed for a specific crop in a situation.

Mechanism of Herbicide Resistance

Dekker and Duke (1995) broadly grouped herbicide in to the following two categories:

Exclusionary Resistance

Resistance is caused in plants due to inaccessibility of the molecule at its site of toxic action. In other words, it is the inability of herbicide molecule to concentrate in right lethal amount at point of action within weed plant. This provides weed a blessed escape from death and avail a sort of herbicide resistance. Such exclusion of herbicide from the site of action can be due to several reasons. It can be differential herbicide uptake due to the morphological barrier on leaves such as extraordinarily increased waxy coating on the cuticle, hairy epidermis and low foliage number and size etc.

It can also be due to differential translocation whereby apoplastic (Xylem tubes) or symplastic path (Phloem cells) restrict or delay movement of right concentration of herbicide at the site of action. (Ozair *et al.*, 1987)

It is likely that herbicide undergoes a sequence of locations before reaching at the site of action *e.g.* some lipophilic herbicides may become immobilized by partitioning into lipid rich glands or oil bodies (Stegink and Vaughn, 1988).

It is also possible that herbicide is rapidly detoxified prior to its reaching at the site of action. The detoxification reaction can either be oxidation, reduction, hydrolysis or conjugation. Resistance gained due to metabolic detoxification involves the role of enzymes. In case of velvetleaf weed *Abutilion theophrasti*, the enzyme Glutathione-s-transferase increases that detoxifies atrazine herbicide, hence the weed gains resistance. Similarly, in *Echinochloa colona* increased contents of enzyme aryl-acylamidase detoxify propanil herbicide. Herbicide metabolism can also be increased due to fast acting Cytochrome P450 monoxygenase with target enzymes such as; AACase (Acetyl-coenzymeA-Ccarboxylase), ALS (Acetolactosynthase) and PS II (Photsynthetic- e- transport pathway).

Site of Action Resistance

Site of action of herbicide in weed biotype is altered in such a way that it is no more vulnerable to be intoxicated by the herbicide. World over, many species were reported to have developed resistance against Sulfonylurea. In *Lactuca sative* biotype ALS enzyme, the site of action of sulfonylurea is modified in such a way that the herbicide can not bind with its site, hence the enzyme is not incapacitated and weed biotype skips the killing effect of herbicide (Eberlin *et.al;* 1999).

Target site-based resistance involves altered binding of herbicides to their target protein. How does the target protein change? A single nucleotide change or mutation in the gene encoding for the protein (mostly the enzyme to which herbicide binds) occurs. Change of one nucleotide changes the amino acid sequence of the protein, hereby destroying the ability of the herbicide to interact with the protein and at the same time do not incapacitate the normal functioning of the enzyme. As such, enzyme keeps normal function in the presence of the herbicide. Mutations leading to herbicide resistance may indirectly alter other unrelated physiological pathways affecting the otherwise growth and development of the resistant biotype. In kochia scopia, mutated resistance against sulfonylurea, reduces or diminishes the ALA (Acetolactate enzyme) sensitivity to normal feedback inhibition patterns. This generates excessive availability of amino acids for cell division and accelerated growth and development. Consequently, sulpfonyl urea resistant biotype of *K. scopia* dominates germination and establishment vigor as compared to its counterpart, susceptible biotype (Tharayil Santhakumar, N. 2003).

It also happens in some cases that the site of action is enlarged or overproduced as a result dilution effect of herbicide occurs. The applied normal rate of herbicide is unable to inactivate the entire amount of enzyme protein produced. Therefore, the extra amount of enzyme produced by the plant biotype can allow it carry on its normal metabolic activities surmounting the lethal effect of the herbicide.

Assessing Herbicide Resistance

In case a herbicide treatment fails to control weeds at a situation, weed resistance may not necessarily be the cause. Before signing a positive statement for weed resistance to herbicide, the following parameters should be tested (Beckie *et al.*, 2000; Gunsolus, 2002; Menalled and Dyer, 2007).

Visual Diagnosis

1. Read the herbicide label carefully. Are other weeds listed on the product label controlled satisfactorily? Chances are only one weed species will show herbicide resistance in any given field situation. Therefore, if several normally susceptible weed species are present, reconsider factors other than herbicide resistance as the cause of the lack of weed control.
2. Also observe, if the uncontrolled weeds exist in patches and each patch contains different species, this case is not of herbicide resistance because it is very unlikely that all species will develop resistance. There could be one of several other reasons.
3. Check if the herbicide is used repeatedly at the same field and is of the same one mode of action. If "yes" chances are that weed is likely evolving resistant biotype.

4. Further survey the area for any previous case of resistant weed reported. Interview growers for resistant-suspect to same herbicide. Did the same herbicide or herbicide with the same site of action fail in the same area of the field in the previous year?
5. Also, investigate if the level of weed control on suspected weed was declining in the past few years.

If the answer to some of the above questions is "yes", chances are that the weed species in question is leading to herbicide resistance.

Bio-Assay

If the above diagnostic survey support that a certain bio-type has likely evolved herbicide resistance, adequate sample of seed/plant material may be collected from the suspect population for the subsequent confirmation tests.

Plant Assay/Seed Collection

For reliable germination and healthy growth of plants during the experiments, healthy plants and stage of seed harvest is essential Particularly the grass weed seeds the best timing to ensure ripened seeds is when plant has shed at least 20 per cent seeds. Collect adequate amount (1000 seeds) from an average area of 50-100 meters by gentle rubbing the inflorescence over paper bag or tray. Label species name, date and location on the bag.

Greenhouse/Plant-Pot Assay

This is preliminary bioassay on mortality, plant vigor check, measurement of whole or fractions of plant biomass. This test will reveal discriminatory spraying effect of herbicide. This test, however, must include a treatment of susceptible set of pots for reference and not for herbicide sensitive or insensitive evaluation. Statistical advice must be sought for appropriate design and replications.

Dose Response Experiments

Use range of doses to formulate a standard response curve. This curve enables quantification of resistance by calculating the dose ratio required to produce the same effect in resistant and susceptible population. Usually the dose required to give a 50-70 reduction in the measured parameter (usually foliage weight or number of surviving plants) relative to the untreated control is determined (Figure 50.6). Ratios of these estimates (variously termed ED50, GR50, LD50 or 150), relative to that of a susceptible population, provide a resistance index (RI) which enables the degree of resistance to be described relatively simply. To obtain a good estimate of ED50 the dose range should be relatively wide and at least six doses are needed. It is usually best that each dose is twice the preceding dose in the range (*e.g.* 10, 20, 40, 80, 160, 320 g a. i./ha). The dose range used should include doses both below and above the field recommended rate as herbicides are normally more active under greenhouse conditions (Moss, 1999).

Single Dose Resistance Assay

After the dose response information is obtained from the above experiment, single or 2-3 discriminating doses can be used for the future screening assays conducted for more population.

Specific Discreet Tests

Several other detailed confirmatory analytical tests can be used such as petri-dish germination assay, chlorophyll fluorescence, leaf disc floating and enzyme specificity/sensitivity assay etc.

Herbicide Resistance Management (Prevention and Delaying Resistance)

Any weed management strategy applied to minimize selection pressure for resistance will block the emergence of resistance. All of the following or combination of some as applicable in a situation may prevent or at least delay the evolution of resistance in weeds.

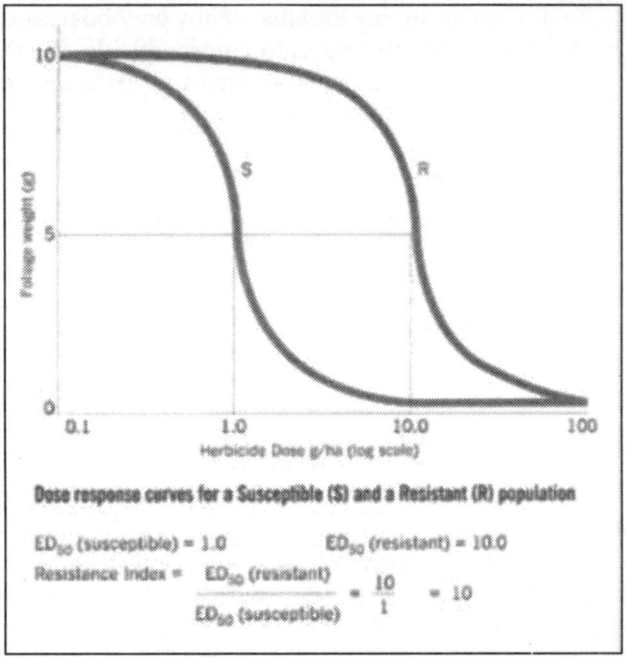

Figure 50.6: Dose Response Curve for Susceptible (S) and Resistant (R) Population (*Source*: Moss, 1999)

Herbicide Rotation

As previously discussed the use of same herbicide or a different herbicide but similar mode of action in consecutive years increases evolution of resistance. Reverse action can be achieved by using different herbicides with different modes of action in the subsequent season(s); this process is called herbicide rotation. Herbicide resistant transgenic crops *e.g.* RR- soybean ready roundup are gradually increasing acreage beyond North America. In this package, single use of herbicide *e.g.* glyphosate in soybean or corn is indispensable. Such a practice will enhance weed resistat culture. Perhaps tank mix application of herbicides could prevent or delay resistance pressure. In Australia when annual rye grass along side railways lines evolved resistance against 10 years persistent use of amitrol and atrazine, a tank mix of glyphosate and sulfometuron-methyl controlled the resistant biotype effectively (Powles and Holtman 1990). However, this combination should subsequently be further rotated in order to delay resistance evolution. Additionally, inclusion of non-chemical control option such as IPM may reduce the potential threat of resistance evolution. Use of the least persistent herbicide reduces the resistance risk. If two herbicides in a tank mix have the same weed control spectrum, it is better to use each separately in seasonal rotation. Do not spray herbicide in one season, as it will allow the susceptible seeds to grow vigorously from the geo-seed-bank reservoir. Consequently, the non resistant biotype will out-compete the fewer resistant biotype. Strong competition with the fewer resistant bio-type. The resistant biotype will experience survival rejection due to natural rejection (survival of the fittest).

Crop Rotation

Growing the same crop every season will invite same inputs including herbicide because of the same ecological culture. Crop rotation allows manipulation of planting time, spectrum of weed infestation, cultivation techniques, choice of herbicide with different mode of action, different stage and different way of application. Using combination of weed control strategies offers a chance to eradicate the so-called resistant biotypes and reduces the chances of their establishment.

Post Treatment Monitoring

While monitoring visual rating for weed control, carefully watch if there is any pattern of weedy patches uncontrolled. Differentiate between resistance and spray application neglect. The inconsistent patches that do not seem to be a spray neglect should be eliminated manually or by superimposed herbicide application that leads to effective killing.

Integrated Cultural Practices

Inclusion of all possible non-chemical weed control methods help more effectively against the weed resistance evolution. In contrast to no-tillage, cultivation practice stirs the soil, buries the early emerged weed seedlings (both susceptible as well as the resistant one) and solairizes the soil. Hand weeding eliminates the weed plants before the seed set, discarding a biotype, 90-100 per cent. Mulching for organic matter will simultaneously debris the weeds before seeding, offering reduced weed population with crop stand.

Herbicide Resistance Management (Post-Evolution)

In order to maintain check over the herbicide resistant biotypes, integrated weed management approaches as discussed above must be incorporated as appropriate. Crop rotation or preferably fallow tillage followed by close cultivation will keep the resistant population down. Extensive manual weed control by effective crew will offer 100 percent eradication of the suspected resistant biotype ensuring leas emergence in the subsequent season. Ensure clean and certified seed is planted each season and clean farm machinery is driven in the farm. It is suggested that power washer be used to clean the machinery from the infested seeds of the obnoxious bio-type.

By the use of these varied weed control practices, farmers have (unconsciously) acted to avoid or greatly delay the emergence of herbicide resistant weed bio-types. Farmers at risk of resistant biotype should be encouraged to maintain this diversity of operations in the control of weeds. The converse is represented by those relatively small numbers of farmers who rely heavily (or exclusively) on the use of selective and non-selective herbicides for weed control and who crop intensively or continuously on some fields. Our observations clearly identify that farmers who practice continuous cropping, or intensive cropping, run a much greater risk of developing resistance. Under such conditions, a consistent selection pressure is placed on the weed population and herbicide resistance is the inevitable result. Farmers who practice such cropping regimes need to be alerted to the probability of resistance appearing under these conditions and should consider modifications to their cropping practices.

Advanced Services for Resistance Management

In the popular scenario of mono-cropping pattern where intensified cropping includes single action broad spectrum ready herbicide with transgenic crop seed, gravity of herbicide resistance is likely to aggravate, and the following extension services both in private and public sectors may offer a prosperous check:

- Research facility must extend window option for DNA finger printing test for resistant and susceptible biotypes.
- Investigation for genetic resistance (one gene or additive gene effect) must be available.
- Development of technologies for remote aerial sensing of fields for clumps areas.
- Arrangement for establishing electronic data-base on regional basis, crop based or herbicide resistance species based for reference facility to the researchers.

Conclusion

It is important that resistance to herbicides is detected timely to implement resistance management strategies before situation grows in to gravity and requires higher cost and human endeavor for launching any extreme strategy. Currently, the progressive growers who grow crops extensively on commercial basis use herbicides as one of the input. There is however, scant recognition by the largely holding growers, the herbicide producers (industry counterparts) and policy making partner (the state or federal governments). This reflection can be assessed from the fact that there is no remedy or precaution on the label that demonstrate any instruction on resistance management. Recognition of the resistance concern is absolutely non-existent in today's marketing plans of the herbicide-producing sector. This is indeed a point to address this issue seriously if they have to keep herbicides as effective agricultural input. All stake holders, particularly the growers, the weed scientist/farm and extension managers in the private and public sectors, and industrial unit (s) must jointly address herbicide resistance. Consorted efforts are required to invest on resistance research and promote action to reduce prolonged or eliminate resistance evolution. So far, as the review of litreature reported in the text reveals, educated and professionally designed strategies can reverse or restrict the extent of resistance. It is, therefore, recommended that farming practices be rethought for actions such as crop rotation, tillage/cultivation practices, herbicide rotation for several factors prescribed in the text of this chapter, incorporation of integrated tools etc. While such reforms are accepted and practiced with the intention to delay or stop resistance management, professional vigilance for likely resistance followed by greenhouse/laboratory testing as discussed for the suspect areas should be monitored.

References

Barnes, J., Johnson, B. and Glenn, N., 2003. Identify glyphosate resistant marestail/horseweed in the fields. *Purdue University Extension Weed Science.*

Beckie, H.J., Heap, I.M., Smeda, R.J. and Hall, L.M., 2000. *Screening of Herbicide Resistance in Weeds*, 14 (2): 428–445.

Dekker, J and Duke, O.S., 1995. Herbicide resistant field crop. *Adv. Agron.*, 54: 69–116.

Delye, C., 2005. Weed resistance to acetyl co-enzyme: A Carboxylase inhibitor: an update. *Weed Science*, 53(5): 728–746.

Duke, S.O., Christy, A.L., Hess, F.D and Holt, Z.S., 1991. *Herbicide Resistant Crops*. Comments from CAST 1991. Council of Ag. Sci. and Tech. Ames, IA.

Eberlin, C.V., Guittieri, M.J., Berger, P.H., Fleming, J.K., Mallory-Smith, C.A., Thil, D.C., Baerg, R.J. and Belkmap, W.R., 1999. Physiological consequences of mutation for ALS-inhibitor resistance. *Weed Science*, 47: 383–392.

Gunsolus, J.L., 2002. *Herbicide Resistant Weeds*. University of Minnesota. North Central Regional Extension. Publication, p. 468.

Heap, I.M., 2007. *International Survey of Herbicide Resistant Weeds online*. Weed Science Society of America. http: //www.weedscience.org/In.asp. Dated 6/7/2008.

Hilton, H.W., 1957. Herbicide tolerance strain of weeds. Hawaiian Sugar Planters Association. *Annual Report*, pp. 69.

Johnson, W.G. and Gibson, K.D., 2006. Glyphosate-resistant weeds and resistant management strategies: An Indiana grower perspective. *Weed Technology*, 20(3): 768–772.

Kawahigshi, H., Hirose, S., Ohkawa, H. and Ohkawa, Y., 2006. Herbicide resistance of Transgenic rice pants expressing human CYP1A1. *Biotechnology Advances*, 25(1): 75–84.

Le Baron, H.M., 1991. Distribution and seriousness of herbicide resistant weed infestations world wide In: *Herbicide Resistance in Weeds and Crops*, (Eds.) Z.C. Casely, G.W. Cussans and R.K. Atkin. Butterwarth-Heinemann, Oxford, pp. 27–55.

Mahajan, G. and Bvar, L.S., 2001. Studies on herbicide resistance in *Phalaris minor* under Punjab conditions. *Indian J. Weed Science*, 33: 1–4.

Malik, R.K. and Brar, L.S., 2001. Studies on herbicide resistance in *Phalaris minor* little seed canry grass resistance to Isoproturon in India. *Weed Technology*, 9: 419–425.

Menalled, F.D. and Dyer, W.E., 2006. Herbicide-resistant weeds. In: *Weed Management Handbook 2006–07*, (Eds.) Uta, Montana, Wyoming. S. Dewey, S. Enole, F. Menalled, S. Miller, R. Whitsides and L. Johnson.

Menalled, F.D and Dyer, W.E., 2007. *Preventing and Managing Herbicide–Resistance in Montana*. MSU Exten. Mont. Guide. MT 200506AG.

Moss, S., 1999. *Detecting Herbicide Resistance*. HRAC Publication http: //www.plantprotection.org/HRAC/Dectecting.htm dated 7/13/2008.

Owen, M.D.K. and Zelaya, I., 2004. Herbicide-resistant crops and weed resistance to herbicides. *Pest Management Science*, 61(3): 301–311.

Ozair, C.A., 2007. Pakistan wheat import likely. Wheat crisis analysis. US farm.com. http: //www.farms.com/swine/freecommentary.asp?commentaryid=5127 dated 11/10/2007.

Ozair, C.A., 2008. The rise and fall of world exports: Global wheat outlook 2008. *Australian Kondinin Group Farming–Ahead*. February Issue, 193: 38–39.

Ozair, C.A., Moshier, L.J. and Werner, G.M., 1987. Absorption, translocation, and metabolism of foliage-applied chloramben in velvetleaf (*Abutilion theophrasti*) and soybean (*Glycine max*). *Weed Science*, 35: 757–762.

Powles, S.B. and Holtum, J.A.M., 1990. *Herbicide Resistant Weeds in Australia*. The Regional Institute Ltd. Farrer Centre. http: /www.regional.org.au/au/roc/1990/roc199001.htm. Dated 2/22/2008.

Prather, T.S., Ditomaso, J.M. and Holt, J.S., 2000. *Herbicide Resistance, Definition and Management Strategies*. Publication of Div. of Ag. and Natural Resources. UC Davis USA, Publication 8012.

Ryan, G.F., 1970. Resistance of common groundsel to simazine and atrazine. *Weed Science*, 18: 614–616.

Stegink, S.J. and Vaughn, K.C., 1988. Norflurazon SAN–9789 reduced abscisic acid levels in cotton seedlings: A glandless isoline is more sensitive than it's glanded counterpart. *Pesticides Biochemistry Physiol.*, 31: 269–275.

Tharayil-Santhakumar, N., 2003. Mechanism of herbicide resistance in weeds. On line book "*Mechanism of Herbicides Resistance in Weeds*". Weed Science Society of America. http: //www.weedscience.org/In.asp. Dated 1/7/2008.

Valverde, B. and Greesel, J., 2006. Dealing with the evolution and spread of *Sorghum helepense* Glyphosate resistance in Argentina. A consultancy Report to SENASA (Confidential).

Vargas, R.N., 2001. *Herbicide Resistance*. Cooperative Exten. Bulletin. UC Davis. USA.

Vargas, R. N. and Wright, S, D., 2004. Herbicide resistance management. In: *Cotton Field Check*. Coop. Extn. Bulletin UC Davis. USA.

WSSA.Com. Herbicide Resistance Committee (Joint consortium of HRAC/NAHRAC/WSSA). http: / /www.weedscience.org/summary/countrysummary.asp. Dated 7/11/2008.

Appendix-I: Herbicide Resistant Weeds by Country and Mode of Action

Sl.No.	Country	Total	A ACCase Inhibitors	B ALS Inhibitor	C1 Triazine	C2 Urea/ Amide	D Bypiri-dilium	K1 Dinitro-aniline	O Synthetic Auxin	Others
1.	Argentina	3	0	1	0	0	0	0	0	2
2.	Australia	51	9	20	5	0	5	2	1	9
3.	Austria	2	0	0	2	0	0	0	0	0
4.	Belgium	18	2	1	7	1	3	1	0	3
5.	Bolivia	3	2	1	0	0	0	0	0	0
6.	Brazil	19	3	8	0	0	0	0	2	6
7.	Bulgaria	4	0	0	2	1	0	1	0	0
8.	Canada	44	3	17	12	3	2	1	4	2
9.	Chile	6	4	1	0	0	0	0	0	1
10.	China	9	0	3	1	2	0	0	0	3
11.	Colombia	4	1	0	0	1	0	0	1	1
12.	Costa Rica	5	1	3	0	1	0	0	0	0
13.	Czech Republic	16	0	2	13	0	0	0	0	1
14.	Denmark	3	1	2	0	0	0	0	0	0
15.	Ecuador	1	0	0	0	1	0	0	0	0
16.	Egypt	1	0	0	0	0	1	0	0	0
17.	El Salvador	1	0	0	0	1	0	0	0	0
18.	Ethiopia	1	1	0	0	0	0	0	0	0
19.	Fiji	1	0	0	0	0	1	0	0	0
20.	France	32	5	2	22	1	0	0	1	1
21.	Germany	19	1	2	13	3	0	0	0	0
22.	Greece	5	1	1	2	1	0	0	0	0
23.	Guatemala	1	0	0	0	1	0	0	0	0
24.	Honduras	1	0	0	0	1	0	0	0	0
25.	Hungary	1	0	0	0	0	0	0	1	0
26.	India	3	1	1	0	1	0	0	0	0
27.	Indonesia	1	0	0	0	0	0	0	1	0
28.	Iran	5	5	0	0	0	0	0	0	0
29.	Ireland	1	0	1	0	0	0	0	0	0
30.	Israel	23	5	6	11	1	0	0	0	0
31.	Italy	16	3	6	4	2	0	0	0	1
32.	Japan	16	0	9	1	0	6	0	0	0
33.	Kenya	1	0	0	0	0	1	0	0	0
34.	Malaysia	16	1	4	0	0	6	0	4	1

Contd...

Appendix-I–Contd...

Sl.No.	Country	Total	A ACCase Inhibitors	B ALS Inhibitor	C1 Triazine	C2 Urea/ Amide	D Bypiri- dilium	K1 Dinitro- aniline	O Synthetic Auxin	Others
35.	Mexico	3	3	0	0	0	0	0	0	0
36.	New Zealand	9	0	1	3	0	0	0	4	1
37.	Nicaragua	1	1	0	0	0	0	0	0	0
38.	Norway	5	0	1	3	1	0	0	0	0
39.	Panama	1	0	0	0	1	0	0	0	0
40.	Paraguay	2	0	1	0	0	0	0	0	1
41.	Philippines	3	0	0	0	1	0	0	1	1
42.	Poland	9	0	1	8	0	0	0	0	0
43.	Portugal	2	0	1	1	0	0	0	0	0
44.	Saudi Arabia	1	1	0	0	0	0	0	0	0
45.	Slovenia	1	0	0	1	0	0	0	0	0
46.	S. Africa	14	3	5	1	0	2	0	0	3
47.	S. Korea	6	0	6	0	0	0	0	0	0
48.	Spain	30	1	3	18	3	0	0	1	4
49.	Sri Lanka	2	0	0	0	1	1	0	0	0
50.	Sweden	2	0	1	0	0	0	0	1	0
51.	Switzerland	14	0	0	11	3	0	0	0	0
52.	Taiwan	1	0	0	0	0	1	0	0	0
53.	Thailand	5	2	0	0	1	0	0	1	1
54.	Netherlands	7	1	0	5	1	0	0	0	0
55.	Tunisia	1	1	0	0	0	0	0	0	0
56.	Turkey	3	2	1	0	0	0	0	0	0
57.	United Kingdom	24	4	4	8	2	2	1	2	1
58.	USA	122	15	38	23	7	4	6	8	21
59.	Venezuela	1	0	0	0	1	0	0	0	0
60.	Yugoslavia	6	0	2	4	0	0	0	0	0

Source: Weedscience.org] http://www.weedscience.org/summary/countrySummary.asp.

Plant Protection, Diversity and Conservation (2011) *Pages* **673–675**
Editors: **D. Prasad & Satya Kumar**
Published by: **BIOTECH BOOKS, NEW DELHI**

Chapter 51
Electronic Waste

Dipak Bagai

Department of Electronics,
Punjab Engineering College (Deemed University), Chandigarh

Our planet is at a critical crossroads. As the high tech revolution evolves at an ever accelerating pace, so do the health and environmental concerns. The "Information Age" has created a society that is global, informed, wired and connected. In the past few years, technological advances in electronics have boosted the economy and improved the general lifestyle of a common man. The ever growing dependence on electronic products has paved the way for an emerging environment concern, called "Electronic Waste". Every year, an estimated 100 million computers and other electronic devices break or become obsolete and are discarded. Electronic waste, or e-waste, is an emerging problem as well as a business opportunity of increasing significance, given the volumes of e-waste being generated and the content of both toxic and valuable materials in them. E-waste is a popular informal name for electronic products nearing the end of their "useful life." Computers, televisions, VCRs, stereos, copiers, mobile phones and fax machines are common electonic products. Many of these products can be reused, refurbished or recycled. Unfortunately, segments of our nation's waste stream. E-waste has become a problem of crisis proportions because of two reasons. Firstly, e-waste is hazardous as the vast amount of computers, televisions, mobile phones and other electronic products that are disposed of every year all contain a variety of toxic substances. When discarded electronics is dumped in landfills, or when the waste is incinerated, contaminants and toxic chemicals are generated and released into the ground or air risking pollution of the environment and toxins entering the food chain are astronomical. Another reason is e-waste being generated at an alarming rate, due to fast obsolescence along with rapidly evolving technology.

Constituents of E-Waste

Electronic and Electrical equipment, consist of multiple components, some having toxic substances that can affect human health and environment, if not properly managed. Generally, these hazards occur on account of improper recycling and disposal methods adopted. There are number of harmful substances (metals) found in e-waste. *"Antimony"* is used primarily in flame proofing, paints, ceramics, alloys, electronics and rubber. Antimony is increasingly used as an alloy that greatly increases leads hardness and strength. Its most important use is as a hardener in lead for storage batteries. Antimony and its compounds in small doses cause headaches, dizziness and depression, while larger doses can cause violent and frequent vomiting. *"Arsenic"* is found in small quantities in the form of gallium arsenide in light emitting diodes. Arsenic is a poisonous metallic element and chronic exposure to it can lead to skin diseases and lung cancer. *"Barium"* is a metallic element used in sparkplugs, fluorescent lamps and in vacuum tubes. It generates poisonous oxides in contact with air. Exposure to barium can lead to muscle weakness, liver and heart problems. *"Brominated Flame Retardants (BFRs)"* is used in the plastic housings of electronic equipment and in circuit boards to prevent flammability. More than 50 per cent of BFR usage in the electronics industry consists of tetrabromo-bis-phenol-(TBBPA), 10 per cent is' polybrominated diphenyl ethers (PBDEs) and less than 1 per cent is polybrominated biphenyls (PBB). *"Beryllium"* found in power supply boxes contain cancer causing agent for lungs. *"Cadmium"* is present in rechargeable NiCd-batteries, CRT screens, printer inks and toners. Cadmium components, absorbed through respiration, seriously affect the kidneys. *"Chromium"*, found in data tapes and floppy disks, is used because of high conductivity and anti corrosive properties. Chromium (VI) compounds are irritating to eyes, skin and mucous membranes. It can also result in DNA damage. *"Lead"* is found in CRT screens, batteries and printed circuit boards. It is also used in solder, lead acid batteries, electronic components and cable sheathing. Exposure to high amount of lead can result in vomiting, diarrohea, appetite loss, abdominal pain, constipation, fatigue and sleeplessness. *"Mercury"* is used extensively in fluorescent lamps, LCDs, alkaline batteries and mercury wetted switches. It is a toxic heavy metal, that bioaccumulates, causing brain and liver damage, if inhaled. Excessive exposure to *"Selenium"* used in photocopying machines, can cause selenosis, leading to hair loss, nail brittleness, and neurological abnormalities. *"Polyvinyl chloride"* (PVC) is an extensively used plastic in electronic appliances. PVC is harmful as it contains about 55 per cent chlorine, which when burned, gives rise to hydrogen chloride gas. This combines with water to produce hydrochloric acid and if inhaled, can lead to respiratory problems.

There are a number of valuable substances in electronic waste. Gold, Silver, Aluminium, plastic etc. are prestigious materials, which recyclers recover from e-waste. *"Silver"* is used mainly in connectors and PWBs to provide conductivity. In mobile phones, it is typically used in the electronics and keypad contacts in the elemental form. *"Gold"* is used in marginal amounts in connectors and PWBs. It is used to provide conductivity and connectivity. Gold has no harmful effects on environment and humans and it can be obtained again after recycling process. *"Aluminum"* is present in almost all electronic products in large quantities. It also provides conductivity. Its less cost, compared to gold and silver and because of its easy recyclability, aluminium is used extensively.

Disposal of E-Waste

The e-waste that is generated can be recycled, reused, and disposed off in landfills or incinerators. The reuse and recycling of outdated electronic products minimize the hazardous effects of electronic waste on the environment. Reuse and recycling also boost energy and resource conservation. After all possibilities for reuse have been exhausted and a computer is slated for disposal, is sent for recycling. By this it is meant that the old raw materials are reclaimed for use in making new products. However, the costs of recycling are still very high due to which most recyclers are not very much willing to take computers for recycling. If the waste cannot be even recycled then it is either sent to landfills or is burnt in incinerators.

Dumping waste in to landfills contaminates the ground water and soil. Toxic chemicals in electronics products leach into the land over time or are released into the atmosphere. These toxic substances can migrate into ground waters, and eventually into lakes, streams, or wells, and raise a potential exposure to humans and other species by entering the food chain.

Burning electronic products into incinerators leads to the formation of toxic gases due to the presence of heavy metals such as lead, cadmium and mercury. Mercury released into the atmosphere can bioaccumulate in the food chain, particularly in fish–the major route of exposure for the humans. Brominated flame retardants generate brominated dioxins and furans when e-waste is burned. If incineration is not at a sufficiently high temperature and sustained for a sufficient time, the plastics and other hydrocarbons may not be completely oxidized to carbon dioxide and water, and may combine with halogens to form, new halogenated hydrocarbons, including dioxins and furans.

E-Waste and Industry

It is a proven fact that, all who produce, distribute, use and dispose of electronic products, have a moral responsibility towards managing electronic waste. Electronic equipment manufacturers should ensure that their products contain lesser toxic constituents, more recycled content and are designed for easy upgradation and disassembly. Initiatives have been taken by some leading companies in this regard.

1. A new packaging technology is already in the offering at Anadigics, which will enhance its product's moisture sensitivity level.
2. National Semiconductor reveals that most of its products are now lead free (except solder). It has also banned the use of cadmium, mercury and chromium in its products.
3. Nokia has created a list of hazardous substance which will not be used in their products and has sent the same to its suppliers.
4. The design of Hewlett-Packard's Office Jet 500 multi-purpose printer has eliminated the need for plastic flame retardants by using a metal chasis and power supply enclosure, utilizes light-emitting diodes (LEDs) instead of a mercury lamp for the scanner, and eliminates the need for batteries by using flash memory technology.
5. Apple's Macintosh Power Books have used longer-life, less toxic rechargeable lithium-ion batteries for the last three product generations, in place of nickel cadmium batteries.
6. The primary plastic resin used in Intel's PCs and servers (ABS + Polycarbonate) have no flame retardants containing PBBs or PBDEs. None of their products contain asbestos, or include lead or cadmium as plastic additives.
7. Panasonic is the first company to apply reflow type lead-free soldering to compact portable minidisc player PC boards. Its video equipment division has been developing a low-cost tin-copper base solder.
8. At Motorola, housings are made of standard engineering plastics. Several Motorola phone models have eliminated the use of brass inserts in their plastic housings.

Finding methods to keep electronic waste out of landfills is a challenge now for all the electronic product manufacturers, recycling and waste management organizations, government agencies and environment management organizations. Indian Govt. should draft legislation in this direction. If all consumers plan to phase out their obsolete computers and other electronic products at the same time, the country may face a tsunami of e-scrap. Ultimately, it is the balance of nature, which always holds supreme.

Plant Protection, Diversity and Conservation (2011) *Pages* **676–682**
Editors: **D. Prasad & Satya Kumar**
Published by: **BIOTECH BOOKS, NEW DELHI**

Chapter 52

Product Development Using Water Hyacinth

Meenu Srivastva and Priaynka Rawat

Department of Textiles and Apparel Designing,
College of Home Science, MPUAT, Udaipur – 313 001

Water hyacinth (*Eichhomia crassipes*) is a noxious aquatic weed which changes its biodiversity with devastating effects on environment by blocking canals and pumps in irrigation projects, interfering with hydro electricity production; and clogging river and canals such that the drainage becomes impossible and there are Hoods. It has posed ecological and economical problems especially in tropical and subtropical 'countries where environmental conditions help grow hyacinth although the year. Water hyacinth is well known for its ability to loose water rapidly through its leaves by transpiration. Thus the plant appears to derive its Hindi name *"samuildra sokh"* which can absorb ocean. In Rajasthan, it is widespread and Udaipur, famous as the city of lakes or Venice of the east; is no exception, with most of its lakes, ponds etc. covered with water hyacinth, badly affecting the beauty of the lakes and in turn tourism.

The massive growth of water hyacinth is a direct result of human activities–We pollute water bodies and the water hyacinth thrives on nutrient rich water. Though there are various physical and chemical methods of controlling aquatic weed, all are painstaking and expensive.

Complete eradication of noxious species has never been possible. Mechanical removal generates large amount of plant biomass which, if not removed, adds to the same problem. Therefore, scientific interests have been directed towards the utilization aspects of water hyacinth for fibre extraction and pulp for paper making in conjunction with manual or mechanical harvesting to recoup some costs.

Products development from water hyacinth requires study of the morphology and physical properties of water hyacinth, fiber and pulp processing and assessment of the usefulness and cost effectiveness of the developed products.

Methodology

Various steps in water hyacinth work are follows:

Collection of Raw Material

The water hyacinth-an aquatic weed was collected from Udaipur city. The stems were used for the extraction of fibre and pulp making.

Extraction of Fibres

Water hyacinth fibres were extracted by chemical retting process. The cut, split stems were air dried for 4-6 hours in the direct sun and then dried stems were immersed in sodium meta-bisulphite/ aqueous KOH in required amount of water.

Testing of Physical Properties of Extracted Fibres

The water hyacinth fibre were conditioned for 48 hours under the testing atmosphere of 65±2 per cent rh and 20±2°C temperature. The fibre were tested for various physical test such as fibre length, diameter/breadth, cell wall thickness, fineness, tensile strength/elongation.

Pulp Making and Development of Handmade Paper Sheet

The cellulosic material obtained as waste after the fibre extraction process were used for making pulp. The pulp was utilized for developing hand made sheet in three different ratios *i.e.*, 1 : 1, 1 : 2, 2 : 3.

Testing of Physical Parameters of Developed Handmade Sheet

The developed handmade sheets of selected different ratio and a varied GSM were analysed for its usefulness. The different papers conditioned for 24 hours kept in a chamber under the testing atmosphere of 65±2 per cent rh and 27±2°C temperature. Sheets were tested for various physical parameters such as brightness of paper, opacity, porosity, bursting strength, tear strength, basis weight, printability.

Brightness of Paper

It indicates the degree of whiteness measured through reflectance by white or near white paper at a single wave length, 457 mm (blue region of visible spectrum). It is a test to measure the effectiveness of bleaching in removing yellowness from pulps and for measuring ageing of paper. For testing the brightness of paper, Technibrite Micro TB instrument was used.

Opacity

Opacity is the ratio of the reflection of a single sheet backed by black body to the reflectance of a single sheet backed by a white body having an absolute effective reflectance of 89 per cent. For testing the opacity of paper, Technibrite Micro TB IC was used.

Porosity

Porosity of containing connected air voids depends upon the number of voids and their distribution in size, shape and orientation. It indicates how a sheet will respond to fluid penetration. The rate of filtration of air through a paper was measured by using Gurley Densometer.

Bursting Strength

Bursting strength of paper was determined using IS : 1060 (Part-I) 1966 test method. It is the ability of a sheet to resist rupture when pressure is applied to one of its side. It is measured by the pressure developed behind a circular rubber diaphragm to burst the paper. For testing the Bursting strength, Messmer bursting tester 1555 instrument was used.

$$\text{Burst Factor} = \frac{\text{Bursting Strength (gm/cm}^2\text{)}}{\text{Substance in gm/m}^2}$$

Tear Strength

Tear strength of the paper can be expressed as an amount of work done in tearing the paper through a specified distance after the tear has been started. For testing the tear strength, ED-20 tear tester was used.

$$\text{Tear Factor} = \frac{\text{Tear resistance}}{\text{Substance in gm/m}^2}$$

Basis Weight

Basis weight of paper is in square meter area in gm/sq. m.

Substance in g/m^2 = 10,000 w/ab

where,

w = Weight in gm of specimen, a = Length in cm of specimen, b = Width in cm of specimen.

Printability

Printability is the property of a paper which yields printing matter of good quality. It is judged by uniformity of colour of the printed area, uniformity of ink transfer, and contrast between the printed and unprinted area, rate of ink setting and drying.

Value Added Products

Value added products like hand made sheet of varied GSM were utilized for developing sixteen different products.

Assessment of Usefulness and Cost Effectiveness of Value Added Products

The assessment .of consumer preferences for the developed products was done by 30 sample subjects/panel of judges. The respondents, 10 from each category were housewives, home scientist, owner of handicraft personnel. A 5 point rating scale penonna was given to 30 respondents for this purpose.

The cost of each products was calculated by determining the cost raw material, labour charges etc. Cost of various accessories used in developing products was also taken in to account to assess the final cost of the product.

Results and Discussion

The fibres extracted with sodium metabisulphate were tested for their physical properties like length, diameter, cell wall thickness, fineness, tensile strength/elongation (Table 52.1).

It was revealed that the strength of water hyacinth fibre is quite low which must be increased by blending with other conventional fibres for its commercial utilization. Further, the pulp obtained could be utilized for handmade sheets as the cell wall thickness contribute towards paper strength.

Table 52.1: Physical Properties of Water Hyacinth Fibre

Sl.No.	Physical Properties	Values
1.	Fibre length (mm)	1.51
2.	Diameter/breadth (mm)	0.020
3.	Cell wall thickness (μ)	3.4
4.	Fineness (μ)	35
5.	Tensile strength/elongation (lbs/mg)	78.98

Good quality handmade paper sheets can be manufactured using cellulosic pulp of unconventional fibre plants in combination with other conventional fibres, cotton rags, recycled paper, etc to improve its strength, opacity and other parameters as per requirement of end products. In the present work. water hyacinth pulp was mixed with cotton rags in shredded form to improve thickness and viscosity of the pulp for manufacturing handmade sheets in three different ratios *i.e.*, 1 : 1, 1 :2 and 2 : 3. The sheets were produced in two different GSM-100 and 200 with different thicknesses. Pure pulp of water hyacinth was also used to develop paper. The results of the testing of physical parameters of developed handmade sheets of 100 GSM have been presented in Table 52.2.

It was found that the paper produced from pure water hyacinth pulp was deficient in most of the parameters except opacity and porosity. Tear and burst strength were found to be high in 1 : 2 *i.e..*, 8.66 mNm^2/gm and 1.94 Kpa. m^2/gm. On the other hand, the 1 : 1 ratio was also found suitable with respect to all the physical parameters tested.

Table 52.2: Testing of Physical Parameter of Handmade Sheet

Sl.No.	Particulars	100 per cent Water Hyacinth	1 : 1 Water Hyacinth : Cotton Rags	1 : 2 Water Hyacinth : Cotton Rags	2 : 3 Cotton Rags : Water Hyacinth
1.	Basis weight (gm/m^2)	50.78	148.23	150.12	152.61
2.	Burst index $Kpa.m^2/gm$	0.49	1.87	1.94	0.86
3.	Tear index (mNm^2/gm)	0.78	7.39	8.66	5.71
4.	Porosity (sec/100 ml)	60.65	–	–	–
5.	Brightness (per cent ISO)	–	26	27.65	14.2
6.	Opacity (per cent)	92.0	98.0	94.21	95.68

Handmade sheets developed were evaluated by the respondents and responses obtained are presented in Table 52.3.

It was found that the uniformity of colour of the printed area was found to be less from pure water hyacinth in case of typing ink and ball pen. The uniformity of colour in 2 : 3 ratio was found to be very good in ink pen and ball pen.

Similarly, the printability of the handmade sheet on uniformity of ink transfer was found to be best in combination of 1 : 2 as well as 2 : 3 in ball pen. Typing ink uniformity on transfer was found to be less in ink pen and typing ink in pure form of water hyacinth.

Table 52.3: Mean Scores Obtained by Uniformity of Colour of the Printed Area

Sl.No.	Parameters	Water Hyacinth	Water Hyacinth : Cotton Rags		Cotton Rags : Water Hyacinth
		100 per cent	1 : 1	1 : 2	2 : 3
1.	Ball pen	1.2	2.4	3.0	4.2
2.	Ink pen	1.8	1.8	2.2	4.6
3.	Gel pen	2.4	1.8	4.2	3.8
4.	Market	2.6	2.2	2.6	3.2
5.	Typing ink	1.4	2.2	3.2	4.0

Again, marker and gel pen were found to be best with 4.2 and 4.6 in 1 : 2 and 2 : 3 ratios on the parameter of contrast between printed and unprinted area. As regards the rate of ink setting and drying ball pen and ink pen scored high (4.2) in combination of 2.3 and were found to be the best.

The result of the physical testing of water hyacinth fibre revealed its unsuitability for developing yarn in pure form. On the other hand, handmade sheets showed bright prospect in developing a variety of handicraft items as per development of different value added articles using variety of colours.

The development of various handicraft items required different GSM (Basis weight). For the light weight delicate articles, sheets of lower GSM (*i.e.* 60–150) and for strong/heavy weight articles, higher GSM (*i.e.*. 200–300) handmade paper sheets are required. Hence, for developing value added articles of two different categories, handmade sheets of 100 and 200 GSM were produced.

Sixteen value added articles have been made from handmade paper sheets. For surface enrichment of developed articles, different accessories were also used like beads, mirrors, *ghungroo*, etc. Table 52.4 lists the articles developed.

Table 52.4: Categories of Developed Value Added Articles

Articles of 100 GSM	Articles of 200 GSM
☆ Book mark	☆ Bandanwar
☆ Bell	☆ Carry bag
☆ Envelope	☆ Clip board
☆ Place mat	☆ Flower pot
☆ Photo frame	☆ Lamp shade
☆ Traditional Barn	☆ Multipurpose box
	☆ Mobile stand
	☆ Photo album
	☆ Telephone diary
	☆ Invitation card

The articles were assessed on the basis of selected parameters *viz.* suitable of design, colour combination, utility of article, overall appearance and marketing potential. It was found that telephone diary achieved the first rank. Photo frame got first rank in design. After telephone diary, the envelope and clip board achieved second and third rank respectively under parameter of utility of article. The clip board was also appreciated.

Effort was made to calculate the cost of developed handmade sheets of various GSM. Table 52.5 shows that the total cost of seven sheets of 100 GSM was found as Rupees 76.09 and sixteen sheets of 200 GSM was found at Rupees 173.92. These costs are less compared to readymade handmade sheets available in the market.

Table 52.5: Cost Estimation of Developed Handmade Sheet

Parameters	Particulars	Amount
Cost of mw material	Water hyacinth	60 kg
	Pulp extracted	12 kg
	Transportation cost	50.00
	Manufacturing charges	200.00
	Total number of developed handmade sheets	23
Cost price of hand made sheet	100 GSM sheet (Total No.7)	76.09
	200 GSM sheet (Total No. 16)	173.92

The cost of each developed handmade sheet was found to be Rupees 10.87 as compared to readymade sheet that sell for about Rupees 12 the market. Hence, it can be concluded that these developed handmade sheets have good market potential. The price of the developed handmade sheet can be lower if manufactured in bulk.

Cost of Developed Value Added Articles

Table 52.6: List of Estimated Price of Different Articles

Sl.No.	Articles	Cost of Handmade Rs. paper .	Cost of Other Rs. Material	Labour Charges (Rs.)	Total Cost (Rs.)
1.	Traditional Bahi	3	2	3	6
2.	Bandanwar	2	2	1	5
3.	Photo frame	12	8	15	35
4.	Clip hoard	10	15	15	40
5.	Multipurpose box	3	5	2	10
6.	Flower pot	12	10	12	34
7.	Place mat	4	3	5	12
8.	Lamp shade	6	35	12	35
9.	Photo album	12	10	10	32
10.	Carry bag	2	2	1	5
11.	Envelopes	1	–	1	2
12.	Book mark	1	1	1	3
13.	Mobile cover	1	2	2	5
14.	Bell	3	5	7	15
15.	Invitation card	1	2	1	4
16.	Telephone diary	13	10	25	48

The labour charges of some of these articles we-re higher due to requirement of creative skill of the handicraft personnel, which can be brought down by increasing the number of articles are made. The costs are not comparable to those of readymade handicraft products available in the market.

For assessment of economic benefits of the value added handicraft items, a Proforma was developed in which total cost and approximate profit percentage ranging from 20-50 per cent and more than 50 per cent were added. This proforma was given to 30 respondents.

The overall opinion was that maximum profit can be gained by the sale of envelopes and traditional Bahi. Flower pots and lamp shade could be sold on 20 per cent profit easily. Multipurpose box could be sold on 50 per cent profit and rest of the items can fetch 20 per cent profit. Thus, it is apparent, that developed articles were found highly acceptable in terms of utility and marketability.

Conclusion

It can be suggested from present study that water hyacinth, a noxious weed can be a potential source of income if cottage industry is established in nearby areas of its occurrence in plentiful amount. If is expected that rural people will be benefited, by way of collecting, drying, supplying thereby adding to their income. It will also help in eradicating this weed simultaneously reducing water pollution. Water hyacinth has great potential in textile and paper industry.

Acknowledgements

The authors are grateful to *Handmade Paper Industry of Vidhya Bhavan Society, Udaipur* (*Rajasthan*) for rendering help in developing handmade paper sheets of water hyacinth and *Kummarappa National Handmade Paper Institute, Jaipur* (*Rajasthan*) for testing and analyzing various physical parameters of the sheets produced by selecting different ratios and also of varied GSM/basis weight.

References

Gopal, B., 1987. *Water Hyacinth*. Elsevier, New York.

Gupta, R.K. and Saikia, S.K., 1998) Ecofriendly handmade paper making. Extraction of jute fibre. Internet: http/nerdata bank.nic.in/nepaper.com.

Muli, J.R., 1996. Environmental problems of Lake Victoria (East Africa): what the international community can do, lakes and reservoirs: Research and management, 2: 47–53.

Sharma, B.K., 2005. *Water Pollution: Environmental Chemistry*, 9th edn. Goel Publishing House, Meerut, p. 32–39.

Visperas, R.V., 1991. *Chronology of handmade paper and paper making*. Internet: http://www.calricestraw.org/Papermaking.html.